Internet
Communication

Steve Jones
General Editor

Vol. 91

The Digital Formations series is part of the Peter Lang Media and Communication list.
Every volume is peer reviewed and meets
the highest quality standards for content and production.

PETER LANG
New York • Bern • Frankfurt • Berlin
Brussels • Vienna • Oxford • Warsaw

James W. Chesebro
David T. McMahan
Preston C. Russett

Internet Communication

PETER LANG

New York • Bern • Frankfurt • Berlin
Brussels • Vienna • Oxford • Warsaw

Library of Congress Cataloging-in-Publication Data

Chesebro, James W.
Internet communication / James W. Chesebro, David T. McMahan, Preston C. Russett.
pages cm — (Digital formations; vol. 91)
Includes bibliographical references and index.
1. Internet—Social aspects. 2. Internet—Social aspects—United States.
3. Internet—Political aspects. 4. Internet—Political aspects—United States.
5. Digital communications. 6. Digital communications—United States.
7. Communication policy. 8. Communication policy—United States.
I. McMahan, David T. II. Russett, Preston C. III. Title.
HM851.C4465   302.23'1—dc23   2014000839
ISBN 978-1-4331-2304-7 (hardcover)
ISBN 978-1-4331-2303-0 (paperback)
ISBN: 978-1-4539-1307-9 (e-book)
ISSN 1526-3169

Bibliographic information published by **Die Deutsche Nationalbibliothek**.
Die Deutsche Nationalbibliothek lists this publication in the "Deutsche
Nationalbibliografie"; detailed bibliographic data are available
on the Internet at http://dnb.d-nb.de/.

© 2014 Peter Lang Publishing, Inc., New York
29 Broadway, 18th floor, New York, NY 10006
www.peterlang.com

Contents

Preface

This volume examines the Internet as a communication system. In this view, the Internet is the single most pervasive, involving, and global communication system ever created by human beings, with a host of political, economic, cognitive, and sociocultural implications. The Internet crosses all cultural boundaries and is the fastest growing global communication system ever witnessed. For example, Facebook is now approaching one billion members, and globally a growing percentage of the world's population in the developed countries surfs the Internet on a daily basis. In this context, on a global level, The Pew Internet & American Life Project reported at the end of 2010 that "although still a relatively young technology, social networking is already a global phenomenon." Based on this survey of 22 nations, Pew (2010; see also *Global Digital Communication*, 2011) reported that social networking is

> especially popular in the United States, where 46% say they use sites like Facebook and MySpace, but other nations are not far behind. At least four-in-ten adults in Poland (43%), Britain (43%) and South Korea (40%) use such sites (respondents were given examples of sites that are popular in their country). And at least a third engage in social networking in France (36%), Spain (34%), Russia (33%), and Brazil (33%). While involvement in social networking is relatively low in many less economically developed

nations, this is largely due to the fact that many in those countries do not go online, rather than disinterest in social networking. When people use the internet in middle and low income countries, they tend to participate in social networking.

By the end of the next year, in its December 2011 survey of 21 nations, Pew (December 20, 2011) reported an increase in these percentages from +4% to +10%.

While social networking is a vivid example of Internet use, it is by no means the most popular use of the Internet. Toward the end of 2011, for example, Purcell reported that sending or reading email was employed by 92% of Internet users. And, all indications are that these emails are displacing face-to-face oral communication as well as other forms of personal and written communication. In terms of search engine activity, she reported that the percentage of Internet users rose from 85% in 2002 to 92% in 2011. Increasingly, the information contained on the Internet defines "what we know" and "what we know to be true." Additionally, some 76% used the Internet to get news online in 2011 (an increase from 68% in 2002), 71% used the Internet to buy products (up from 61% in 2002), and some 65% used the Internet for social networking (up from 11% in 2005) (Purcell, 2011). Further, this mounting popularity and the seemingly endless appeal of the Internet has led some countries, such as China and Korea, to begin treating problematic Web use as a national health crisis (Dokoupil, 2012).

Among those most likely to enter undergraduate communication classes, media use is even greater and has been increasing for the last 10 years. Specifically, Lenhart (April 2011) reported in 2011 that both 93% of teens aged 12 to 17 and 93% of "young adults" (those 18-to-29-years-of-age) are regular Internet users (compared to 81% of 30-to-49-year-olds, 70% of 50-to-64-year-olds, and 38% of those 65 and older). Many of these users, over one-third of smartphone owners, get online before getting out of bed in the morning (Dokoupil, 2012). More generally, the Kaiser Family Foundation (2010) reported that total daily media use for those 8 to 18 years of age has increased from 6 hours and 19 minutes in 1999 to 6 hours and 22 minutes in 2004 and 7 hours and 38 minutes in 2009. Moreover, when the amount of time using more than one medium at a time is considered, Kaiser reported that, "today's youth pack a total of 10 hours and 45 minutes worth of media content into those daily 7 ½ hours—an increase of 3 hours and 15 minutes media exposure per day over the past five years" (The Henry J. Kaiser Family Foundation, January 2010).

Focal Points of This Book

In its broadest conception, this volume specifically explores the ways in which the technology of the Internet itself, beyond its specific content, possesses its own message-generating capabilities that dramatically and decisively affect its users. At the same time, this book focuses on the power of media theories to explain, describe, interpret, and evaluate the Internet in insightful, useful, and thoughtful ways. Ultimately, the practical applications are used to test theories, and theories are used to explain practice. Specifically, the actual concepts, processes, functions, and outcomes of the Internet as a global communication technology are used as a way of testing the validity and reliability of media theories, and media theories are also used as a way of identifying the powers and limitations of the Internet as a communication system.

More specifically, four focal points define the approach to communication employed in this textbook:

1. ***Research Driven.*** The Internet is now one of the most frequent objects of study for researchers. For example, the Pew Research Center, especially through its Pew Internet & American Life Project, has been publishing research findings—predominantly surveys—regarding the social uses of virtually all dimensions of the Internet since the year 2000. The Pew Research Center has been specifically committed to tracking recent developments in the use and meanings attributed to new computer technologies. Additionally, new research findings regarding the "new media" now regularly appear in academic journals. It is, therefore, now more than reasonable to let research findings establish a foundation for the analysis of Internet functions and uses.

2. ***Symbolic Perspective.*** A symbolic perspective is also employed to define and explain the Internet. The Internet can be viewed from any number of perspectives. It is a political system; it is a cultural system; it is a social system. In this book, we will examine the ways in which the Internet uniquely formats content and selectively highlights certain content rather than other, and how, in these ways, the Internet ultimately functions as a message-generating system.

3. ***User Perspective.*** As a communication system, the Internet can additionally be viewed as a series of technological systems that require mastery. While recognizing the importance of mastering these technological skills, this volume remains user-centered. We focus on the ways in which

the Internet affects human choices, human processes and interactions, and human outcomes.

4. ***Critical Perspective.*** The word *criticism* can refer to judgments that are predominantly negative, but the word can also identify evaluations designed to promote understanding of the objects under review. In terms of the Internet, we need to ask, not only what the Internet is doing to the human symbolic processes, but also ***if*** those effects are beneficial and/or harmful, and ***in what ways*** the processes and effects set off by the Internet are beneficial and/or harmful. In this sense, as critics, we will seek to ***describe*** how the Internet functions as a communication system, identify or ***interpret*** the different ways in which the Internet can be understood, and to render judgments about or ***evaluate*** both the benefits and disadvantages of the Internet as a communication system. In every case, however, we need to go beyond saying what we like and dislike, for such claims are only statements of our personal preferences; they are not criticism.

Criticism is deliberative; it is political; it seeks to go beyond personal preferences; or, in short, it seeks to enhance the quality of life for the people within a society (Levy, 2001, pp. 211–217). Criticism is also a reason-giving activity; a critic provides reasons why the critic's descriptions, interpretations, and evaluations are appropriate. These reasons are accordingly verifiable; the reasons can and should be sustained and confirmed with reasonable research; the reasons offered by the critic can be demonstrated to be more true than false given the conditions being explored.

Objectives of This Book

1. ***Understand and appreciate the role that technology can play as part of a larger communication process***. In different ways, one purpose of this book is to identify the cognitive function of technology. Every technology isolates certain stimuli, features those stimuli, and constructs a view of reality based upon this selective presentation of some—rather than all—of the available stimuli. We are particularly concerned about how technology affects our cognitive and emotional reactions to ourselves, other people, and to the world in which we exist.

2. ***Acquire the ability to function more effectively as a consumer of the information and understandings generated by the Internet.*** This volume provides multiple ways or sets of constructs for describing and interpreting Internet content.

3. ***Provide a foundation for assessing the benefits and limitations of Internet concepts and processes.*** Specifically, this book will provide a range of strategies and intervention techniques for dealing more effectively with, and judging the usefulness and value of, Internet content.

4. ***Provide exposure to new understandings and research regarding the nature and characteristics of the Internet as a communication system and how this new mode of communication may be altering and affecting other modes of communication.*** The Internet is explicitly interacting with all of the other modes of communication. This book carefully examines these interactions.

5. ***Develop a critical perspective of the Internet and how it is affecting human communication.*** This volume enables readers to think critically and be able to verbally articulate the sociocultural issues involved in the transformation toward digital technologies such as the Internet.

Pedagogical Features of This Book

Numerous pedagogical features have been crafted for this volume, many of which can be easily integrated into the classroom in addition to their use as independent learning tools for students. These pedagogical tools encompass three fundamental characteristics of the Internet and its study. First, profound *transformation* continuously takes place in the Internet's development, understanding, influence, and use. Accordingly, as with the volume itself, the pedagogical tools approach the Internet and its study as continually evolving and ever changing. Second, social and academic approaches to the Internet are quite *controversial.* There exist many alternative approaches and disagreements concerning the study of the Internet and its influence on society. Therefore, the pedagogical tools engage the controversial nature of the Internet by offering and encouraging the exploration of dissimilar and contentious approaches to the material. Finally, the Internet impacts *everyday life.* Indeed, the Internet's impact now touches every aspect of a person's life and daily experiences. Consequently, many of the pedagogical tools either directly focus attention on the everyday impact of the Internet or indirectly lead students to reflect on their personal use and understanding of the Internet.

Chapters are structured to fully guide student exploration of the material. Every chapter begins with an **introduction** that establishes and outlines the material to be examined and is then followed by a list of **focus questions**. These focus questions are purposefully embedded within the chapter rather than offered before the chapter begins to increase the chances that they will actually be used by students and to provide students with some awareness of the material prior to giving them a list of questions to consider. **Key terms** are bolded throughout chapters to emphasize important material for students and are listed at the very end of the chapters to reinforce their recognition by students. Multiple **diagrams** and **charts** are included to visually synthesize the material for students as well as to guide their reading of chapters. The **conclusion and future possibilities** section of each chapter also serves as a pedagogical tool for students. In recognition of the transformative nature of the Internet, rather than simply rehashing what was just provided, these conclusions review the material with an eye to the future. Future trends and changes are offered, while prompting students to speculate about what transformations they believe will take place.

Five types of pedagogical features appear in every chapter—as needed and in terms of the appropriateness of the content of each chapter—to assist understanding and prompt students to further contemplate the material being examined.

Alternative View boxes can appear numerous times throughout each chapter to provide students with counter arguments to those offered by the authors and to examine multiple perspectives of an issue. We find these "alternative view" boxes particularly important from an educational perspective because they foster critical thinking. The "alternative view" boxes specifically ask students not to accept all that they read, but to anticipate that even claims made with tremendous confidence may be concealing issues. We hope these "alternative view" boxes serve this critical thinking function.

Current Issues boxes encourage students to go beyond the material provided by searching for recent innovations and events connected to particular topics. For example, in the American Politics chapter, students are urged to search for news and reports regarding the use of Twitter by political candidates.

Debate It boxes prompt students to consider two contradictory statements by providing support for and support against both statements. These boxes compel students to fully consider multiple sides of an issue and can be used to stimulate classroom discussion. Within the Pragmatics of Human Communication chapter, for instance, students are asked to debate the merits of searching for healthcare information online.

Ethical Issues boxes engage students in ethical questions surrounding the materials examined. Students are asked in the Social Networking chapter, for example, if it is ethical for companies to use social networking sites to understand information about applicants or existing employees.

Everyday Impact boxes urge students to ponder the use and influence of the Internet in their everyday lives. In the Business of the Internet chapter, for instance, students are asked to consider how their purchasing decisions are influenced by their online activities.

Finally, **Future Focus** boxes offer students predictions about the future of the Internet. For example, in the Pragmatics of Human Communication chapter, we predict that activities such as shopping, travel, and health care will become increasing based in virtual realities. Students are then asked whether they agree with this prediction and are encouraged to offer their own predictions regarding the development of these activities.

We cannot claim that there is evidence drawn from classroom studies that demonstrate that each of these types of pedagogical features can actually do what they are designed to do. For example, "Ethical Issues" may not engage all students in the ethical questions surrounding the material examined in various chapters. But, it is our hope that these boxes may catch the attention of some students and encourage them to at least consider what they have read from the perspective encouraged by each pedagogical box. And, we do think that the content of these boxes could also be a stimulus for some classroom discussions.

In all, we hope you enjoy and benefit from this book in any number of social and personal ways. We have enjoyed producing this book.

Your textbook authors,

James W. Chesebro, David T. McMahan,
and Preston Russett

August 2013

1

Power and Meanings of the Internet as a Communication System

The Internet is now the largest and most pervasive global communication system ever created in the history of human beings, affecting more people in more ways than any other human invention. In greater detail, Cantoni and Tardini (2006) aptly noted:

> As a matter of fact, the internet is one of the newest and most powerful communication technologies, and is rapidly spreading worldwide. It is no doubt changing in depth the way we interact and communicate; both in everyday life and in our professional activities; it is changing our social life; the way we conceive of ourselves and our relationships with others; the way we learn; the way we buy; the way we take care of our health; the way we interact with civil services, and so on. (p. 2)

In this volume, we literally explore each of these uses, and how the Internet is now affecting virtually all of these dimensions of our lives.

Indeed, we now have an amazing diversity of devices to connect us to the Internet, such as laptops, desktops, mobile phones, tablets, touch pads, e-readers, netbooks, and web-enabled televisions. For example, 66% of 18-to-29-year-olds in the United States own and have a smartphone with them 24 hours a day and seven days a week (Rainie, September 11, 2012). And, with all of these devices, the Pew Research Center's Internet & American Life

Project has tracked 55 activities that people carry out on the Internet, from searching to find information (91% of all adult U.S. Internet users), all the way down to the 55th use, with 4% of adult users visiting virtual worlds such as Second Life ("Trend Data (Adults)," August 2012).

Moreover, Internet use is global. While not yet universal, Internet usage is rapidly spreading. The International Telecommunications Union (2013) reported that 77% of those in the "developed world" are Internet users; 31% of those in the "developing world" are Internet users; and, more generally, 39% of those in the "world globally" are Internet users. But, we do expect a fully global participation in the Internet, especially if mobile phone subscriptions are any indication. In the first quarter of 2013, mobile phone subscriptions have increased globally from 2,205 million to 6,835 million, or 96.2% of the international population (Internet World Statistics, 2013). And, as we argue shortly, as an increasing percentage of the world's population are Millennials, the use of the Internet will become commonplace. Indeed, in rather dramatic and precise terms, Google's Eric Schmidt (Internet Stats Today, April 15, 2013) has "predicted that everyone in the world will be online by [the year] 2020."

Additionally, in terms of age, Internet use is dramatically increasing (Lenhart, 2011, p. 6). Some 93% of those aged 12 to 17 and 18 to 29 access the Internet regularly. Eighty-one percent of adults between the ages of 30 and 49 access the Internet regularly, with 70% of adults between the ages of 50 and 65 accessing the Internet regularly. While only 38% of adults over the age of 65 access the Internet regularly, every year an increasing percentage of this group accesses the Internet. Certainly, the age group that will dominate the American culture in the years to come are intensive Internet users. Perhaps Lenhart (2011, p. 5) captured the point simply and directly when she proclaimed that, "Everyone uses the Internet."

Questions Previewing This Chapter

1. *What activities or functions do people carry out or use the Internet for?*
2. *What sociocultural context and what future is the Internet fostering?*
3. *How should the notion of digital and the Internet be defined, especially within a communication context?*

Preview

In this chapter, we explore three issues: (1) **Uses of the Internet**: We are particularly intrigued by growing use of the Internet since 1995 as well as the particular functions carried out on the Internet today. (2) **Sociocultural Context and Future Internet Driver**: An assessment of the sociocultural factors that have stimulated the growth of the Internet, especially in terms of the sociocultural grouping that we think will drive the future growth and development of the Internet. (3) **Central Digital and Internet Definitions**: While much computer terminology can simply be clarified by way of examples, formal definitions of some of these concepts can be useful, especially when going beyond this volume and linking the findings reported here to the analyses provided by others.

Uses of the Internet

For all practical purposes, the Internet became a popular sociocultural device in 1995. Certainly, universities, the federal government, and corporations had been fashioning links among their computer systems for years, especially since World War II, which so dramatically underscored the use of computer systems in all kinds of military operations. We will explore these military uses as well as a host of other factors shaping the emergence of the computer and Internet age in Chapter 3. But for now, it is particularly important to note that the Internet itself became a popular device by 1995. As Figure 1.1 indicates, by mid-June of 1995, only 14% of the American population was using the Internet. Given that now over 80% of the population in the United States regularly uses the Internet, by usage, the growth from 14% to 81% in 2012 can be viewed as nothing less than spectacular in terms of sociocultural developments. Essentially, the Internet became an indispensable part of the American culture in less than 15 years. Few changes or technologies have taken hold so dramatically or quickly.

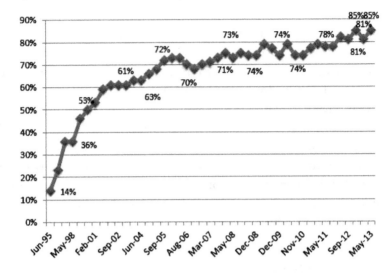

% of American adults who use the internet, over time

Source: Pew Internet & American Life Project Surveys, March 2000-May 2013. All surveys prior to March 2000 were conducted by the Pew Research Center for People & the Press.

Figure 1.1. Internet Adoption, 1995–2013.

And, as we shall suggest shortly in this chapter, when we consider the development of the Internet in the years to come, the same dramatic use in the Internet is like to occur in all developed and developing countries.

A key question involved in the growth of the Internet turns on how it has been used by people. We are especially intrigued by this question because it identifies the functions of the Internet. As we mentioned earlier in this chapter, the Pew Research Center's Internet & American Life Project has tracked some 55 distinct uses that have been made of the Internet. Table 1.1 provides an overview of these uses.

Table 1.1. Internet Uses by U.S. Adults.

Rank	Use	% of Adult Internet Users in the United States Who Do This Online	Month and Year of Pew Internet Survey
1	Use a search engine to find information	91%	2/1/2012
2	Send or read email	91%	8/1/2011

Table 1.1. *Continued*

Rank	Use	% of Adult Internet Users in the United States Who Do This Online	Month and Year of Pew Internet Survey
3	Look for info on a hobby or interest	84%	8/1/2011
4	Search for a map or driving directions	84%	8/1/2011
5	Check the weather	81%	5/1/2010
6	Look for health/medical info	80%	9/1/2010
7	Look for information online about a service or product you are thinking of buying	78%	9/1/2010
8	Get news	76%	5/1/2011
9	Go online just for fun or to pass the time	74%	8/1/2011
10	Buy a product	71%	5/1/2011
11	Watch a video on a video-sharing site such as YouTube or Vimeo	71%	5/1/2011
12	Search for info about someone you know or might meet	69%	9/1/2009
13	Look for "how-to," "do-it-yourself," or repair information	68%	8/1/2011
14	Visit a local, state, or federal government website	67%	5/1/2011
15	Use a social-networking site such as Facebook, LinkedIn, or Google+	66%	2/1/2012
16	Buy or make a reservation for travel	65%	5/1/2011
17	Do any banking online	61%	5/1/2011
18	Look online for news or information about politics	61%	8/1/2011
19	Look online for info about a job	56%	5/1/2011

Table 1.1. *Continued*

Rank	Use	% of Adult Internet Users in the United States Who Do This Online	Month and Year of Pew Internet Survey
20	Look for information on Wikipedia	53%	5/1/2010
21	Use online classified ads or sites such as Craigslist	53%	5/1/2010
22	Get news or information on sports	52%	1/1/2010
23	Take a virtual tour of a location online	52%	8/1/2011
24	Do any type of research for your job	51%	3/1/2007
25	Upload photos to a website so you can share them with others online	46%	11/1/2010
26	Send instant messages	46%	12/1/2010
27	Pay to access or download digital content online	43%	8/1/2008
28	Look for info about a place to live	39%	8/1/2006
29	Download music files to your computer	37%	12/1/2007
30	Get financial info online, such as stock quotes or mortgage interest rates	37%	5/1/12010
31	Rate a product, service, or person using an online rating system	37%	5/1/2011
31	Play online games	36%	9/1/2010
33	Categorize or tag online content such as a photo, news story, or blog post	33%	12/1/2008
33	Read someone else's online journal or blog	32%	5/1/2010

Table 1.1. *Continued*

Rank	Use	% of Adult Internet Users in the United States Who Do This Online	Month and Year of Pew Internet Survey
34	Look for religious/spiritual info	32%	9/1/2010
35	Post a comment or review online about a product you bought or services you received	32%	9/1/2009
36	Post comments to an online news group, website, blog, or photo site	32%	9/1/2010
37	Share something online that you created yourself	30%	9/1/2010
38	Research your family's history or genealogy online	27%	9/1/2009
39	Download video files to your computer	27%	12/1/2007
40	Participate in an online auction	26%	9/1/2010
41	Make a donation to a charity online	25%	9/1/2011
42	Make a phone call online, using a service such as Skype or Vonage	25%	8/1/2011
43	Participate in an online discussion, a listserv or other online group forum that helps people with personal issues or health problems	22%	12/1/2006
44	Download a podcast so you can listen to it or view it later	21%	9/1/2010
45	View live images online of a remote location or person, using webcam	17%	9/1/2009
46	Create or work on web pages or blogs for others, including friends, groups you belong to, or for work	15%	9/1/2009

Table 1.1. *Continued*

Rank	Use	% of Adult Internet Users in the United States Who Do This Online	Month and Year of Pew Internet Survey
47	Take material you find online—such as songs, text, or images—and remix it into your own artistic creation	15%	5/1/2008
48	Download or share files using peer-to-peer file-sharing networks, such as Bit Torrent or LimeWire	15%	8/1/2006
49	Sell something online	15%	9/1/2009
50	Use Twitter	15%	2/1/2012
51	Create or work on your own webpage	14%	1/1/2010
52	Create or work on your own online journal or blog	14%	5/1/2011
53	Buy or sell stocks, bonds, or mutual funds	11%	9/1/2009
54	Use an online dating website	8%	9/1/2009
55	Visit virtual worlds such as Second Life	4%	9/1/2009

The range of uses or functions carried out on the Internet is extensive, and it suggests that virtually all dimensions of life can be and actually are handled on the Internet. Yet, Table 1.1 is limited. It does not give us the shades of meaning we might need to be aware of if we are to make proper use of the findings it reports.

Everyday Impact: Select an evening when you know you will be on the Internet for an hour or more. Keep a log of which sites you are on and for how long you are on them. Keep track of how many people you interact with on each site. Compare your log with friends of yours in this class. Are face-to-face interactions and online interactions related or linked in any way? When do you trust your face-to-face interactions more than what you find online? When do you trust your online interactions more than your face-to-face interactions? If you have a chance, chat about these findings in your class.

For example, demographics are not emphasized. As a case in point, we know that those 55 years and older are more likely to seek health information on the Internet or that relatives and friends of older people are likely to seek health information with the older person in mind. Likewise, this table reflects only what those 21 years of age and older do on the Internet. More than 36% of those under 21 years of age go on the Internet to play games, and indeed, for this group, these games can be viewed as an entire set of virtual realities that can dominate and control the lives of these younger users. In all, demographics can make a tremendous difference in how we interpret the findings reported on Table 1.1.

Additionally, Table 1.1 does not tell us how long people actually stay online or how involved they are when they are completing a function on the Internet. For example, while 76% of people get their news on the Internet, the amount of time each person devotes to news gathering can vary tremendously, and also vary tremendously in terms of reasons and motives. Likewise, while 51% of people use the Internet to do research for their job, the amount of time a person must spend on the Internet for such research and the significance of these research searches can vary tremendously.

Hence, while the 55 uses of the Internet identified on Table 1.1 are extremely informative, especially when we are seeking a sense of the scope of Internet use in people's lives, these functions must be discussed carefully and precisely. Table 1.1 simply does not reflect demographic differences among people nor does it reveal the amount of time that different people might engage in these uses, or their motives for doing so. Nonetheless, we can draw a major conclusion: The Internet is a massive set of interactions devoted to virtually all of life's major activities. And, if we recall the findings reported on Table 1.1, the Internet is playing an increasing role in the lives of more and more Internet users every year. Perhaps displaying too much confidence in the power of the Internet, some have suggested that the Internet could play a role in the lives of virtually everyone in the world by the year 2020. While we cannot confidently predict that all computer users will be on the Internet by the year 2020, we are sure that for the vast majority of computers—in all of their various forms—the content, forms, and patterns of Internet use are extremely like to influence, if not determine, the communication systems of all computer users.

Sociocultural Context and Future Internet Driver

In Chapter 3 of this volume, "The Past—The Development and Evolution of Digital Technologies," we examine some of the founders of the digital revolution, some of the technologies that contribute to the information age, and make a placement of the digital revolution among the historical technological innovations that have defined the American experience. While Chapter 3 actually provides a more complete foundation for our observations on these matters, at this juncture we want to make a preliminary conception and placement of the current sociocultural system and identify the force we believe is currently driving the nature of this digital revolution.

The Sociocultural Context

Our point of departure here begins with the typical American home in the year 2005, its basic components, and what a child growing up in the typical America home would encounter as a "normal" part of his or her daily existence. *The Gallup Poll* provides a description of this context. Some 10 years ago, in the December 23, 2005 issue of *The Gallup Poll*, Joseph Carroll listed in "Americans Inventory Their Gadgets" the "most popular gadgets," "owned by at least half of Americans," and the percentage of American homes with these gadgets:

> VCR – 88%
> DVD – 83%
> Cellular Phone –78%
> Cable TV – 68%
> Desktop Computer – 65%
> Digital Camera – 50%

Beyond these gadgets that were in at least 50% of American homes, another set of gadgets was also present. While they did not meet the 50% standard, we now know they constituted the future of the American home. These gadgets included:

> High-speed Internet – 48%
> Video Game System – 36%
> Satellite Television – 30%
> Laptop Computer – 30%
> Portable DVD Player – 30%

Indeed, by the year 2006, Rainie reported that the "typical media profile" of a 21-year-old entering the labor force showed that they had already:

- Played video games for 5,000 hours
- Exchanged 250,000 emails, instant messages, and phone text messages

- Interacted for 10,000 hours on cell phones
- Spend 3,500 hours online

Of all of these set of behaviors, Prensky (in Rainie, 2006) concluded that, "The preference" of the typical 21-year-old "is for sharing, staying connected, instantaneity, multi-tasking, assembling random information into patterns, and using technology in new ways."

Today, these technologies have evolved, and variations on these 2005 devices are now evident. Sabino (2012) reported that children under the age of 13 (ages 6 to 12) now live in households with, or have access to:

Laptop Computer – 81%
Desktop Computer – 80%
Mobile Phone – 69%
Tablet – 37%
iPod Touch – 30%
E-Reader – 26%
Netbook – 17%
Web-Enabled Television – 16%

In this regard, the Kaiser Family Foundation (2010, p. 2) reported that those between the ages of 8 and 18 years have increasingly invested their time in these technologies. Comparing the overall media use of young people from 1999 through 2004 and into 2009, the Kaiser Family Foundation reported that media use has increased from 6 hours and 19 minutes per day in the year 1999, to 6 hours and 21 minutes in the year 2004, and finally to 7 hours and 38 minutes per day in the year 2009. In this context, the "amount of time spent" with computers "in a typical day" has increased from 27 minutes a day in 1999 to 1 hour and 2 minutes a day in 2004, and 1 hour and 29 minutes a day in 2009. Video-game time was recorded separately but roughly paralleled what young people did with their computers (26 minutes in 1999, 49 minutes in 2004, and 1 hour and 13 minutes in 2009). Music/audio content (often downloaded from computer systems) took up three times as much as computer time in the year 1999 but declined to one-and- a-half times as much as computer time by the year 2009.

In all of these senses, the sociocultural context of the younger American normally includes communication technologies as a matter of routine, which implicitly conveys the message that scientific innovation has enhanced and created a better life for typical Americans. Accordingly, while it is true that a major "rhetoric of science" (e.g., Gross, 1996, and Prelli, 1989) has dominated American life, the power of this rhetoric stems not from an understanding and use of scientific methods and procedures, but from the vast array of communication technologies produced by the scientific and technology American industries and the fact that the communication devices of these industries are now in the typical American home.

The Future Internet Driver

The evolution of the Internet has certainly been shaped by the founders of the digital revolution, the various technologies that contribute to the information age, and the historical forces that gave rise to the digital revolution among a wide array of the historical technological innovations. Additionally, the nature of the Internet is also influenced by how we use it today. In this regard, some groups within the American culture use the Internet more than others, and these specialized and repetitive uses can strongly affect the identity of the Internet. In this regard, these potential sociocultural groupings now dominate the use of the Internet in the United States. Each of these three groups is age specific (influenced by events of their time) and possesses a unique governing core set of values, patterns of behavior, and specific actions. These grouping are more readily identified as generations, and we employ the works of Strauss and Howe (1991) for our analysis of generations and Internet use here.

In *Generations: The History of America's Future 1584 to 2069*, William Strauss and Neil Howe (1991) proposed a scheme for distinguishing some 18 generations—in terms of continuity, age location in history, social role, cycles, and examples—that constitute the sociocultural evolution of the United States. Since the mid-2000s, the Pew Internet & American Life Project has repeatedly confirmed the statistical uniqueness of each generation in terms of media access, uses, and functions (see Jones & Fox, for example). The last three of these generations included the Baby Boomers, Thirteenth Generation, and Millennials. Each of these generations has been characterized by unique social experiences and personalities that influence how members of these respective generations view the world and influence the roles they perform within society at given moments in their life span. Table 1.2 details the unique traits that Howe and Strauss (2000, pp. 43–44) attributed to the Millennials as a generation.

Table 1.2. Seven Distinguishing Traits of the Millennial Generation.

"This Millennial persona has seven distinguishing traits. Into the Oh-Oh decade, America can expect to see more evidence that Millennials are:

—*Special.* From precious-baby movies of the early '80s to the effusive rhetoric surrounding the high school Class of 2000, older generations have inculcated in Millennials the sense that they are, collectively, vital to the nation and to their parents' sense of purpose.

—*Sheltered.* Starting with the early-'80s child-abuse frenzy, continuing through the explosion of kid safety rules and devices, and now continuing with a post-Columbine lockdown of public schools, Millennials are the focus of the most sweeping youth safety movement in American history.

—*Confident.* With high levels of truth and optimism—and newly felt connections to parents and future—Millennial teens are beginning to equate good news for themselves with good news for their country. They often boast about their generation's power and potential.

—*Team-oriented.* From Barney and soccer to school uniforms and a new classroom emphasis on group learning, Millennials are developing strong team instincts and tight peer bonds.

—*Achieving.* With accountability and higher school standards rising to the very top of America's political agenda, Millennials are on track to become the best-educated and best-behaved adults in the nation's history.

—*Pressured.* Pushed to study hard, avoid personal risks, and take full advantage of the collective opportunities adults are offering them, Millennials feel a 'trophy kid' pressure to excel.

—*Conventional.* Taking pride in their improving behavior and more comfortable with their parents' values than any other generation in living memory, Millennials support convention—the idea that social rules can help."

The specific argument we make here is that the Millennials are particularly intrigued by the Internet, are constantly expanding its uses, avoid policies that might potentially hamper the Internet (e.g., Brown, June 20, 2013), and are indeed a global sociocultural grouping that has transcended national boundaries.

By way of background, it is convenient to examine how Strauss and Howe characterized each of these generations in a way that enables us to compare these generations. Table 1.3 summarizes some of the distinguishing features of each generation as well as illustrating how each generation can be understood.

Table 1.3. Comparison of Baby Boomer, Thirteenth, and Millennial Generations.*

Characteristics	Baby Boomers	Thirteenth Generation	Millennial Generation
Birth Years	1943 through 1960	1961 through 1981	1982 through 2002 or 2003
Total Number	79 million with 69 million still alive	93 million with 79 million still alive	76 million
Representative Members	Oliver North, Janis Joplin, Joe Namath, Angela Davis, Steve Martin, Donald Trump, David Letterman, and Oprah Winfrey	Michael J. Fox, Eddie Murphy, Jon Bon Jovi, Tom Cruise, Michael Jordon, Whitney Houston, Mike Tyson, Gary Coleman, and Jennifer Capriati	Jessica McClure
Psychological Attributes	Inner-Driven and Judgmental	Pragmatism and Cynicism	Protected, Civic Virtue, and Happy
Perspective on Work	Career	Job	Teamwork, achievement, modesty, and good conduct
Communication	Diplomatic	Blunt	Respectful
Approval	Seek validation	Indifferent	Collaboration
Policies and Procedures	Protective	Mistrustful	Ruler followers
Work Ethic	Driven	Balanced	Technology and future-oriented

Note. See: William Strauss and Neil Howe, *Generations: History of America's Future 1584 to 2069* (William Morrow and Company, 1991); Neil Howe and Bill Strauss, *13th Gen: Abort, Retry, Ignore, Fail?* (Vintage Books, 1993); William Strauss and Neil Howe, *The Fourth Turning: An American Prophecy* (Broadway Books, 1997); and Neil Howe and William Strauss, *Millennials Rising: The Next Great Generation* (Vintage Books, 2000).

We should initially note that Millennials—as a generation—use technology, and specifically computers and the Internet (89% to 93% for 18-to-24-year-olds and 12-to-17-year-olds), more than any other group, and they have since they were first included in the studies completed by the Pew Internet & American Life

Project (see, e.g., Jones & Fox, January 28, 2009, p. 2). As Jones and Fox concluded, "young people dominate the online population" (p. 2) and in general "are the most likely. . .to use the Internet" (p. 3).

Future Focus: Using Strauss and Howe's system for classifying generations, what generation are you in: "Baby Boomers," "Thirteenth Generation," or "Millennials"? As an Internet user, how has your use of the Internet changed? Are you shifting generations? What is the most powerful factor in your environment fostering this change? What is the most powerful factor in your environment hampering this change? With which friend or group of friends do you talk about computer use most? How is this friend or group of friends affecting your computer use and choices?

As we examine Table 1.3, of all of the rich characteristics we might explore about these three generations, we are particularly intrigued by Strauss and Howe's observation that the work ethic of the Millennials is strongly influenced by technology and the future it might create. Indeed, this attribute is decisively underscored by Spanish telecom company Telefónica, which, in partnership with the Financial Times, commissioned 12,171 online quantitative interviews of Millennials, aged 18–30, across 27 countries in six regions, including North America, Latin America, Western Europe, Central and Eastern Europe, Asia, and Middle East and Africa. Penn Schoen Berland conducted the 190-question survey from January 11 to February 4, 2013. Millennials from Argentina, Australia, Brazil, Canada, Chile, China, Colombia, Czech Republic, Egypt, France, Germany, India, Israel, Italy, Japan, Kingdom of Saudi Arabia, Korea, Mexico, Peru, Poland, Russia, South Africa, Spain, Turkey, United Kingdom, United States, and Venezuela were surveyed. Country sample sizes represented in the global number are weighted by the percent of the population in each country with access to the Internet. The global margin of error is +/-.9%.

The findings as reported by Telefónica are particularly relevant to our discussion of the role of Millennials in terms of the Internet. It should initially be noted that the Millennials in this survey were more likely to own a technological device (82% owned a smartphone, 80% owned a laptop, 66% owned a desktop, and 46% owned a tablet). Specifically, this generation said they were "very optimistic about their future" (61%), "think optimistically about their country's future" (79%), and "believe they can make a global difference" (74%). Additionally, for this global generation, 55% of the Millennials surveyed believe that the

"Internet, including social media is the best resource for" "developing a news story or crisis," "providing credible news coverage," and for "entertainment." Of course, we should note—as Strauss and Howe already did—that Millennials are just beginning to emerge as adults and will not really reach "midlife" (ages 44 to 65) and elderhood (ages 65 and older) for at least a decade, a time when we expect the values and orientations of the Millennials will dramatically shape and control not only public policies and the dominant lifestyle of Americans but also virtually all of the major countries of the world.

In our view, we are still only at the inception of the digital revolution. We fully expect that the Internet and digital communication will only increase in use and will dominate how people throughout the world communicate and understand. We are left with the prophetic power of a claim made by Chesebro (2000, p. 8) some 15 years ago: "The Internet is the single most pervasive, involving and global communication system ever created by human beings, with a host of untapped and unknown political, economic, and sociocultural implications."

With these notions regarding the power of the Internet in mind, we close this chapter by considering formal definitions for many of the terms we use throughout this volume.

Central Digital and Internet Definitions

Essential notions behind digital communication technologies do require attention. At the same time, it should be noted that in everyday parlance, digital communication technologies are simply defined by example. Digital or interactive technology can also be defined by example, and they include any or all of the following: The Internet, CD-ROMs, video games, home video and DVD, interactive TV, cell telephone, and/or ATM or automated teller machine.[1]

All of these examples are unified by the fact that they process and convey information with the same format. In this regard, we like the directness of the *North Carolina Echo* (Exploring Cultural Heritage Online) when they defined *digitization* as "the conversion from printed paper, film, or other media formats to an electronic format where the object is represented as either black and white dots, color or grayscale pixels, or 1s and 0s." While precise for some, such a definition also contains elements of potential confusion regarding what this process actually is and how black and white dots or the numbers 1 and 0 can represent all of the content of other media.

Structural Features of Interactive Technology

When we define the nature of interactive technologies here, we are examining computer mediated communication, CD-ROMs, computer games, and so forth. All of the technologies we examine in this unit emphasize the interactivity between individuals and the technologies they are using. In this conception, the individual is no longer a passive entity, and the introduction of interactive or computer technologies virtually requires that the individual be viewed and understood as more active and self-determining than ever.

In terms of its scope, interactive technologies include the Internet, websites, computer multimedia, computer games, CD-ROMs, DVDs, computer graphics, and virtual reality.

As a point of departure, Manovich (2001) aptly argued:

> This new revolution is arguably more profound than the previous ones, and we are just beginning to register its initial effects. Indeed, the introduction of the printing press affected only one stage of cultural communication—the distribution of media. Similarly, the introduction of photography affected only one type of cultural communication—still images. In contrast, the computer media revolution affects all stages of communication, including acquisition, manipulation, storage, and distribution; it also affects all types of media—texts, still images, moving images, sound, and spatial constructions. (p. 19)

As the individual moves through each of these MUDS (multi-user domains or websites), different virtual environments are encountered. And, because no two of these searches are likely to be the same, the foundations for knowledge will vary for each person. In these ways, knowledge is a personal construction, designed to suit individual needs.

How do interactive or computer technologies shape information? Some six structural features can be isolated.

One key structural feature is provided by Manovich (2001, pp. 19–46), who gave us the first of six structural features of interactive technology:

1. **Numeric Representations**—all programs transmit information in a digital form. All communication is becoming digital. How interactive we make it once we get it into a digital form is up to us—virtually all communicative messages (from oral to music to printed to telecommunications) can probably be formatted in a digital code. Beyond using numeric representations, Manovich also suggested that (in descending order of importance and significance) interactive technologies also employ modularity, automation, variability, and transcoding.

The second through sixth structural features of interactive technology are drawn from Sheizaf Rafaeli in a dialogue held in the online *Journal of Computer-Mediated Communication* (http://www.ascusc.org/jcmc/vol1/issue4/rafaeli.html).

2. **Multimedia**: The Internet combines all of the other sensory channels in one Place. What is unique is that it is integrated into one place.

3. **Hypertextuality**: All data bases can be linked to each other.

4. **Packet Switching**: Packet switching can be most easily defined by comparing it to circuit switching. Circuit switching is the opposite of packet switching. By way of example, telephone lines employ circuit switching. On a telephone circuit, a message being sent to you moves along a line, you receive it immediately—when it's sent—and you receive it with all of the intonation and pauses in the original message. Computers don't use circuit switching; computers employ packet switching. The messages are grouped, and they are sent as space becomes available and in terms of whatever priorities govern the use of the line.

5. **Synchronicity**: There are multiple forms of feedback. Some feedback systems send messages and only after the initial message has been received and internally recognized is a message sent that indicates the original message was sent (asynchronistic feedback). Other feedback systems send and receive messages at the same time. These are synchronistic feedback systems. Functioning as a classical synchronistic system, our face-to-face verbal interactions to people are constantly sending both verbal and nonverbal messages at the same time. Computer systems differ from human face-to-face verbal and nonverbal interactions. Functioning as an asynchronistic feedback system, a computer message is sent to a human or another computer; once it is received, the message is encoded and formally received; then a response can be sent back.

6. **Interactivity**: Rafaeli maintained that, "By interactivity we mean the extent to which communication reflects back on itself, feeds on and responds to the past" (p. 8).

The Internet and Other Traditional Media

In a noteworthy recent essay, Cai (2004) asked if our various media can be substituted for each other. What substitutes for computer interaction? What substitutions are made for what computers are doing with other media? The computer is a unique medium because it combines in a unique way all of the other attributes of other media. When people were deprived of the computer they didn't turn to the TV, they felt a void…there was simply nothing available that could do what a computer interaction did.

More generally, Cai's (2004) analysis suggests that a computer interaction may indeed generate a new kind of experience that cannot be equaled by another communication. This new kind of experience may begin to show us new things, reveal new information and phenomena that we never felt or understood before. In other words, a computer interaction can reveal a new kind of reality for us.

In one sense, Cai's (2004) analysis is about the nature of reality. Cai implicitly suggested that something cognitively happens when you use a computer that doesn't happen when you use any other technology. Your brain goes different. Cognitively you know certain things through interactive technologies that you don't with other media.

Virtual Reality

Over two decades ago, the *Journal of Communication* provided a special issue dealing with "Virtual Reality: A Communication Perspective." One of the essays, by Jonathan Steuer, was outstanding. Steuer (1992) provided four definitions and observations that detailed the notion of computer interactive realities that are worthy of consideration here.

Presence

> The key to defining virtual reality in terms of human experience rather than technological hardware is the concept of *presence*. Presence can be thought of as the experience of one's physical environment; it refers not to one's surroundings as they exist in the physical world, but to the perception of those surroundings as mediated by both automatic and controlled mental processes. (p. 75)

> *"Presence is defined as the sense of being in an environment"* (p. 75).

Telepresence

> When perception is mediated by a communication technology, one is forced to perceive *two* separate environments simultaneously: the physical environment in which one is actually present and the environment presented via the medium. The term *telepresence* can be used to describe the precedence of the latter experience in favor of the former; that is, telepresence is the extent to which one feels present in the mediated environment, rather than in the immediate physical environment. (pp. 75–76)

"Telepresence is defined as the experience of presence in an environment by means of a communication medium. In other words, *presence* refers to the *natural* perception of an environment, and *telepresence* refers to the *mediated* perception of an environment" (p. 76).

Virtual reality

"A virtual reality is defined as a real or simulated environment in which a perceiver experiences telepresence" (pp. 76–77).

Two major dimensions across which communication technologies vary are discussed here as determinants of telepresence. The first, *vividness*, refers to the ability of a technology to produce a sensorially rich mediated environment. The second, *interactivity*, refers to the degree to which users of a medium can influence the form or content of the mediated environment. (p. 80)

Vividness is affected by the "breadth" and "depth" of the "richness" of the mediated environment. (p. 81)

Interactivity

Interactivity is *"the extent to which users can participate in modifying the form and content of a mediated environment in real time"* (p. 84). Interactivity is specifically affected by "speed," "range," and "mapping" (p. 81).

Conclusion and Predictions

Since 1995 the Internet has become an increasingly growing force within the United States and the world. We have more and more devices to access the Internet, and globally some 77% of those in developed companies, 31% in developing countries, and overall roughly one-third of the world population regularly accesses the Internet. In the United States, when the age of users is considered, some 93% of those aged 12 to 17 regularly access the Internet and 70% of those between 50 and 65 regularly access the Internet. And, use of the Internet is profound. People can and do use the Internet for some 55 different functions, with these functions virtually paralleling what people do in everyday life. While a host of factors affect Internet use, certainly what people want and decide they need affects how they use the Internet. In our view, the major cultural driver for Internet use and expansion are the members of the Millennial generation (born between 1982 and 2002/3), and this generation can appropriately be viewed as functioning at a global level. We have concluded that the Internet will continue to expand and grow in use and in terms of popular access. All of these computer systems are unified by their use of digital systems, a method of converting and transmitting information through interactive systems. These digital systems ultimately create a new reality, a *virtual reality*, which functions for many as a reality system equal in power to everyday face-to-face reality. In all, we find the creation and extension of these digital systems to be both challenging and amazingly exciting in terms of what they are doing to human beings and what human beings are doing to them.

Key Terms

Analog Communication

Digital Communication

Desktop Computers

Devices:

E-readers

Laptops

Mobile Phones

Tablets

Touch Pads

Web-Enabled Television

Generations as Communication Systems:

Baby Boomers

Thirteenth Generation

Millennials

Packet Switching

Synchronicity

Telepresence

Virtual Reality

Chapter Exercises

Current Issues: Are computers replacing face-to-face communication? Is it easier and more efficient—and perhaps even more comfortable—to leave people emails or text messages rather than chatting to them face-to-face? Are computers slowly controlling how we interact with other people? Every year, do we use computers and smart phones to carry out more and more of the activities we used to do with friends in face-to-face situations?

Debate It: Divide the class into two groups to debate the resolution, "Resolved that the shifts in society toward a computer-dominated social system is more fair, equal, and effective."

Ethical Consideration: Some people are more comfortable interacting through computerized media while others find face-to-face communication more efficient and effective. When or under what conditions should societal systems require computer competence? What happens to people who cannot effectively master computer requirement?

Everyday Impact: When are face-to-face modes of interaction required? Are online relationships viable and are such activities as sexual interactions ever appropriate for digital systems?

Future Focus: Will people increase or decrease their use of digital systems? Are work environments increasingly relying on digital systems? Is it true that online dating systems have more success matching people in terms of dating, marriage, and interpersonal satisfaction? If so, will people use these systems more and more in the future?

Endnote

1. For some, ATM = *Asynchronous Transfer Mode* in voice, video, and data modes are combined into very large-scale integration and data transfer networks.

2

Traditional Media and the Rise of Digital Communication Technologies

The way we access, deliver, and store our media is constantly changing. Our vinyl records became cassette tapes before becoming CDs. Now, all music is being converted to the current standard, the **MP3**. The CD format slowly fades from memory as records and cassette tapes are sold to hobbyists at flea markets. The mainstream relies on iTunes, Amazon, and Torrents for MP3s. Our VHS tapes were rendered obsolete by the quality of the DVD before the DVD in turn, and its successor, the Blu-Ray, became bulky compared to the **MPEG-4**, a digital file viewable on our computers and our phones. Our bookstores are closing as books become downloadable from virtual shops and viewable on compact touch screens as **e-books**. Our news is now available online for free instead of a dollar fifty from the nearest street vendor. Our magazines and other mail have also transcended the streets, no longer tethered to post offices and postage stamps. E-mail circumvents our mail carriers and delivers birthday cards and love letters direct and instantaneously.

Today, our media are digital, streamed from and stored in virtual clouds and on hard drives. Wireless Internet and Internet-ready devices are increasingly ubiquitous. Consumers are almost always connected or seconds away from being connected. Connected consumers can seamlessly engage and surf vast collections of entertainment and information through software built into their mobile phones, touch tablets, e-readers, and laptops. Digital eyeglasses are on the horizon that promise to enable users to wear the Internet 24 hours a day. As a result, consumers no longer need to enter via swinging doors with clanging bells and theft detectors to access, rent, or purchase this or that media experience.

Motivated by the Internet and other complementary digital innovations, the media industry and traditional media experiences are in a rapid state of conversion and **convergence**. Industries built on traditional media systems and analogue content find they must adapt to compete or disband in defeat. Digital communication technologies and their peddlers are becoming the new default while the traditional analog experience slowly becomes more and more niche.

> ### *Are you a member of "The Analogue Revolution"?*
>
> *Across various fringe markets, the analogue experience has been repurposed as sales gimmick, exploited by entrepreneurs and hobbyists and targeted at niche audiences. (For example, see http://www.lomography.com/) Why might a person or group be interested in revisiting the analogue experience? Have you ever or would you ever deliberately buy analogue instead of digital?*

Preview: The Information Technology Revolution

In the early 1980s, before the commercialization of the World Wide Web and the smartphone, researchers were already discussing the effects of computers on traditional communication mediums. In 1985, Forester declared "the dramatic reduction in the cost of computing power made possible by microelectronics" has spawned a "'convergence' of electronics, computing, and telecommunications and the unleashing of a tidal wave of technological innovation which scientists are now calling the 'Information Technology Revolution'" (p. xiii). The revolution described by Forester has continued to reshape our media and communication landscapes. In this chapter, we will discuss in detail five communication media (books, newspapers, music, film, and television) that have been dramatically impacted by this snowballing Information Technology Revolution and its endless wave of technological innovations. First, we present an overview of each medium, focusing on how that medium and industry are responding to, and being reimagined by, digital communication technologies. No industry has remained untouched or unaltered. Second, we discuss the effects and new relationships with media that have resulted from our adoption of digital communication systems. A new media environment is forming. Third, and finally, we propose potential future outcomes for current observable trends. Ultimately, this chapter will answer the following questions:

Questions Previewing This Chapter

1. *What did popular media look like before the Internet?*
2. *How are particular communication mediums adapting and responding to the challenges and promises of digital technologies and online systems?*
3. *How do traditional media industries and rising online media industries interact, collaborate, and compete?*
4. *How have digital communication technologies created a new media environment?*

Books and the Book Industry

Since the 19th century, books have been printed and print books have prospered as a mass medium and major component of our communication system. As many have argued,

> It is difficult to imagine what Western culture could be, or indeed the culture of any major civilization anywhere in the world today, without the wealth or resources that are preserved, disseminated and handed down from one generation to the next in the form of the book. (Thompson, 2005, p. 1)

Books offer readers entertainment and knowledge. For many scholars, being educated has commonly been defined in terms of how many books a person has read (see Grimes, 2008).

Despite the perceived value of books, the book publishing industry has never achieved dominance among the mass media (Potter, 2008). As a medium, the book industry has responded to competition from other mass media by being niche-oriented. As noted by Potter (2008), "Many publishers sell books to only one niche, such as college-level science texts, library reference books, religious books, children's mystery novels, and so on" (p. 373). This trend has continued today. Many believe this trend is essential to the survival and repackaging of traditional models. In *The Long Tail: Why the Future of Business Is Selling Less of More, Wired* editor-in-chief Chris Anderson (2006) promoted the exploitation of the niche as the future of all successful business in the digital age.

A Steady Decrease in Print Book Consumption

Across 2012, the number of people reading printed books fell from 72% of the population ages 16 and older to 67% (Raineie & Duggan, 2012). Today, with the embrace of the Internet, reading books appears to be increasingly unpopular, aside from the occasional anomalous spike in mainstream consumption habits. These spikes often occur in conjunction with the unveiling and popular reception of a film or television adaptation (i.e., Rowling's *Harry Potter* series, Martin's *A Song of Fire and Ice* series, and Collins's *The Hunger Games* series). Despite the pervasive coverage and embrace of particular collections and authors, the Association of American Publishers continue to report a steady decline in sales of U.S. print books since 2005 (Keller, 2011). Of further note, while popular fiction is widely consumed, *The Wall Street Journal* found contemporary readers routinely quit, drop, or abandon works of nonfiction before completion (Alter, 2012).

Digital natives are increasingly dismissing and/or avoiding nonfiction and print books, especially as they grow older. The National Assessment of Educational Progress reported, "the percentage of kids who said they read for fun almost every day dropped from 43 percent in fourth grade to 19 percent in eighth grade" (Rich, 2007, p. A15). In *To Read or Not to Read: A Question of National Consequence*, the National Endowment for the Arts (2007) found young adults reading fewer books in general, reporting, "less than one-third of 13-year-olds are daily readers," "of those 15 to 18-years-old, only 34% read a book for at least 5 minutes a day and only 26% of this group read a book for at least 30 minutes a day," and "nearly half of all Americans ages 18 to 24 read no books for pleasure."

The decrease in print book consumption can and does have social and economic consequences. First, reading less means reading less well. The National Endowment for the Arts found, "poor reading skills correlate heavily with lack of employment, lower wages, and fewer opportunities for advancement" (2007, p. 5). Second, libraries were once architectural shrines sporting massive book collections. Now, computer screens increasingly dominate libraries with books buried in the peripherals or transferred online. The undergraduate libraries at the University of Texas in Austin and the University of Southern California in Los Angeles have replaced all books with computer screens (see Canon, 2006, and Eberhart, 2008). Third, the book industry, predominantly sustained through print services, is struggling to adapt to stay in business and stay relevant. The trajectories of both Borders and Barnes & Noble, two previously dominant booksellers, show us that traditional models and expectations must be reconfigured to stay afloat in the digital age. In the next section, we take a closer look at how the book industry is being impacted by digital innovations and the declining interest in printed books.

Alternative View: Shifting Habits and Collateral Damage

Younger audiences may avoid reading print books front cover to back cover, but undoubtedly read content daily in sips and gulps when surfing online, hopping from hyperlink to hyperlink, post to e-book. Indeed, e-book readers are growing in number. Across 2012, those who read e-books "increased from 16% of all Americans ages 16 and older to 23%" (Rainie & Duggan, 2012). These percentages represent a relatively small chunk of Americans (less than a quarter), but the steady rise is significant.

What we read and how we read it will continue to impact our cognitive performance and relationships with information. Frequent Internet browsers master essential skills in searching, assessing quality, and synthesizing information. Communication experts, however, voice fears that a sole diet of "quick-fix information nuggets" may lead to an inability to undertake "deep, critical analysis of issues and challenging information" (Pew Research Center, 2012, p. 4).

Transitions: Imploding Bookstores and Exploding e-Book Sales

Transitions often create collateral damage. The fit survive through adaptation. Due to its inability to adapt and America's declining interest in reading print books, Borders, a major bookstore chain founded in 1971, filed for bankruptcy in 2011. The Borders's bankruptcy erased over 200 bookstores and 20,000 jobs. Retail buildings that once offered consumers a colorful plethora of print books turned dark and ominous. Barnes and Noble, a Borders contemporary, felt the boom in business and the increasing pressure to respond to what Borders could not—new media and the digital revolution. At the dawn of 2012, *The New York Times* declared Barnes & Noble the last player standing between "traditional book publishers and oblivion" (Bosman, 2012, p. BU1).

Touch tablets and e-books

For Barnes & Noble, and other threatened print book-based industries, hope and profits were found in developing touch tablets, e-Readers, and **e-Books**. Touch tablets and e-Readers are mobile, compact, touch-activated and wireless-ready. Users of these devices can seamlessly connect to content and retail stores hosted online. Each store offers users the ability to download a wide range of constantly updated content for their devices, including electronic books, or e-Books.

When an e-book is purchased, it is stored on the device for as long as the user desires. Multiple e-books and other media can be simultaneously stored on the device.

Since their release, the use of e-readers, and e-book sales, are steadily increasing. Sales more than doubled from $4 million in the second quarter of 2006 to $8.1 million in the second quarter of 2007 (Wayner, 2007). By March of 2011, e-book sales reached $69 million, increasing 146% from the year before (Miller & Bosman, 2011). Use of e-readers increased from 8% to 15% from 2010 to 2011. Use and ownership of one of these devices nearly doubled again from December 2011 to January 2012 (see Table 2.01). At the onset of 2012, The Pew Research Center reported that 29% of adults in the United States now owned an e-Reader or touch tablet (Rainie, Zichuhr, Purcell, Madden, & Brenner, 2012). Furthermore, in the first quarter of 2012, *The Wall Street Journal* announced e-book sales had surpassed print book sales, generating $282 million in sales versus $230 million for print (Alter, 2012). As one market for publishers was declining, a new market with new competition, new expectations for profits, and new ideas was rising.

A Case Study in Conversion: New and Old Battle for an Emerging Marketplace

At the start of the e-Book explosion, Amazon quickly became one of the forerunners in sales of e-readers and e-books following the late 2007 release of its Kindle e-reader. Following the release and popularity of the Kindle, Amazon developed and released several more e-reader devices, including the Amazon Kindle Fire. These devices, combined with their well-established online presence, earned Amazon notable financial successes in the e-book marketplace across 2008 and into 2011. In 2008, Amazon reported that 125,000 book titles were available as e-books with e-book sales accounting for over 6% of total book sales (Bezos, 2008, p. R3). After 2008, e-books steadily increased in accessibility and popularity. Within three years, Amazon's catalog of e-books increased from 125,000to over 950,000 with e-book sales accounting for approximately 14% of all general consumer fiction and nonfiction booksales (Miller & Bosman, 2011). Amazon, however, was not the only capitalist interested in reinventing the book and turning e-books and e-readers into profits.

Alternative View: Market Monopoly

Despite its successes enlivening the book market, many industry insiders and publishers believe Amazon is devaluing the book, establishing a dangerous monopoly in the publishing industry, and pushing unprecedented low prices. For instance, Amazon insists on a $9.99 price for new e-book releases despite protests from publishers and authors. Several prominent publishers have repeatedly and unsuccessfully combated unchecked discounted prices on Amazon. In February 2010, Macmillan, a publishing group founded in the 1800s and active in dozens of countries, saw Amazon remove the buy buttons from all their active titles on the site after disputes over e-book prices (Reid, 2010). Publishers Weekly reported that when content developers work with Amazon, it's not a dialogue, but more a dictation of terms (Deahl & Milliot, 2011).

Barnes & Noble: The Nook

In 2009, with the release of the Nook, a floundering Barnes & Noble entered the e-reader and e-book market that Amazon had been creating and controlling for two years. Barnes & Noble believed e-readers and e-books could help save its declining stock and show a continuing relevancy in a changing marketplace. The last player standing between traditional bookstores and oblivion was not going to fade without a fight. The Nook provided moderate financial success, but, as of 2012, has not been able to re-establish Barnes & Noble as a dominant bookseller. In 2012, Barnes & Noble began discussing the economic and social viability of splitting the Nook and e-book business from its traditional bookstore business (Trachtenberg & Peers, 2012). Barnes & Noble's interest in condensing its business and foci reflect an ongoing trend cited by both Potter (2008) and Anderson (2006): security can be found in being niche-oriented.

Apple: The iPad and iBooks

In competition with both Amazon's Kindle and Barnes & Noble's Nook are Apple's iPad devices. The first iPad was released in 2010. The iPad can act as an e-reader, like the Kindle or Nook, but also comes equipped with a webcam, an Internet browser, and other seemingly endless applications. Like the Amazon tablet, the Apple tablet devices offer seamless access to an established online store overflowing with content.

Championing convergence and textbook reform, in early 2012, Apple released an iPad application ("APP") called **iBooks 2**. Apple did not wish to simply enter the e-book marketplace, Apple wanted to reinvent that marketplace with its own unique signature. iBooks 2 promised to change the education landscape with low-cost and low-weight textbooks. With iBooks 2, Apple planned to compete with sales at Barnes & Noble and Amazon by offering a new type book-reading experience that implemented video graphics, note-taking capabilities, and advanced interactivity. In a telling collaboration of old and new, Apple recruited several major established print publishers to help facilitate the growth of iBooks 2. These publishers included Pearson, McGraw-Hill, and Houghton Mifflin Harcourt.

Self-publication

Accompanying this announcement, Apple also released **iBooks Author**. iBooks Author enables users to create and self-publish their own unique content for use and purchase through iBooks 2. On the shoulders of the Internet, self-publishing is increasing in popularity for both new and established authors. One such budding author, Darcie Chan, released her debut novel, *The Mill River Recluse*, digitally after 100 literary agents rejected the text. The novel became a best-seller without ever being printed or handled by a publishing team (Alter, 2011). Through iBooks Author, Apple hoped to attract budding entrepreneurs like Chan, authors interested in circumnavigating traditional publishing systems and releasing content directly to the public.

The emerging trends of self-publication, self-distribution, and shopping online reflect an ongoing shift from mass communication systems to individually controlled digital communication systems. Our digital communication systems are more interactive and person-centered than previous models. Where the detached media industry once dictated consumers' options, today's consumers and entrepreneurs have more control over what content is produced, consumed, and distributed.

Newspapers, Magazines, and the Post Office

Newspapers first emerged as a purely persuasive vehicle for political parties in the early 18th century. These newspapers had a limited and local reach with a primary interest in influencing political opinions. Across the 19th century and into the 20th century, newspapers evolved into a mass medium covering topics both inside and outside politics. The newspaper became a primary and dominant source for both information and entertainment. By the early 1900s, 93% of all U.S. households subscribed to newspapers (Potter, 2008, p. 375).

With the commercial adoption of radio and television in the 1930s and 1940s, newspapers became less and less dominant as a mass medium in America. In *The Postwar Decline of American Newspapers, 1945–1965*, Davies (2006) tracked the beginnings of a steady and ongoing decline in newspaper readership and circulation. The communicative power attributed to the medium declined with its dwindling popularity.

Internet Prompts Sharp(er) Declines

Despite a steady decline in readership and circulation, in 1993, according to the Pew Research Center for the People & the Press (2006), 53% of Americans were still getting their news from the newspaper. However, as the Internet became more accessible for more Americans across the late 1990s and into the 2000s, newspaper readership declined significantly. By 2006, only 34% of Americans were reading print newspapers daily, with younger audiences bypassing the newspaper completely (The Pew Research Center for the People & the Press). In 2004, only 23% of young adults aged 18–49 were daily newspaper readers (Keeter, 2006). Accordingly, Vass (2006) projected, as "a purely text-based medium [newspapers] will struggle in the future" with newspaper readers "heading into the cemetery" and non-readers "just getting out of college."

Americans, especially younger Americans, increasingly circumvent newspapers in favor of 24-hour news coverage on television and online. News is readily available for free online and endlessly updated. Consumers can instantly access information without ever needing to leave their homes to fetch the paper. The migration of readers to the Internet has also motivated advertisers to migrate online, further endangering an already limping industry.

Siphoning Newspaper Revenue: The Rise of Online Advertising

According to the Newspaper Association of America, in 2000, "classifieds contributed $19.6 billion or 40% of total newspaper revenues of $48.7 billion" (Sass, 2011). With the rise of the Internet, however, websites like Craigslist became a popular alternative for advertisers. The Internet offered advertisers cheap, endless space for promotions with a wider potential scope than the newspaper. Indeed, Craigslist offered free online classifieds for 570 cities worldwide and low-priced classifieds for 18 major American cities, from New York to San Francisco. Additionally, any user from any location could access and search classifieds from any region. Classifieds began appearing online in droves. During this period, the newspaper industry saw revenues decline as online profits for organizations like Craigslist swelled.

In its early stages in 2004, Craigslist was estimated to have brought in $9 million in revenues (Stone, 2009). By 2009, online revenues for Craigslist were projected to reach $100 million, a 23% increase from 2008 (Stone, 2009). In 2011, *The Wall Street Journal* reported the Internet as the second largest ad-supported medium (Steel, 2011, p. B1). Meanwhile, in contrast, newspaper advertising in the United States steadily declined from 2004 to 2010 (see Sass, 2011). From 2005–2007, total classified revenues dropped 18% from $17.3 billion to $14.2 billion (Sass, 2011). From 2007–2009, classified advertising in newspapers in the United States had dropped 56% to $6.2 billion (Sass, 2011).

Seeking Unique Visitors: Newspapers Shift to the Internet

The Internet has siphoned, and continues to siphon, advertisers and readers from print newspapers. As a result, the Internet grows while the newspaper industry continues to shrink at an accelerated rate. *The New York Times*, *The Wall Street Journal*, and *The Washington Post* have all been forced to downsize and cut employees (see Peters, 2012; Plambeck, 2010). Newspapers have responded to competition by becoming smaller, flashier, thinner, more local, and flooded with more advertisements (Perez-Pena, June 9, 2008, p. C1). In addition, and most important to their ongoing relevance and survival, newspapers have gone online. More and more newspapers are creating a digital presence to replace or compliment their print editions. Some newspapers, such as *The Wall Street Journal* and *The New York Times*, also offer online subscription packages or downloadable editions for purchase on touch tablets.

Newspapers can also be seen increasingly quantifying their success based not on circulation numbers, but on how many unique visitors their website can garner monthly (see Peters, 2012). Indeed, the Internet has become a critical factor determining today's successful newspapers. In June 2010, comScore reported, "more than 123 million Americans visited newspaper sites in May, representing 57 percent of the total U.S. Internet audience." *The New York Times* was the top visited newspaper website across 2010 averaging over 30 million unique visitors per month. In May 2011, however, comScore reported that *The*

Huffington Post, "the Internet Newspaper," a six-year-old newspaper born online and only available online, topped *The New York Times* with 35.6 million unique visitors (compared to 33.6 for NYT.com). *The Huffington Post* has proved itself a dominant player in the competition for newsreaders in the digital age. Here, new media outshined a long-established king of print news distribution.

Magazine Plight Echoes the Newspaper Narrative

Mirroring a similar trajectory to newspapers, contemporary print magazines in America are witnessing unimpeded declines in sales. For over three years, magazine subscriptions and newsstand sales have been steadily dropping (Clifford, 2010; Haughney, 2012; Vega, 2012). In mid-2012, magazine newsstand sales were down nearly 10% (Haughney, 2012). *Vanity Fair*, *The New Yorker*, *People*, *Us Weekly*, *Vogue*, *Cosmopolitan*, and many other popular magazines all showed various steep declines in sales (Haughney, 2012). Celebrity magazines, specifically, were cited as suffering because of similar content readily available online for free.

Some magazine publishers sought renewed profits through embracing popular electronic platforms to showcase and distribute their wares. *The Atlantic*, a popular monthly magazine over a century old, reported their digital advertising revenue exceeded print revenue for the first time in October 2011 (Peters, 2011). The digital rise of *The Atlantic* may be a unique case, but from 2011 to 2012, digital sales for all magazines increased notably with distributors selling more than five million digital copies (Haughney, 2012). Yet, insiders accustomed to the previous profit standard set by the print model express concerns over current digital sales (Haughney, 2012). Sales expectations set by print for many decades, however, might need to be revaluated based on the new consumption habits and expectations of digital consumers.

> ### *Alternative View: Digital Promotes Less Physical Waste*
>
> *A shift toward providing content primarily through electronic platforms would indirectly respond to environmental concerns about forest exploitation and material waste. Publishing directly to an electronic platform reduces the need to harvest physical materials to print physical copies every release cycle (monthly, quarterly, annually, etc.) for every single magazine. Some electronic magazines, such as the* Food Network Magazine, *even allow users to create digital shopping lists for recipes found inside their magazine with a tap of the screen, further eliminating a need for paper.*

The U.S. Post Office: An Endangered Icon

For decades, the U.S. Post Office was a trusted and enduring national monument. Since 2006, however, the volume of physical mail being sent and received in America has declined some 25% (Venteicher, 2013). As magazines, letters, bills and other personal and professional communicative transactions gravitate towards the purely digital, the necessity, appeal, and profits for terrestrial mail diminish. Further, while profits fade, losses rise. In 2012, the U.S. Postal Service reported $15.9 billion in losses (Venteicher, 2013; Nixon, 2013). By 2013, *The Los Angeles Times* declared, the "Postal Service is on its last legs, with little help in sight" (Venteicher).

Today, struggling with a dissolving relevancy and mainstream appeal, desperation has descended on the terrestrial mail industry. In a controversial attempt to reduce its mounting losses, the Post Office declared its intent to end Saturday delivery services (Nixon, 2013). The reduction would save the industry $2 billion per year. U.S. Congress, however, rejected the proposal, leaving the Post Office floundering in multiplying losses while consumers continue delivering paperless correspondences instantaneously via email.

Music, Radio, and the Music Industry

Music has been, and continues to be, an enduring component of all cultures across the decades. As observed by Zimmer (2010), "Anthropologists have yet to discover a single human culture without its own form of music" (p. 28). The ways in which we receive and access this music today, however, has changed. As witnessed in and across other popular communication industries, digital technologies have significantly transformed the ways in which persons access, receive, and distribute music and other traditional radio content.

Music began as a live spectacle accessed on street corners, in pubs, and in theaters. In the late 1800s, the gramophone, phonograph, and record player shifted music to a shared home entertainment experience. With records, consumers could listen to music as a physical commodity at any time without attending a live scheduled event. In the early 1900s, radio broadcasting further tapped this market, distributing both music and other entertainment directly to the home through radio receivers. By 1936, as reported by Potter (2008), "there was an average of one receiver per household" and "people were spending more time with this medium than any other medium" (pp. 379–380). The radio surpassed the newspaper as the most popular mass medium of the time. With the commercialization of the television, the radio saw notable declines in revenue, listeners, and its dominance as a mass medium (Potter, 2008). Television surpassed the radio.

Repackaging Music for Convenience and Mobility

Despite the allure of television, music and talk radio continued to have appeal across the 1900s and into the present. The music industry adapted its products to meet the needs of this changing market. Bulky records were converted to pocket-sized cassette tapes. Unlike the record, these tapes were small and could be played at home or in the car. Cassette tapes made music more mobile and gave the consumer more control over the music they could play.

Music quickly became digitized when the CD (compact disc) launched in the mid-1980s. By the late 1980s, hundreds of millions of CDs were being manufactured for popular consumption. The CD had the same mobility of the cassette tape with a higher sound quality and more space for more songs. Tapes and records had to be flipped in order to hear an entire collection of songs, but the CD enabled the listener to conveniently play entire albums without interruption.

Across the '80s, '90s, and into 2000, music was predominantly a physical commodity heard through CDs and CD players (Potter, 2008). The music industry was booming. However, the axioms of mobility and convenience created efficient and appealing, but economically destructive, new technologies and habits. The industry was redefined and deflated by the rise of online sharing and the **MP3**. MP3 is a mobile and versatile digital file storable on recordable CDs, computers, and an endless variety of other electronic devices (or MP3 players).

The Crippler: File Sharing Online

Between 1999 and 2005, CD sales fell approximately 25%, according to the Recording Industry Association of America (Williams, 2006). When Napster, an early online file sharing service, exploded in popularity in the late 1990s, it offered users endless and convenient access to downloadable MP3s. Younger consumers began favoring downloading music online for free over trekking to the record store to unload their allowances.

Many posit that Napster destroyed the market, introducing the notion that music should and could be free online (Mitchell, 2008; Williams, 2006). After the record labels helped shut down Napster, a trend was already established. A new market emerged that wanted music to be available online in digital form. When Napster went dark, users found hundreds of open-armed alternatives eager to meet the growing demand for downloadable MP3s.

Music packaged as an MP3 was more versatile than music on a record, a cassette, or a CD. The consumer was given more control over their music listening experience. Additionally, consumers were given more options than ever before. National and international content was no longer tethered to certain physical regions. Persons could pick and choose specific individual songs, remixes, and artists to download, creating personal and eclectic libraries to store on computers and mobile, battery-operated MP3 players. By 2009, 45% of American adults reported owning an MP3 player, up from 11% in 2005 (Rainie et al., 2012).

The popularity of downloading and sharing MP3s online through services such as Napster, eMule, and Limewire, prompted record stores to close in waves. By late 2003, approximately 900 independent record stores had closed nationwide. In 2004 and 2006, national music store operators Musicland and Tower Records filed for bankruptcy protection (Williams, 2006; Kafka, 2006). Online downloading, and their struggle to properly respond to it, were cited as causes for their financial misfortunes.

Seeking Profits Online: Peddling MP3s, Online Radio, Podcasts, and Streaming Services

From 2008 to 2009, as reported online by Nielsen's SoundScan service, CD sales plummeted another 20% in the United States (Smith, 2009). Record stores responded by becoming increasingly niche, targeting their products at aging audiences still accustomed to perusing offline shops and purchasing content as a physical commodity. The music industry was forced to respond to profit losses by embracing the world that was crippling them, the Internet and online markets. Online sharing had created new expectations and new business models were needed to meet these expectations. The Internet offered the music industry and entrepreneurs a new vehicle for tapping profits from modern consumers and digital natives. Three approaches the music industry has taken to adapt to the Internet are outlined here.

Alternative Views: Redefining Distribution

Not all artists have suffered under online file sharing. Some artists, comedians, and musicians, are creating novel distribution models online to discourage illegal downloading of their content. These artists exploit the Internet to independently publish and market downloadable content on their personal websites and social networks. In October 2007, the platinum-selling and multi-Grammy-award-winning band Radiohead helped popularize this trend when they announced their album, In Rainbows, *would be released only online through their personal website. The cost of the download was up to each individual downloader who could pay anywhere between nothing and $200. Approximately 1.2 million customers downloaded and paid for the album in its first three days available on-line (Buskirk, 2007). Another 240 thousand people reportedly downloaded the album for free through illegal file sharing services (Buskirk, 2007). Comedian Louis C. K. has successfully released content to millions of buyers in a similar fashion. Will this "pay-what-you-want" model grow in popularity in the future? Is it a viable model for artists, both established and aspiring? Would you be more willing to pay for content if it was distributed with this model?*

First, more and more music has been made available for download online through iTunes, Amazon.com, and bands' own websites. Users can download entire albums or purchase individual songs of their choosing as MP3s. Some artists have started offering content for free online in hopes of building buzz and attracting fans to attend live performances. Live performances have become a heavily relied on commodity while offering free MP3s online has become an audition for new and old fans (Caramanica, 2011). Second, music is increasingly streamed online, creating online radio supported by advertisements and/or subscriptions. Websites such as Rhapsody, Pandora, and Spotify offer any consumer with an Internet connection instant access to massive music libraries.

Third, exploiting wireless Internet and the versatility of the MP3, radio and talk-based programming are also more readily available online. SiriusXM Radio now offers its 20 million some subscribers an application for the Android, iPhone, iPad, BlackBerry, and other mobile electronic devices. Listeners to SiriusXM are no longer tethered to a radio receiver built into their car or living room.

Podcasts, invented in late 2004, also offer consumers more mobility and control over their individual listening experiences. Podcasts are seemingly infinite in their foci and breadth, with Podcasts devoted to discussing entertainment, technology, education, politics, and much more. Podcasts are created and published by established celebrities, rising amateurs, and loquacious neighbors.

Podcasts can be released daily, weekly, and monthly and all are downloadable as MP3s online. The consumer accesses these podcasts at any time of their choosing after their release. Today's listener can listen when they can instead of listening when they must. As a result, finite terrestrial radio is becoming outdated as Internet radio and podcasts are more widely embraced.

Film and the Film Industry

The cinema has been a popular feature of the American landscape for over 100 years. For Americans in the 20th century, watching films started as a novelty, but quickly became a habit (Potter, 2008). When feature films first became a widely embraced mode of communication in 1912, they could only be accessed and viewed inside theater houses. Fueled by the invention of Hollywood and the Hollywood star machine, audiences, however, became increasingly captivated by this medium and ventured out to these theaters in their droves. As Potter (2008) observed, "In 1927, an average of 60 million people attended motion pictures every week. By 1929, this figure was more than 110 million people" (p. 378). By the 1940s, there were already some 20,000 theaters with over 11.1 million movie seats available (Potter, 2008).

When the television became commercially available in the late 1940s, audiences were given the alternative of staying at home instead of venturing to theaters to see visual entertainment. People responded by staying home more. With the television, movies started to become a home entertainment event monitored by the individual. Here, a trend slowly emerged that has continued to balloon today with rippling effects on the film industry.

The Gravitation Toward On-Demand Entertainment

The economic and social effects of this home viewing trend became more explicit with the advent of videotapes and video rentals. With videotapes and rental stores, consumers were given more options and more control. Staying at home became a more accessible and dynamic alternative. Again, people responded by staying home more, directly affecting the box office and the film industry. Across the 21st century, the preference for consuming at home has only grown in popularity. The rise of DVDs and online **VOD services** such as Netflix has given consumers more methods for viewing current films at home with home theater systems that simulate the theater experience. New releases can be available on VOD services some 45 days after their release in theaters (Barnes, 2010). As a result, in part, of these growing alternatives, in mid-2011, Hollywood reported its fourth consecutive summer of declining attendance at the theater, with sales down 4% for the year (Barnes, 2010, p. B1).

Alternative Views: Hosts Respond to Declines with Personalization

A 2012 study commissioned by the Hollywood Reporter *found over half of young people ages 18–34 believed the use of social media while watching television and/or film enhances the experience. Various industry entrepreneurs are adapting in response to these types of evolving trends. In addition to offering VOD services, groups are offering consumers more control and more social alternatives to the once static and impersonal theater experience. The Alamo Drafthouses in Texas, Colorado, and Virginia offer alcohol during film screenings and promote a more active and community-flavored experience. Also, online, we see services such as Tugg.com emerging (in Beta form as of 2012). The website partnered with AMC Theaters, Cinemark Theaters, Rave Cinemas, Regal Cinemas, and other major theater chains to launch a service allowing consumers to request, promote, and vote for individual films to be screened at their local theaters. Here, viewers are given more ownership and control over their theater experience, as they become event coordinators and promoters for a screening of their choosing (or design).*

In 2012, instead of going to the theater or purchasing a movie, most consumers rent movies on DVD for a dollar from their local Redbox or stream new releases online instantly through VOD services. Redbox, a kiosk the size of a bulky refrigerator, often located outside grocery stores or restaurants, boasts over 29,000 locations nationwide with over 18 million discs in circulation. In 2012, Netflix.com had over 23 million active subscribers streaming content through the Internet on computers, gaming consoles, mobile phones, and television sets. Across 2011 and 2012, Netflix began to expand internationally, offering services in Latin America and Europe. Netflix's multiple successful expansions hint at a global demand for VOD services.

Home Viewing, Hollywood, and the Cable Company

Today, audiences in North America show a preference for consuming films digitally, at home, on the move, and on their own terms. In 2006, the Pew Research Center reported, "Three-quarters [71%] of all adults say they would prefer watching movies at home rather than in a theater" (Taylor, Funk, & Craighill). As a result, we see a continuing decline in box office income in the

United States with an increased reliance on foreign markets to earn profits (Holson, 2006). In fact, Hollywood has started producing more films with global narratives intent on enticing international audiences. As *The New York Times* observed, "the Oscar for best picture, for three consecutive years, has gone to films—'Slumdog Millionaire,' 'The Hurt Locker,' and 'The King's Speech'— that used globe-spanning financial networks to create stories aimed at global audiences" (Kulish & Cieply, 2011, p. A1).

In a telling sign for the direction of home entertainment, in February 2012 the Weinstein Co. circumvented cable providers completely, signing an exclusive deal with Netflix.com to stream their films (Fritz & James, 2012). In 2012, the multiple Oscar winning *The Artist* would not be seen on cable, but would be available to stream on demand on Netflix.com.

Comcast, America's largest cable provider, responded to online competition by launching a streaming service, XfinityStreampix, in early 2012. During this same period, Verizon teamed with Redbox to announce the launch of an online streaming service to compete directly with Netflix and others. Each of these blossoming streaming services offer the film industry a variety of new outlets for making profits off their content.

Current digital technologies appear to complement and fuel an ongoing preference for home entertainment. The film industry has adapted by exploring global markets and VOD services to stay active financially. However, digital technologies have introduced a second insatiable threat to the traditional Hollywood system: **piracy**. The accessibility, widespread dissemination, and consumption of pirated entertainment online are some of the largest threats endangering the entertainment industry today.

Responding to Piracy: The MPAA, the Pirate Bay, and the Kim Dotcom Takedown

Free pirated copies of films and television programs abound. Typing a few select keywords into a favorite web browser can reaffirm the abundance of free content online. In 2009, the Motion Picture Association of America (MPAA) reported, "illegal downloads and streams are now responsible for about 40 percent of the revenue the industry loses annually as a result of piracy" (Stelter & Stone, 2009, p. A19). Indeed, HBO's *Game of Thrones* saw an estimated 4.3 million illegal downloads, costing the company some $170.7 million in revenue in 2012 (Friedman, 2013). In response, the MPAA and the entertainment industry have employed several strategies to combat piracy online. Three of these strategies are discussed here.

First, the entertainment industry has made more content available online for streaming. Film studios, such as the Weinstein Co., are signing deals with VOD services to make films and television episodes digitally available on demand quickly after their release. Consumers have grown accustomed to accessing content instantaneously through the Internet. If content is not available legally through Hulu.com, Netflix.com, or other alternatives, then users are more apt to access content illegally through filesharing websites such as The Pirate Bay.

Second, the entertainment industry and the U.S. government have started publicly persecuting specific individuals maintaining popular filesharing websites offering massive libraries of pirated content. Instead of persecuting the guests, they are persecuting the hosts (for now). In 2009, four persons, each a cofounder of the filesharing website The Pirate Bay, were convicted of violating copyright law, sentenced to one year in prison, and ordered to pay $3.6 million in damages (Pfanner, 2009). The conviction was a significant victory for the music and movie industries. Piracy online, however, continued.

Beheading Megaupload and the Online Backlash

In 2012, the movie and music industries responded to online piracy again, targeting another high-profile filesharing website, Megaupload.com. For years, Megaupload allowed users to trade, stream, and download content, often pirated. Seven key operators connected with the website were charged with five counts of copyright infringement and conspiracy (Sisario, 2012). Together, these operators have been accused of causing "$500 million in damages to copyright owners and of making $175 million through selling ads and premium subscriptions" (Sisario, 2012, p. B1).

These prosecutions were widely covered internationally coordinated events. As a result, Peter Sunde, cofounder of The Pirate Bay, and Kim Dotcom, founder of Megaupload, became Internet celebrities and martyrs. Governments were accused of interfering with an open web and citizens of the Internet used cyber warfare to voice their disapproval. After Kim Dotcom's arrest in early 2012, the "hacktivist" group Anonymous responded by launching a series of cyber attacks against the Department of Justice, the FBI, and Universal Music Group (Williams, 2012).

In early 2012, The Pirate Bay co-founder Peter Sunde borrowed the pages of Wired.com to speak directly to filesharing supporters and the entertainment industry. In the article, titled "It's Evolution, Stupid," Sunde lashed out against traditional media, saying,

> The problem here is that we're allowing this dying industry to dictate the terms of our democracy. We allow them to dictate new laws (ACTA, SOPA, PIPA, IPRED, IPRED2, TPP, TRIPS, to name a few recent ones) that forbid evolution. If you don't give up before you're sued, they corrupt the legal system....The Internet is being controlled by a corrupt industry. We need to stop it. (Sunde, 2012)

Sunde's remarks and the actions of Anonymous are signs of an ideological and economic war brewing online. Battle lines have been drawn between digital "entrepreneurs," modern consumers, and the entertainment industry.

New Economy Versus Old Economy: The Rise and Fall of SOPA

The third strategy employed by the MPAA to combat piracy, as mentioned by Sunde above, is the development of antipiracy legislation, legislation such as SOPA (the Stop Online Privacy Act). With the rise and (potentially temporary) fall of SOPA, we see a unique collision between the Internet and the government that should be explored in deeper detail.

SOPA was developed to combat piracy online in national and foreign markets. In 2012, when SOPA was preparing to be passed by Congress, the Internet exploded with coverage and backlash against the proposed bill. SOPA put forward legislation that would allow governments to shut down websites associated with copyright infringement without trial or a traditional court hearing. Journalists, activists, and entrepreneurs found the proposed legislation by Congress to be loosely defined. Many popular websites believed the bill, if passed, threatened the open web and would permit online censorship (see Barrett, 2012). In late 2011, Internet giants Google, Facebook, Twitter, Yahoo, eBay, LinkedIn, AOL, and others used *The New York Times* to publish an open letter titled, "We stand together to protect innovation." The letter spoke out directly against SOPA, highlighting the flaws implicit across the bill's stated goals (see Figure P.01). (An online copy of the full letter can be found at http://boingboing.net/2011/11/16/internet-giants-place-full-pag.html.)

Across late 2011 and into 2012, an anti-SOPA campaign grew online culminating on January 18, 2012, when numerous influential and high-traffic websites, from Reddit.com to Google.com to Wikipedia.org, voiced their disapproval with website blackouts and/or SOPA banners on their homepages. *The New York Times* interpreted the event as the new economy rising against the old, saying, on Wednesday this formidable old guard was forced to make way for the new as Web powerhouses backed by Internet activists rallied opposition to the legislation through Internet blackouts and cascading criticism, sending an unmistakable message to lawmakers grappling with new media issues: Don't mess with the Internet. (Weisman, 2012, p. A1)

The anti-SOPA campaign, funded and led by online entrepreneurs and activists, succeeded in delaying the bill. As of 2012, Congress has not passed SOPA (or PIPA). The "new economy" and new media scored a momentary victory. The online response to SOPA has shown the MPAA and the U.S. government that they must design a more sensitive approach to combating piracy nationally and internationally if they hope to prevent ongoing declines in profits.

Television and the Television Industry

Television captivated the United States in the late 1940s and quickly began siphoning audiences from all other available popular mass mediums. Across the 20th century and into the 21st century, television ruled as the dominant medium for entertainment and information. Television initially offered only a handful of broadcast channels, but the rise of cable television created hundreds of channels and a wide range of eclectic content for the choosey viewer. Accordingly, it became increasingly uncommon for a single broadcast to reach a mass audience (see Chesebro, 2003). Prime time network programming gradually reached less people while viewership spread itself across hundreds of channels.

Despite becoming increasingly niche-oriented, viewing television programming remained a massively popular leisure activity for the majority of Americans. As reported by the Bureau of Labor Statistics, on an average day in 2004, men and women watched approximately three or more hours of television. Watching television was the leading leisure activity for those surveyed. Socializing was the second most common leisure activity accounting for under an hour each day. Persons were spending more time with their television sets than talking with their friends and family. In 2010, according to Nielsen, more than half of all American homes had three or more television sets (Mindlin, 2010, p. B2).

Enter the Internet: Television's Supremacy Wobbles as TV Viewers Go Online

As seen throughout this chapter, no industry has evaded the transformative influence and allure of the Internet. In mid-2010, the pew Research Center uncovered an increasing decline in the perceived necessity of the television set. Taylor and Wang (2010) announced, "after occupying center stage in the American household for much of the 20th century, two of the grand old luminaries of consumer technology—the television set and the landline telephone—are suffering from a sharp decline in public perception that they are necessities of life" (p. 1). In 2010, 42% of Americans considered the television set to be a necessity, down from 52% in 2009 and 64% in 2006 (Taylor & Wang, 2010, p. 1).

Research suggests other devices, such as computers and smartphones, are becoming substitutes and replacements for the television set (Mindlin, 2010). In 2011, the Nielsen Company released research echoing this trend, reporting a 2.2% decline in the percentage of American households now owning television sets (Stelter, 2011). The decline was attributed to an increase in poverty and changing preferences in younger viewers. Young people, according to Nielsen, are watching more television online without a need for traditional television sets and cable television providers (Stelter, 2011). In April 2012, "some 181 million Internet users watched nearly 37 billion online content videos" (Sullivan, 2012).

In 2002, 82% of Americans claimed television as their main source for national and international news, according to the Pew Research Center (2011). As of 2011, 66% of Americans claim television as their main source for national and international news while 41% now get their news online. For people under 30, "the internet has surpassed television as the main source of national and international news" (Pew Research Center, January 4, 2011). A shift is occurring.

Television Industry Mirroring Newspaper Industry

Nielsen echoes the findings of the Pew Research Center, noting a steady decline in viewership of broadcast news since 2002 (see Schechner, 2009; Arango & Carter, 2009). In 1980, newscasts amassed 50 million some viewers, but, by 2009, the same newscasts amass only 22 million some viewers (Stelter & Carter, 2010, p. B5). The once boastful big three broadcast networks are a thin shadow of their dominant pre-1980s figure. Accordingly, to survive financially, television news staffs are being cut in scores (Stelter & Carter, 2010). *The New York Times* ominously compared the struggles of the television news industry to those faced previously by the newspaper industry (Stelter & Carter, 2010).

The migration of advertising revenue and viewership online has threatened the security and longstanding dominance of another industry. However, despite these well-documented declines and struggles, advertisers continue to perceive television as a dominant and widely accessed medium. As of 2011, the $59.7 billion U.S. television-advertising market continues to grow steadily (Steel, 2011, p. B1). Major marketers continue to funnel money into television advertisements, citing the persuasive value of video storytelling and the ability to efficiently reach a mass audience. Online, it can be difficult for advertisers to reach a mass audience, needing to plant adverts across thousands of appropriate websites (Steel, 2011, p. B2). Yet, with the explosion of available television online, the industry, like those before it, will be forced to find effective models to profit from Internet programming.

Television Programming Goes Online

Established broadcast and cable companies and rising online companies are responding aggressively to the growing market for on-demand television programming. NBC, CBS, ABC, MTV, HBO, Comcast, Hulu, Google, Apple, Amazon, Netflix, YouTube, and more are all trading, streaming, and delivering entertainment through the Internet. Each has developed, or is developing, specific systems or deals for delivering content online. Television programs will air live through traditional cable providers before becoming available online mere hours after broadcast. Today, television programming is less dependent on cable boxes and can now be accessed on mobile phones, computers, tablets, and gaming consoles.

In 2008, Netflix partnered with Roku to release the Roku DVP, a small black box that connects to the television set, enabling consumers to stream television programs and films from Netflix to their living rooms. Additionally, Netflix offers applications for the Xbox, the PlayStation 3, the Nintendo Wii, the iPad, and smartphones. In response to the competition, other companies are offering more applications and services for wireless streaming and beaming on a variety of devices. In late 2011, HBO and Time Warner Cable announced they would begin streaming programming over the Internet (Schechner, 2011). Also, Google and Apple have teased releasing their own self-contained advanced television sets capable of streaming programming and more (Bloomberg News, 2011; Miller, 2011). Nielsen has already begun measuring how users view entertainment online and through iPhones, iPads, and other computers.

Alternative Views

While television networks such as Fox and NBC produce online content to complement and drive viewers to their offline content, other entities circumvent the television and cable boxes completely. These groups view the Internet as a profitable and widely accessible distribution platform. In 2012, Netflix released its first exclusive original series, Lilyhammer. *The entire show was made available online only through its streaming service. Award-winning director David Fincher and actor Kevin Spacey have also signed deals to release original content through Netflix. Additionally, other online streaming services, such as Hulu and YouTube, have begun announcing and releasing original programming (*Battle-ground, Spoilers, Up to Speed, *and* We Got Next*) for online streaming only. How will original programming released online and made available online only affect advertising and the terrestrial television industry? Will Netflix and other online streaming services usurp Comcast and other cable companies as the dominant means for viewing television programming?*

In 2012, Hulu, AOL, Microsoft Advertising, Digitas, Yahoo!, and Google/YouTube hosted the first "Digital Content New Fronts" (DCNF). Here, online programmers showcased digital content for advertisers and industry representatives. Similar to the "upfronts" held for traditional television programming, this event was an opportunity for brands to invest in upcoming and incoming online content. The website for DCNF declared, "consumers are shifting. Things don't 'go' digital anymore. They start digital."

The future landscape of television online and its effects on advertising and television offline are only just beginning to take shape. However, despite the changes in packaging and distribution, television content continues to be a widely demanded and dominant source of entertainment.

Digital Communication Technologies as the New Media Environment

Digital communication technologies have created a new media environment. As covered throughout this chapter, the methods in which we interact with, distribute, and receive information and entertainment have changed dramatically on the shoulders of the Internet. In this section, we summarize six key features and phenomena related to the new media environment.

Individual Centered

Technologies have shifted from a focus on mass-communication systems to individual-centered digital communication systems. These new digital-communication systems are more interactive, person centered, and individually controlled. Where the media industry once dictated consumer's options, audiences today have more control over what messages are produced, consumed, and distributed.

Untethered

Media is increasingly untethered to a predetermined time and place. With the lightweight and low-power demand of digital technologies, media have become endlessly mobile (Chesebro, 2007). The individual chooses when and where to access their personally selected media experience. The individual can seamlessly enter, stop, pause, exit, and re-enter media on their own terms in a space of their choosing.

Interactive

Interactivity abounds online as a controlling norm and is an essential feature of popular online sites such as Wikipedia and Myspace (Chesebro, 2007). Popular media, such as television, music, and books, can now be seen tapping the popularity of interactivity to increase profits and captivate audiences enamored with interactivity. From voting on Facebook for a favorite contestant to following artists on Twitter, media producers are promoting online components that compliment offline content. These online interactions are intended to extend a particular media experience and/or enhance a viewer's relationship with that media experience. Interactivity can produce a more intimate link between the audience and the content and/or artists.

Our offline media is becoming increasingly connected to online communities and social media sites. Today, audiences gather online (as opposed to offline) to discuss their favorite shows with others, often while they are watching them live. According to Bluefin Labs, there were some 13 million "social media comments" during the CBS broadcast of the 2012 Grammy Awards (Stross, 2012, p. BU5). Twitter, Facebook, and blogs have become the new water cooler where audiences discuss and react to this and that media experience.

Concurrent Media Exposure

Many individuals are responding to the abundance of available options by consuming multiple media simultaneously. When an individual surfs the web on their laptop or tablet while watching Netflix on their television set, this is an instance of **concurrent media exposure**. With a host of available content and limited time in the day, audiences do not restrict themselves to passively consuming one medium at one single moment (Chesebro, 2007). The Ball State University *Middletown Media Studies* reported over 30% of media use involves "exposure to two or more media" at one time (2005, pp. 16–21). Nielsen echoed these findings in 2012, reporting, "40 percent of smartphone and tablet owners in the United States self-reported that they used their devices daily while watching TV" (Stross, 2012, p. BU5). The rise in popularity of **concurrent media exposure** should encourage us to rethink our approach to understanding the effects of one particular media experience on the individual as the individual is often exposing themselves to multiple media experiences simultaneously.

Multifaceted Simultaneous Consumption

The rise in popularity of interactivity and concurrent media exposure has also introduced a rise in **multitasking**. Multitasking involves the simultaneous pursuit of multiple activities (Chesebro, 2007). Today's youth report multitasking most of the time, if not all the time (Foehr, 2006, p. 6). In *Media Multitasking Among American Youth: Prevalence, Predictors and Pairings,* one 17-year-old encapsulated today's multitasking youth, saying, "I multitask every single second I am online. At this moment, I am watching TV, checking my email every two minutes, reading a newsgroup about who shot JFK, burning some music to a CD and writing this message" (Foehr, 2006, p. 1). With the spreading of consumer attention across multiple media simultaneously, both scholars and marketers will need to conduct further research into the effects and cognitive limitations of this practice.

Convergence

The Information Technology Revolution has inspired and encouraged **convergence**. Here, convergence should be understood as the blurring of boundaries between media channels (Chesebro, 2007). For example, one can look at the smartphone. On one single device, the smartphone, users can access television, photography, comics, music, geographical positioning systems, and much more.

Multiple media experiences have been concentrated into one single, dynamic package. The device enables users to navigate multiple and different types of media experiences.

Conclusion and Future Possibilities

Media producers and packaging and distribution agencies are in an unavoidable and exciting flux. The Internet has redesigned, and continues to redesign, how we access and deliver news and entertainment. In this final section, we end this chapter by offering a collection of predictions concerning the future of the media industry.

The media industry will continue to be transformed by digital technologies and online systems. What has been covered in this chapter is only the beginning of the bottom of the slope. Every day, new innovations and ideas emerge demanding adaptation and imposing immediate change, from augmented reality to touch- and voice-activated interfaces. The predictions we offer now are based on current observable trends and should be received not as finite, but as open-ended musings. Here, we offer four predictions for the incoming media environment.

Mass Digitization

First, we believe physical media will become artifacts and antiques, harvested by sentimental collectors for museums of nostalgia while libraries and consumers convert all their collections into digital archives accessible online through clouds (or online lockers). One man, Brewster Kahle, has already begun amassing and preserving hundreds of thousands of physical texts and films in one single location, the Physical Archive of the Internet Archive. Kahle's archive is intended as a physical copy of its digital counterpart, the Internet Archive. As of 2012, the Internet Archive has digitized two million books and Kahle's Physical Archive of the Internet Archive is on course to preserve 10 million texts (Streitfeld, 2012).

Television Online

Second, television programming will no longer be tethered to cable boxes and narrow time slots, but will be released online daily and/or weekly for streaming and download (similar to today's distribution model for Podcasts and Netflix's distribution model for their original content). As younger generations accustomed to going online for content grow older and the Internet becomes the dominant host for content, the cable cord will be cut. Cable companies will become reliant on selling Internet subscriptions to stay afloat and relevant. Internet subscriptions, however, will no longer be flat rates, but will be based on month-to-month usage figures. Users downloading or streaming mass amounts of content will be charged accordingly.

Infinite Global Libraries

Third, our access to vast amounts of national and international media and information will increase exponentially across the next decade. Our ability to access seemingly infinite amounts of global entertainment and news will further compartmentalize consumers momentarily into niche pockets. Social networks and online recommendation systems will be relied on to create cohesion and common experiences in relationships. Large audiences circling around a single program at a single moment in time will become increasingly rare and reserved for live events.

Additionally, more options will affect the consumer emotionally and cognitively. In "The Paradox of Choice," Schwartz (2005) argued our growing abundance of options contributes to consumer paralysis and heightened levels of consumer anxiety, dissatisfaction, and regret.[1] Consumers and producers of media will develop advanced, and more invasive, methods for qualifying and synthesizing information, as well as aiding, targeting, and directing choice in an expanding sea of limitless information.

Social Entertainment

Fourth and finally, the media viewing experience will continue to become more interactive and social. The boundaries between consumer, producer, and distributer will continue to appear to dissolve. Artists (or their publicists) will interact directly more and more with their fans via online social portals. The silent and passive consumption of entertainment and news will become increasingly niche as new generations show preferences for multitasking, immediate commentary, sharing, and interactivity. In this type of environment, at the neighborhood cinema, instead of being asked to shut off phones and tablets, patrons will be asked to simply keep their screens dim while in use.

Classroom Exercises

Debate It:

Resolved: Online shopping is superior to offline shopping.

Resolved: Offline shopping is superior to online shopping.

Divide students into two opposing groups and debate the advantages and disadvantages of shopping online and offline. Consider the impact of online shopping on offline industries, such as bookstores, music shops, and other retail stores. Also, consider the impact of offline shopping on online industries, such as Amazon. How do environment, interface, and other variables impact the consumer experience and influence preference?

Current Issues: For this exercise, instruct students to search print media and digital media for news stories and events discussing the same current event. Preferably, select a story or event related to the media industry. How do these stories compare across platforms and across sources? What are the differences and similarities between print media and digital media? Who reported on the event first? Is the first report superior to other reports? Which source is most trustworthy?

Everyday Impact: As discussed above, music dissemination online has transformed music distribution and expected profits. For this exercise, encourage students to explore how their own daily behaviors and choices might be impacting the music industry. Do students download music illegally online? Why do persons download music for free online? How does this trend impact aspiring musicians and established musicians? What should the music industry do to respond to this enduring trend? Should all music be free?

Current Issues: The rise of multitasking amongst youth can be a controversial and dividing phenomenon. The impact of multitasking on personal and professional contexts is a ripe research field. For this exercise, instruct students to search for articles discussing the impact of multitasking on professional or personal life. How is the popularity of multitasking affecting the modern workplace? How does multitasking impact social environments?

Key Terms

MPEG-4	Piracy
MP3	SOPA
E-book	Concurrent Media Exposure
Digitize	Multitasking
Podcast	Convergence
iBooks	VOD Services

Endnote

1. For a summary of Schwartz's arguments, see his TED talk here: http://www.ted.com/talks/barry_schwartz_on_the_paradox_of_choice.html

3

The Past—The Development and Evolution of Digital Technologies

As we noted at the outset of Chapter 1, we remain convinced that the Internet is "now the largest and most pervasive global communication system ever created in the history of human beings, affecting more people in more ways than any other human invention."[1] Yet, it is difficult to identify all of those people and the related technologies that contributed to the development of the Internet.

We are not computer scientists nor are we professional historians. We are communication specialists focusing on the Internet as an extremely profound communication invention and technology. While some may be intrigued by the technical features of the Internet, such matters affect us only if they affect how people transmit and receive messages through the Internet. Certainly, we cannot remain indifferent to these "technical features," for far too often they affect and determine how human beings can communicate with the Internet. Likewise, there is no question that different historical eras are more or less receptive to using technologies to communicate with other human beings. Indeed, we find such analyses particularly fascinating and intriguing.

Yet, here we want to discuss the development and evolution of the Internet as an emerging and developing communication device and technology. While some of the individual inventors affected the technical effectiveness of the Internet, we specifically focus on those who directly impacted how, when, where, and even to whom and why people send and receive messages through the Internet. Indeed, from our perspective, the innovators we discuss here focused on digital technologies while also making a major contribution to communication theory, and we believe these contributions are increasingly being recognized and becoming part of the history of the discipline of communication.

Questions Previewing This Chapter

I. Who Were the Founders of the Digital Revolution? Seven Who Made a Difference:
1. Alan Mathison Turing.
2. Norbert Wiener.
3. Harold Adams Innis.
4. Jack S. Kilby and Robert Norton Noyce.
5. Herbert Marshall McLuhan.
6. Steven Jobs.
7. Tim Berners-Lee.
II. What Seven Technologies Contributed to the Information Age?
1. Printing Press.
2. Calculating Machine.
3. Analytical Machine.
4. Technologies Generated by Electricity.
5. Electromechanical Calculating Machines.
6. ENIAC or the Electronic Numerical Integrator and Calculator.
7. Single Silicon Integrated Circuit.

Founders of the Digital Revolution: Seven[2] Who Made a Difference

At the outset, we must formally admit our fascination with those founders and innovators of the digital revolution. We find these individuals particularly intriguing especially in terms of their ability to conceive of what did not exist and imagine how these omissions might be filled. The creativity and resourcefulness of the early innovators and founders we survey here are particularly worthy of attention. In many cases, we find what they accomplished to be not just novel and functional but also poetic in the sense that the creation even goes beyond what is feasible and creates an imaginative awareness of the possible futures of technologies. In this regard, we begin our survey of the seven computer innovators with Alan Turing.

Alan Mathison Turing (June 23, 1912 to June 7, 1954)[3]

Of all of the early digital innovators, Alan Turing is perhaps one of the most recognized names among the founders of the digital revolution. Because of the personal and actually tragic attributes of Turing (which we shall mention shortly), several popular conceptions volumes about Turing have been published (e.g., Dyson, 2012). Additionally, Turing was the object of study in a 1986–1988 play, *Breaking the Code*, which ran both in London and on Broadway. There was also a 1996 BBC television production (broadcast in the United States by PBS). In all three performances Turing was played by Derek Jacobi. The Broadway production was nominated for three Tony Awards including Best Actor in a Play, Best Featured Actor in a Play, and Best Director of a Play, and for two Drama Desk Awards, for Best Actor and Best Featured Actor. And, some dramatic stories of Turing's life have continued to be portrayed through 2012 (Pease, 2012).

In terms of Turing's immediate contributions to the digital revolution, we appropriately begin by noting that Turing received his PhD from Princeton University in the United States. Officially, during World War II, he worked in the British Foreign Office in the Code and Cypher School; for his efforts in this capacity—helping to break the cryptic messages used by the Germans—he was made an officer of the Order of the British Empire. And, professionally, he was awarded with a Fellow of the Royal Society in 1951.

Briefly stated, Turing is associated with three contributions related to the digital revolution.

First, Turing was one of the original founders of contemporary computer theory. As Pask and Curran (1982, p. 19) reported, Turing established the "fundamental principles of computation" and as Hofstadter (1983, p. 1) argued, Turing was "in large part responsible" for the "concept of computers as we know them, as well as for incisive theorems about the power of computers."

Second, Turing was associated with the development of the contemporary computer architecture and stored computer programs. Carpenter and Doran (1977, p. 269) reported that Turing had a central role in the development of "the first complete design of a stored program computer architecture." Randell (1973, p. 350) argued that Turing's wartime contact with John von Neumann may have "played some part in the development of the practical stored program concept."

Third, Turing was one of the originators of the concept of artificial intelligence. As Evans (1979, p. 24) put it, Turing was "the first person to face up to" the problem of whether or not the ability to "think" is a "unique" feature of human beings. Indeed, as Feigenbaum and McCorduck (1983, p. 155) put it, "The first inkling that a computer might be capable of intelligent behavior came from the brilliant Cambridge logician Alan Turing."

Turing's contributions generally emerged from his applied and practical experiences. Especially at the outset of his career, Turing used various computer systems to help him solve immediate and real problems.

During World War II, Turing's task was to decode and translate and then transmit the Germans' everyday war plans and specific strategies. The Germans had employed a random system to code their military plans, and these coded plans were transmitted by radio to other Germans. Another machine, also containing the Enigma code, was used to decode the transmitted message. Turing created a machine capable of deciphering the Germans' Enigma coding machine. By shifting to an "electronic" technology, his division was "responsible" for "cracking the German cipher code during World War II. Breaking that code was an intellectual feat that is now widely recognized as a key element in the Allies' eventual triumph over the Nazi war machine."

Following World War II, Turing joined the National Physical Laboratory (NPL) where he designed and developed an electronic computer. He later quit the NPL to head the Computing Machine Laboratory where he designed the Ferranti Mark I, which was the first electronic digital computer to be commercially available. He was also a member of the mathematics faculty at Manchester University, which gave him access to one of the world's only computers. In this context, Turing was continually intrigued by the meaning that should be assigned to the potential cognitive power of computer systems. Indeed, he predicted that the processing system of computers would be able to produce an outcome that would be virtually indistinguishable from the output generated by the human thinking process. Towards this end, Turing proposed what has now come to be called the "Turing test." The famous Turing test was developed for determining if and when computers achieved the equivalent of human intelligence. In developing the test, Turing noted that there is "An unwillingness to admit the possibility" that human beings "can have any rivals in intellectual power" (in Meltzer & Michie, 1969, p. 3).

In 1950, Turing actually established a procedure for determining when machine intelligence equaled human intelligence. Employing the basics of an "imitation game," Turing suggested that two rooms would be connected only by teleprinters. In the first room, humans could read the messages from one of two teleprinters, they could type questions into either one of the teleprinters, and they would receive messages back from either of the two teleprinters. In the second room, one teleprinter would be controlled by a human being, and the other teleprinter would be connected to and controlled solely by a computer. Turing (1950) predicted that by the year 2000, the

> average interrogator will not have more than a 70 percent chance of making the right identification [of the computer and human being] after five minutes of questioning. I believe that at the end of the century the use of words and general education opinion will have altered so much that one will be able to speak of machines thinking without expecting to be contradicted. (p. 412)

In all of these ways, in terms of the development of concepts and mechanisms relevant to the digital revolution, Turing was exceptional. Google computer specialist Vint Cert (in "Google doodles Alan Turing's 100th b'day," IBNLive. com, 2009) described Alan Turing as "The man who challenged everyone's thinking" as well as the "father of computing" (in "Who was Alan Turing?" IBNLive. com, 2009). More specifically, Campbell-Kelly (2009, p. 62) placed Turing in the center of the digital revolution when he argued that, "The information age began with the realization that machines could emulate the power of minds." Campbell-Kelly goes on to argue that, "the modern electronic computers that came out of World War II gave rise to the notion of the universal computer—a machine capable of any kind of information processing, even including manipulations of its own programs."

Turing is noteworthy, however, for more than these contributions to the digital revolution. His personal life has been a source of intrigue for many for more than 50 years and his personal life story is part of the drama that makes Turing such a powerful figure in this history of the digital revolution.

One event from 1952 captures Turing's personal issues. As Copeland (in Pease, 2012) tells it, after:

> reporting a petty burglary, Turing found himself being investigated for "acts of gross indecency" after he revealed he had had a male lover in his house. Faced with the prospect of imprisonment, and perhaps the loss of the mathematics post he held at Manchester University which gave him access to one of the world's only computers, Turing accepted the alternative of "chemical castration"—hormone treatment that was supposed to suppress his sexual urges.

While some maintained that this "treatment" ultimately resulted in Turing's suicide in 1954 by cyanide poisoning with an apple (a mime of the fairy tale of Snow White and the Seven Dwarfs), in 2012, Copeland (in Pease, 2012) argued that the "evidence that was presented [at Turing inquest]" would "not today be accepted as sufficient to establish a suicide verdict" and he has argued that the "death may equally probably have been an accident." After reviewing all of the available evidence, Copeland concluded that, "The exact circumstances of Turing's death will probably always be unclear. Perhaps we should just shrug our shoulders, and focus on Turing's life and extraordinary work."

But, the issues involved in Turing's death did not end in 1954. In 2009, the British Prime Minister Gordon Brown (in Whiteman, 2009) issued a "posthumous apology for the 'appalling' treatment of Alan Turing, the British code-breaker who was chemically castrated for being gay."

As Whiteman (2009) detailed the apology,

> In a statement on the British Government Web site, Prime Minister Gordon Brown acknowledged Turing's "outstanding" contribution during World War II. "He truly was one of those individuals we can point to whose unique contribution helped to turn the tide of war," he wrote, adding, "The debt of gratitude he is owed makes it all the more horrifying, therefore, that he was treated so inhumanely."

For over 50 years now, Turing has been recognized as significant. His contributions to the digital revolution are noteworthy in several respects, and the attention that his personal life has received has continued to make Turing intriguing. We are not sure that Turing should be identified and named the "father of computing," but we are confident that his applied use of computers during World War II and his conceptions of artificial intelligence were original and deserve attention even today.

Current Issues: Are scientific inventions and even revolutions affected by the culture in which they emerge? Should we judge the scientific creativity by the type of culture that fosters and encourages scientific creativity? What kind of culture is best suited for the development of scientific inventions?

Norbert Wiener (November 26, 1894 to March 18, 1964)

By virtually any standard, Norbert Wiener was a classical scholar and educator. He received his PhD at Harvard in mathematical logic, and he became a professor of mathematics at MIT. Like Turing, Wiener's major contribution to the digital revolution first emerged during World War II. Wiener was one of the original scientists who worked on ballistics on the first computer, ENIAC (Electronic Numerical Integrator and Calculator, unveiled in 1946 by the Moore School of Engineering at the University of Pennsylvania), at the Aberdeen Proving Grounds in Aberdeen, Maryland. For Wiener, while a practical question related to successful military efforts, he also raised—as we shall see—a profoundly important question defining the meaning of computer-human communication when he asked how human beings could achieve gunfire control.

As a communication specialist intrigued by the issues governing communication theory, Wiener coined the word *cybernetics* and developed the discipline of cybernetics. In his view, cybernetics explored questions of control. Control is ultimately a function of message-sending and feedback. As Wiener (in Conway & Siegelman, 2005, p. ix) put it, "if my control is to be effective I must take cognizance of any messages from him [another person] which may indicate that an order is understood and has been obeyed." Ultimately, cybernetics involved five essential principles:

1. Cybernetics became that branch of the discipline of developing communication systems that enabled human beings to interact with machines. In *The Human Use of Human Beings*, Wiener (1950, p. 16) outlined his major claim:

   It is the thesis of this book that society can only be understood through the study of the messages and the communication facilities which belong to it; and that in the future development of these messages and communication facilities, messages between man and machines, between machines and man, and between machine and machine, are destined to play an ever-increasing part. (p. 16)

2. Communication is conceived as a process and system that decays. Entropy governs communication systems—feedback systems must be enhanced consistently to keep communication systems functional.

3. Coined and integrated a host of key terms in communication that describe and determine how all digital communication technologies function, including *feedback, information, control, input, output, stability, homeostasis, prediction,* and *filtering*.

4. Designed the first human-machine interactions, creating the first bionic arm controlled by the user's own thoughts.

5. Forecast the massive changes in technology to come, altering all areas of society. Ethically and morally, Wiener worried that people would surrender to machines, reducing their own choices, and Wiener "feared for humanity's future."

In sum, Conway and Siegelman (2005) aptly captured the meaning of Weiner and his conceptions for the digital revolution:

> He is the father of the information age. His work has shaped the lives of billions of people. His discoveries have transformed the world's economies and cultures. He was one of the most brilliant minds of the twentieth century, a child prodigy who became a world-class genius and visionary thinker, an absentminded professor whose eccentricities assume mythical proportions, a best-selling author whose name was a household word during America's first heyday of high technology. His footprints are everywhere today, etched in silicon, wandering in cyberspace, and in every corner of daily life. (p. ix)

Harold Adams Innis (November 5, 1894 to November 8, 1952)

With tongue in cheek, some have ironically speculated that Canada has only produced two major communication theorists who contributed to the digital revolution. If true, these two theorists were truly outstanding. One was Harold Innis.

Innis was a professor of political economy at the University of Toronto. His major contributions were published in two volumes at the beginning of the 1950s. The first was *Empire and Communications*, published in 1950, and the second was *The Bias of Communication* in 1951.

These two volumes contain a host of major concepts that an investigation of the history of the digital revolution would find especially relevant.

In the most general of terms, it should be initially noted that Innis approached communication in a unique fashion. Rather than focusing on the content of communication as a central key to its meaning, Innis drew attention to the medium of every message. He reasoned that when messages employed the same medium (such as orality, written/printed, telecommunication systems such as radio and television, or digital) they might be grouped together and the common features of these groups might be extracted and used to define the medium of that mode of communication. Traditionally, communication theorists had focused on the idea expressed in a message, and they slighted the energy that would be devoted to the medium of a communication. In this regard, Innis's focus represented a major shift in how communication itself was to be defined, characterized, and understood.

Within the context of the medium of communication, three of Innis's contributions are outstanding.

The first of these conceptions involves Innis's belief that every medium of communication contains its own use and sense of time, a use and sense of time that is unique to a specific medium and ultimately defines the medium in a major way.

When speaking to others face-to-face, for example, the immediate time when an utterance is made is part of the utterance itself. The utterance must be understood within the context of the time when it was uttered. In this sense, for Innis, oral communication was frequently most effective in a small community for long periods of time. In contrast, the temporal context or time constraints of hand-written notes are extremely different than the time constraints of oral communication. Hand-written notes are affected by the culture in which they are written; in other cultures, a hand-written note might be read in completely different ways. Hand-written notes then could be understood to exist within a single person or even multiple communities. In all, this notion meant that media contain a *time-binding* characteristic. Time was used by Innis as a way of identify a unique feature of each medium.

Second, for Innis, a medium of communication also involved a *space-binding* attribute. Accordingly, we know that radio, television, and newspapers are particularly effective in large communities for short periods of time. Such media are especially understood and interpreted within a commercial context, if not—Innis argued—as a form of imperialism.

Third, Innis provided us with a new vocabulary of basic terms for analyzing communication. He suggested that the analysis of communication must begin by recognizing:

1. Types of media;
2. Each medium functions within a particular temporal system;
3. Each medium functions with a particular spatial context;
4. Media are designed for different kinds of communities of different sizes; and,
5. Each medium is appropriate for and supports a different kind of social system—depending on the medium, the sociocultural outcome will vary along a continuum from ideation to various forms of profit/dominance.

In terms of our specific interest in the digital revolution, Innis certainly must be recognized for his formal statement that all communication systems are socializing systems that create and reinforce a specific and particular type of culture. Within this context, accordingly, every communication system governs and controls time, space, and therefore human perception, apprehension, and understanding. Appropriately, it is possible to read Innis and conclude that Innis is arguing for the notion, viability, and usefulness of a *technological culture*.

In terms of the digital revolution, Innis can be read to imply that digital communication technologies are creating their own cultural arenas with their own sense of time and space that determines human perception, apprehension, and understanding. Given Innis' contributions, we find it more than reasonable to understand how digital communication technologies can be cast as systems creating their own realities—*virtual realities*, which might have the means and power to compete with the everyday reality or realities created by oral, written, and telecommunication systems.

Jack S. Kilby (November 8, 1923 to June 21, 2005) and Robert Norton Noyce (December 12, 1927 to June 3, 1990)

We treat Jack Kilby and Robert Noyce as one inventor for they both created the same technological device at virtually the same time.

Jack S. Kilby designed the first integrated circuit after arriving at Texas Instruments in 1958.[4] He also performed the first successful laboratory demonstration of the simple microchip which took place on September 12, 1958. Ultimately, this integrated silicon chip replaced a wide array of component parts such as transistors, resistors, and capacitors. Among other factors such as cost and ease of transporting, in practical terms, the development of the integrated silicon chip and its universal replacement in the computer industry meant that people would no longer walk into computers, but that computers would become handheld devices. In 2000, Kilby received the Nobel Prize in Physics.

Robert Norton Noyce was nicknamed "the Mayor of Silicon Valley," because he cofounded Fairchild Semiconductor in 1957 and Intel in 1968. He is also credited (along with Jack Kilby) with the invention of the integrated circuit or microchip. Berlin (2005) provided many of the rich details regarding Noyce's life:

Bob Noyce's nickname was the "Mayor of Silicon Valley." He was one of the very first scientists to work in the area—long before the stretch of California had earned the Silicon name—and he ran two of the companies that had the greatest impact on the silicon industry: Fairchild Semiconductor and Intel. He also invented the integrated chip, one of the stepping stones along the way to the microprocessors in today's computers.

Noyce, the son of a preacher, grew up in Grinnell, Iowa. He was a physics major at Grinnell College, and exhibited while there an almost baffling amount of confidence. He was always the leader of the crowd. This could turn against him occasionally—the local farmers didn't approve of him and weren't likely to forgive quickly when he did something like steal a pig for a college luau. The prank nearly got Noyce expelled, even though the only reason the farmer knew about it was because Noyce had confessed and offered to pay for it.

While in college, Noyce's physics professor Grant Gale got hold of two of the very first transistors ever to come out of Bell Labs. Gale showed them off to his class and Noyce was hooked. The field was young, though, so when Noyce went to MIT in 1948 for his PhD, he found he knew more about transistors than many of his professors.

After a brief stint making transistors for the electronics firm Philco, Noyce decided he wanted to work at Shockley Semiconductor. In a single day, he flew with his wife and two kids to California, bought a house, and went to visit Shockley to ask for a job—in that order.

As it was, Shockley and Noyce's scientific vision—and egos—clashed. When seven of the young researchers at Shockley semiconductor got together to consider leaving the company, they realized they needed a leader. All seven thought Noyce, aged 29 but full of confidence, was the natural choice. So Noyce became the eighth in the group that left Shockley in 1957 and founded Fairchild Semiconductor.

Noyce was the general manager of the company and while there invented the integrated chip—a chip of silicon with many transistors all etched into it at once. That was the first time he revolutionized the semiconductor industry. He stayed with Fairchild until 1968, when he left with Gordon Moore to found Intel. At Intel he oversaw Ted Hoff's invention of the microprocessor—that was his second revolution.

At both companies, Noyce introduced a very casual working atmosphere, the kind of atmosphere that has become a cultural stereotype of how California companies work. But along with that open atmosphere came responsibility. Noyce learned from Shockley's mistakes and he gave his young, bright employees phenomenal room to accomplish what they wished, in many ways defining the Silicon Valley working style was his third revolution.

Noyce died of heart failure in 1990, at the age of 62.

In all, Kirby and Noyce are noteworthy for one invention, but an invention that made all the difference to the computer industry, its viability as a commercial entity, and the ability of the computer industry to make the computer part of millions of homes throughout the United States and the world. The integrated silicon chip—and its inventors—are truly remarkable—the integrated silicon chip literally revolutionized the digital industry.

Herbert Marshall McLuhan (July 21, 1911 to December 31, 1980)

Marshall McLuhan, as he was known by virtually all of this colleagues and the public, was the second, but perhaps the most famous, of the Canadian communication theorists and innovators. His contributions were, however, profoundly conceptual, and he virtually taught the entire world to welcome and receive digital technologies as well as how to think about, understand, and incorporate digital technologies into their personal lives.

McLuhan was a professor at the University of Toronto and he was Director of the Centre for Culture and Technology. Within these roles, he formulated five notions that made all the difference in how people understood media and how ready people were to adopt and embrace the digital revolution.[5]

His first and perhaps most powerful notion was that the *medium is the message*. Distinguishing between form and context (as Innis had), McLuhan formulated his notion in terms that caught the imagination of many, and was repeated again and again. In greater detail, when explaining "the medium is the message," McLuhan argued that the type of technology used affects what the message is. In another context, McLuhan had argued that, "we shape our tools and they in turn shape us." This conception is now viewed, by some, as a form of technological determinism. But, the notion might also be said to possess an ironic formulation, for "The medium is the massage" can be understood as a formulation that emphasizes, not how people adjust to technologies, but how technology "adjusts" people to environments and stimuli.

Second, McLuhan argued that technology is culture-creating. McLuhan argued that each communication technology creates its own sociocultural norms and systems. For example, of the book, McLuhan argued: "Printing, a ditto device, confirmed and extended the new visual stress. It created the portable book, which men could read in privacy and in isolation from others."

Third, in McLuhan's view, technologies are extensions of the human senses. McLuhan created an alignment in which each human sensory system was understood to be extended by contemporary technologies. Hence, the following equations dominate McLuhan's thinking:

1. Books extend the eye;
2. Television and electricity extend the nervous system; and,
3. Pencils and pens extend the sense of touch.

For McLuhan, these equations can be influenced by culture, and different ratios of senses exist in cultures with different degrees of technology use and/or domination.

Fourth, McLuhan held that the electronic media were creating a global village. In it, "intimacy" is more like a village than like a removed sense created by the vastness of the world. In this sense, the electronic media will unify and retribalize the human race. For many, digital technologies only seem to underscore the significance and power of McLuhan's view at this point.

Fifth, McLuhan proposed the use of a *hot and cool media* dichotomy, at least for conceptual understanding and for critical use especially in terms of cultural characterization. In this sense, the dichotomy is probably best viewed as a continuum from hot to cool with the possibility that some cultures may swing between hot and cool media or use both simultaneously. Thus, roughly speaking, in McLuhan's view, a medium could be understood as *hot* if it stimulated only one sensory system, while a *cool* medium affected several sensory systems simultaneously. Hence, because it affects predominantly sight, a newspaper is a hot medium. In contrast, television is a cool medium because it affects two of the human sensory systems: sight and sound. Accordingly, a culture that conveys its rules and guidelines predominantly through print, such as the Bible, would be viewed as a "hot culture," while a culture that conveyed its societal norms through television would be a more cool culture. Of course, this hot-cool dichotomy is value-oriented. Particularly, in the American culture, the use of a "cool culture" would be more positive for some than the conception of a hot culture that could only be understood through one medium or perspective. In this sense, the hot-cool dichotomy might serve as a resource for a critic or critics.

As a founder of the digital revolution and especially from the perspective of those in the discipline of communication, we think McLuhan should be viewed in particularly selective ways. We fully admit that these "selective ways" will add qualifiers, for some, to how McLuhan's contributions can be viewed and valued. So be it. Nonetheless, it is within this context that McLuhan has served four particular roles.

First, he popularized communication in general and electronic communication in particular. In his day, McLuhan made many television appearances, and his writings generated a host of popular reactions. Indeed, McLuhan can be said to have popularized communication and technology from 1965 through 1980. During this period, more than any other person, McLuhan can be said to have made communication popular. According to Rogers, "During his lifetime McLuhan did more than any other individual to interest the general public in communication study."

Second, McLuhan employed a host of rhetorical and stylistic devices, techniques of exaggerated humor and rebellion. Gary Wolf identified McLuhan as "a critic and an academic rebel," prone to incredible pronouncements and humorous quips. For example, to underscore the power of television, McLuhan predicted it would destroy the book. McLuhan's exaggerated humor and rebellious nature was one of the ways in which he popularized communication and technology. Ambiguity was a critical and central strategy to popularize his ideas. In all, McLuhan knew how to keep his notions in the public eye. Among a host of notions, he made the word *communication* a popular and everyday concept, but he also changed the meanings associated with communication media. He underscored, for example, the cognitive power of television at a time when it was predominantly viewed as "mere entertainment." In all, McLuhan cast technology as an active agent capable of generating culture and identity for entire populations.

Third, McLuhan cast communication itself as a shocking and overwhelming tool. McLuhan shocked, and his shocks always stemmed from what he claimed communication and technology were doing. For example, McLuhan proclaimed that the invention of the alphabet gave primacy to sight over hearing and therefore emphasized space over sound. Ultimately, knowledge and understandings are equated with what could be seen and later read.

Fourth, McLuhan—almost paradoxically—has been and will always be understood as an established scholar. Indeed, McLuhan was a Cambridge-educated scholar, trained in the classics, aware of Modernists such as James Joyce and Ezra Pound, and aware of findings in and about antiquity. Dozens of universities awarded McLuhan honorary degrees and he secured the Schweitzer Chair in the Humanities at Fordham University. Visitors at the University of Toronto Centre for Culture and Technology included an unending stream of celebrities such as John Lennon and Yoko Ono.

In all, McLuhan knew and understood propaganda and how it worked, especially in the contemporary popular culture. But, he also proposed notions that transformed the popular culture, extended the appreciation of the links among technology and communication, and provided people with tools and concepts for analyzing and debating the various roles which technology and communication were having on people. He certainly has generated mixed reactions and understandings. Humor, radicalism, and exaggeration are all risky strategies. They can individually affect groups of people in different ways, but as a group, they might also suggest that McLuhan had no enduring idea or contribution. The 1988 publication of Marshall and Eric McLuhan's *Laws of Media: The New Science*, formally authored by Marshall (eight years after his death) and his son Eric McLuhan, has done little to make people think of McLuhan's original notions as part of a scientific framework. Indeed, the original concepts offered by McLuhan continue to intrigue many. For example, at the Seventh Annual International Seminar on Social Communication in Porto Alegre, Brazil, in August 2003, seven international communication scholars were invited, and representatives from the United States, France, and England were predictably present. But, in addition, two of the seven international scholars were from McLuhan's Centre for Culture and Technology and they argued for the centrality of McLuhan in the 21st century.

> **Debate It**: Rather than focus on McLuhan's specific concepts about digital technologies, McLuhan should be understood and appreciated as an advocate of sociocultural change (e.g., as a society, we benefit more if we define and understand McLuhan's techniques or strategies for encouraging and stimulating technical and cultural adaption rather than his specific ideas about how technology has changed society).

Steven Jobs (February 24, 1955 to October 5, 2011)

At the time of his death, Jobs was best known as the cofounder (along with Steve Wozniak), chair, and CEO of Apple, Inc. He had supervised the production and development of the iMAC, iTunes, iPod, iPhone, and iPad. But, equally important, he was perceived, defined, and understood to be charismatic. For example, at the time of his death, he was variously described as a "visionary who transformed the digital age" (Markoff, October 2, 2011), the "wizard and the mortal: two sides of genius" (Stross, 2011), "Steve Jobs and the coolest show on Earth" (Gelernter, 2011), "the genius of Jobs" (Isaacson, 2011), and "father of the digital revolution" (*People and Lifestyle*, October 14, 2012).

Certainly, as an entrepreneur, his achievement is extremely difficult to equal. After a power struggle with the Apple board of directors in 1985, he left Apple and founded NeXT, a computer platform development company specializing in business and higher education. Also, after leaving Apple, Jobs began his work with Lucasfilm (*Toy Story* in 1995) and Pixar (CEO and majority shareholder which he sold to Disney in 2006). In the late 1990s, Jobs returned to Apple as an adviser and took control of the company as an interim CEO. As Kane (2012) and Isaacson (2012) reported, Jobs brought Apple from near bankruptcy to profitability by 1998. In 2011, Apple was the world's most valuable publicly traded company, one of the "greatest turnarounds in business history" (Wikipedia, 2013, p. 2).

When Jobs died of respiratory arrest related to his metastatic tumor (pancreas neuroendocrine tumor) on October 5, 2011, Deutschman (2011, September 5,

p. 30) captured a sentiment held by many at the time: "Exit the King." More specifically, Theroux (2011, September 5, p. 36) noted Jobs had "revolutionized our world" and that "Steve Job's dazzling inventions have forever changed us."

Tim Berners-Lee (June 8, 1955)

Tim Berners-Lee is now a Senior Research Scientist within the "Lab." The Lab merged with the AI lab to became "CSAIL," the Computer Science and Artificial Intelligence Laboratory at MIT.

On his webpage, Tim Berners-Lee's work with the World Wide Web is described in these words:

> In 1989, he proposed a global hypertext project, to be known as the World Wide Web. Based on the earlier "Enquire" work, it was designed to allow people to work together by combining their knowledge in a web of hypertext documents. He wrote the first World Wide Web server, "*httpd*," and the first client, "*WorldWideWeb*" a what-you-see-is-what-you-get hypertext browser/editor which ran in the NeXTStep environment. This work was started in October 1990, and the program "WorldWideWeb" first made available within CERN in December, and on the Internet at large in the summer of 1991.

Through 1991 and 1993, Berners-Lee continued working on the design of the Web, coordinating feedback from users across the Internet. His initial specifications of URLs, HTTP and HTML were refined and discussed in larger circles as the Web technology spread.

In 1994, Berners-Lee founded the World Wide Web Consortium at the Laboratory for Computer Science (LCS) at the Massachusetts Institute of Technology (MIT). Since that time he has served as the Director of the World Wide Web Consortium, which coordinates Web development worldwide, with teams at MIT, at INRIA in France, and at Keio University in Japan. The Consortium takes as its goal to lead the Web to its full potential, ensuring its stability through rapid evolution and revolutionary transformations of its usage. The Consortium can be found at http://www.w3.org/.

On April 15, 2004, Berners-Lee was awarded the first ever Millennium Technology Prize from the Finnish Technology Award Foundation, established in 2002 by eight Finnish high-technology organizations. The prize carried an award of one million euros.

In November 2006, he launched Web Science as a formal discipline at the University of Massachusetts Institute of Technology and University of Southampton in England. Leading researchers from 16 of the world's top universities have since expanded on that effort (see, e.g., Hendler et al., 2008; Anderson & Rainie, 2010).

In 2009, then British Prime Minister Gordon Brown announced Berners-Lee would work with the UK Government to help make data more open and accessible on the Web, building on the work of the Power of Information Task Force. Berners-Lee commented that,

> The changes signal a wider cultural change in Government based on an assumption that information should be in the public domain unless there is a good reason not to—not the other way around…. Greater openness, accountability and transparency in Government will give people greater choice and make it easier for individuals to get more directly involved in issues that matter to them. ("Tim Berners-Lee," June 29, 2013, Wikipedia)

Since 2009, Berners-Lee has continued to received awards and distinctions such as an honorary degree from Harvard University, and he was inducted into the IEEE Intelligent Systems' AI Hall of Fame and the Internet Hall of Fame by the Internet Society. He was recognized for the invention of the World Wide Web at the 2012 Summer Olympics opening ceremony in London.

In our view, while we believe that Berners-Lee will continue to be active and his efforts globally recognized, we expect that his program creating the Web—one of the most impressive global communication systems ever made—and his efforts to create Web Science as an academic area will continue to dominate his work and reinforce the notion of Berners-Lee as one of the founders of digital communication technologies.

Seven Technologies Contributing to the Creation of the Information Age[7]

Inventions tempt us; they can intrigue us. They can tell us something about how those before us were thinking, the resources they used to create, and how they might have intended the inventions to function. Indeed, as *Webster's* (1981, pp. 602–603) tells us, an invention is "a product of the imagination" and "a device" or "process originated after study and experiment." When we examine inventions, we begin to understand how something emerged from nothing, for inventing is the decision to "produce something for the first time through the use of the imagination or of ingenious thinking and experiment."

Computers and their massive formatting and delivery systems have fascinated human beings for thousands of years. In 2008, for example, *New York Times* reporter John Noble Wilford reported that "scientists found" a "surviving marvel of ancient Greek technology known as the Antikythera Mechanism" (see Freeth,

2009). The device was used to predict "solar eclipses and also organized the calendar in the colonies of Corinth on Sicily. The scientists said this implied a likely connection with Archimedes," who lived and died in 212 BC. The Antikythera Mechanism is sometimes "called the first analog computer," reported Wilford.

If true, the notion of computers—especially analog computers—fascinated humans even before electricity was discovered and computers became digital. Additionally, the scientists who examined the Antikythera Mechanism also seemed to be functioning rhetorically, perhaps seeking to overcome the fear that some have of new technologies, and therefore proclaiming that "computers have always been with us."

Toward this end, we begin our survey of relevant technologies with the printing press, an analogic and non-computerized device that nonetheless created part of the cultural and social need for today's computers.

Printing Press

Invented by Johannes Gutenberg in 1450, the printing press produced a respect for the culture of information, pride in the accumulation of information, and, ultimately, even a belief in the right to possess and have all information. The printing press also suggested that information could be made available to all—it could be mass-produced and mass-distributed (Ong, 1982). Without a doubt, the printing press created a host of "revolutions." Sometimes these revolutions could be marked by sheer volume. For example, by the beginning of the 1970s, some 3,000 notices and forms were manufactured for every man, woman, and child in the United States (Chesebro & Bonsall, 1989, p. 16).

Two major social institutions are associated with the emergence and use of the printing press.

First, the printing press affected how learning could be achieved. While we might trace these changes from the earliest grades through high school, the impact of the printing press is most dramatic at the university level. Indeed, as we understand it today, the printing press fostered a rapid and profound expansion of the modern university. Haskins (1957) described the early universities, noting that the first universities emerged between 1119 and 1158, at the University of Paris and University of Bologna. Students would band together and select a topic they wanted to explore. They would then find an appropriate "instructor," pool their funds to pay for their instructor, and these instructors gradually became professionals who were indeed paid to "profess" what they felt required attention and internalization. While the classroom itself continued to employ an oral give-and-take, the manufacturing and distribution of textbooks encouraged each student to read and understand on his/her own. In this sense, the mass production of textbooks promoted a sense that what we know and how we know it is essentially a private matter. And, knowledge slowly moved from what the community knew and discussed to what each individual read and internalized. And, of course, there were differences in what different people read and know. Hence, the printing press provided a sense that knowledge could be transferred directly from a book to an individual student. And, if the authors of the textbook were not known, the book itself could easily be perceived as the source of knowledge. In all of these ways, the printing press compartmentalized knowledge into books, made learning a private matter, and began to encourage the development of individualization ("We become what we read").

Second, the printing press also encouraged and fostered the rise of the liberal democracies, especially in the United States and France. The United States and France represented the first commitments to freedom of expression and the right to information. In this regard, Thomas Paine's 1776 *Common Sense* is a powerful example of the popular power of the appeal to the right to information early in the development of popular and representative democracies. The publication and distribution of Paine's notions required the printing press, and in this regard, the mass production of books, magazines, and newspapers created a foundation for liberal democracy, the importance of a free press as well as a host of related rights associated with freedom of expression.

Calculating Machines

Blaise Pascal created the first calculating machine at the age of 21 in the year 1642. Fifty of the machines he manufactured are in existence today. Pascal's invention demonstrated that machines could function and carry out some intellectual activities faster and more accurately than human beings.

Analytical Machine

Establishing the basic principles of computing (programs in memory, punch cards to retain and process data, and arithmetic functions), Charles Babbage (born in 1792) sketched—for some 40 years—the engines he thought would be necessary to carry out computing activities. His machines were not built in his day. If they had been, they would have covered a football field and been run by six steam engines. Nonetheless, his design possessed equivalents of a modern logic center, memory, control unit, mathematical center, and operating system that could be changed or reprogrammed at will.

Technologies Generated by Electricity

Beginning in the 1870s, Thomas Edison's inventions—such as the phonograph, the light bulb, and the electric utility—were creative, but his ultimate test was that they should be practical and profitable. Using the southern section of Manhattan as his extended example, Edison even demonstrated how electricity might be a source of power in a viable and real community. Electricity was also to become, of course, the energy source for the contemporary computer. As Cortada (2002, p. 28) put it, Edison can be "seen as an early player in the world of information."

Electromechanical Calculating Machines

In 1880, William S. Burroughs completed the design for a electromechanical calculating machine, and he moved his invention into mass production. The mass distribution of these machines made computers part of the everyday environment. While these machines fostered some fear of computers, for most people calculators were tools and devices that could be used to make life easier.

ENIAC or the Electronic Numerical Integrator and Calculator

As we noted that the outset of this unit, ENIAC was unveiled in 1946 by the Moore School of Engineering at the University of Pennsylvania. It was installed at the Aberdeen weapons-proving grounds in Maryland. It compiled ballistic tables for new guns and missiles, and could perform some 5,000 addition and subtraction computations per second. The machine was massive and unreliable. It was composed of 17,000 vacuum tubes, and had 70,000 resistors, 10,000 capacitors, and 6,000 switches. It filled a huge room, weighed 30 tons, required a tremendous amount of electricity, frequently overheated, required 24-hour maintenance, and had to be rewired to change its program. While the size of the ENIAC seems beyond control today, at the time, the size of the ENIAC also underscored the significance of computers (i.e., a kind of size = power relationship). Indeed, one literally walked inside a computer. And, the size of the ENIAC also made computer engineers appear to be highly trained specialists, hence enhancing the credibility and ethos of computers and implicitly suggesting that that computers should be viewed and treated as very special devices. However, while it was the first modern computer, and it was impressive in many ways, the ENIAC was also limited by its core component, the vacuum tube.

Single Silicon Integrated Circuit

Developed independently in the mid-1950s by two firms—Texas Instruments and Fairchild Semiconductor—the single silicon integrated circuit eliminated hand-wiring by "printing" circuits on a board. In 1971, Intel Corporation produced the first integrated circuit on a tiny silicon chip, thereby establishing the procedure for the mass production of the microprocessor found in the current personal computer. These first chips were the size of a postage stamp. By 1975, Altair produced for mass-market distribution a computer kit for under $700, using the principles embodied in the integrated circuit. In 1977, Steve Wozniak and Steven Jobs introduced Apple II, the first preassembled desktop personal computer.

While each of these devices deserves special attention and examination, we must admit that we see them as more unified, with one device contributing to the development and evolution of the next. Figure 3.1, "The Evolution of Computer Hardware," provided an overview of the interrelationships we see among computer devices

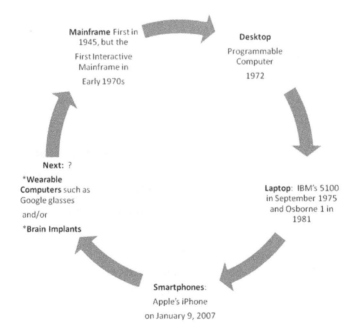

Figure 3.1. The evolution of digital hardware.

Conclusion and Future Possibilities

The history of the contemporary digital revolution might be approached by any number of ways. Some have viewed major historical events, such as the launching of Sputnik, as key to the birth and development of the contemporary digital revolution. Or, major technological eras might be viewed sequentially as the pattern that gave rise to the Internet. We encourage all of these approaches. However, from our perspective, the history of the contemporary digital revolution features the unique contributions of specific innovators and specific devices. In our view, these innovators and devices affected how people interacted with each other, understood themselves and others, and could manipulate information creatively towards intriguing outcomes.

Key Terms

Analytical Machine
Artificial Intelligence
Cybernetics
ENIAC
Silicon Integrated Circuit
Technological Culture

Chapter Exercises

Current Issues: Do existing programs allow digital systems to duplicate human emotions and reactions? During a telephone conversation, when or how do you know that you are talking to a computer rather than a human person? Should additional progress be made to make digital systems "more human" if they will help people deal with complex computer systems?

Debate It: As Watson demonstrated on the television show *Jeopardy* and a host of computer chess programs have illustrated, resolved that digital systems are recognized as potential uses of information that is more effective and efficient than human information processing for several tasks.

Ethical Consideration: Even if digital systems can process information more efficiently and effectively than human beings, many people believe that human beings should always retain decision-making powers when major sociocultural issues are at stake. Others think that decision-making decision and information effectiveness and efficiently should be linked and the primary foundation for action even if other systems are viewed as important, if not more important, than human decisions. Which is the most ethical system for decision-making?

Everyday Impact: The typical new car now has some 100 computer chips embedded in the automotive engine. These chips have increasingly determined the effectiveness and efficiency of the contemporary automobile and will probably continue to exert more and more influence on how devices such as cars operate. Indeed, automotive mechanics now find they must employ complex testing devices when checking a car's potential problems. If more and more human devices and machinery are regulated by digital systems, will you feel like you have lost control over the machinery in your world?

Future Focus: If you were to pick the best future forecaster, who might that person be? For example, given this key role in creating the code that allows the Internet to function and his work on advanced digital systems, would Tim Berners-Lee be your choice? We end the last chapter of this volume suggesting such a choice. Given the founders discussed in this chapter, is there an alternative founder who should be more strongly recognized?

Endnotes

1. Chapter 1, page 1, of this volume.
2. We have identified seven founders who made a difference in terms of the digital communication revolution. In terms of our area of specialization, each of these seven have also made a major contribution to the discipline of communication at the theoretical and applied levels. We think each of these seven deserve the attention we have given them here. At the same time, we also recognize that several others made contributions that we would also want to recognize here; there are outstanding volumes, in this regard, such as Simon Lavington's *Early British Computers* (Digital Press, 1980), that can and should be consulted. Others who have made contributions to both computer innovation and communication include, in alphabetical order, the following: Paul Baran (the pioneer who formulated the notion of "message blocks" for bundling information on computer systems and one of the engineers who developed the ARPANET which led to the development of the Internet); Bryce Bayer (creator of the checkerboard-like filter that enabled digital cameras to capture vivid color images); William Duckworth (composer, author, and performer of interactive

and electronic musical projects); Douglas C. Engelbart ("father of the computer mouse" who created the mouse in 1964 and demonstrated it in 1968 as well as refined word processing, bit-mapped computer displays, and navigating online using links); William Henry "Bill" Gates III (cofounder with Paul Allen of Microsoft in April 1975, an aggressive and effective businessman who made Microsoft the world's largest software maker measured by revenues); William Ford Gibson (an American-Canadian speculative fiction novelist who has been called the "noir prophet" of the cyberpunk subgenre; he coined the term "cyberspace" in his short story "Burning Chrome" (1982) and later popularized the concept in his debut novel, *Neuromancer* (1984); Jacob Goldman (a research and development laboratory specialist—at Ford and later at Xerox—who encouraged corporate investment in computer innovations for which some have identified him as "a Father of Computing" [Markoff, 2011]); John E. Karlin (an electrical engineer who focused on the behavioral science implications of keypad designs in terms of how humans and machines interact); J. C. R. Licklider (Licklider's development to the Internet consists of ideas, not inventions; he foresaw the need for networked computers with easy user interfaces including graphical computing, point-and-click interfaces, digital libraries, e-commerce, and online banking; working for several years at ARPA where he set the stage for the creation of the ARPANET; and in his 1960 volume, *Man Computer Symbiosis*, he developed the idea that computers should be developed with the goal "to enable men and computers to cooperate in making decisions and controlling complex situations without inflexible dependence on predetermined programs"); John McCarthy (one of the leading engineers involved in the creation of artificial intelligence, or AI, as well as a major formulator of computer time-sharing); Daniel D. McCracker (one of the first "serious and systematic computer educators" for the larger society and culture in the 1950s and 1960s, especially for businesses); Peter G. Neumann (a pioneering engineering computer scientist at SRI International who has repeatedly predicted security and reliability problems for the computer industry); Larry Page and Sergey Brin (cofounders of Google 1998, developed powerful search systems, and digitized everyday environments into virtual realities); Jon Postel (established standards for the Internet, administered the numeric system for integrating diverse computer systems, and administered the infrastructure of the Internet address system under a federal government contract for 30 years); Dennis Ritchie (longtime Bell Labs research engineer who invented the language underlying Microsoft Windows and much of the other software running on computers around the world); Jack Tramiel (founder of the Commodore 64, which became the best-selling single-personal computer, selling 17 million copies, and later, as owner of Atari, introduced "Pong" in the early 1970s); Anthony J. Wiener (coauthor with Herman Kahn of *The Year 2000: A Framework for Speculation on the Next Thirty-three Years*); Steve Wozniak (invented the Apple I computer and cofounded, with Steven Jobs, the Apple corporation); and Mark

Zuckerberg (creative and dedicated business leader who developed social media and especially Facebook as a global communication system).

3. For a more detailed treatment of Turing, his ideas, and the ways in which he linked computer science and communication, see Chesebro (1993, Winter).

4. Recognition for the design of the first integrated circuit was also given to Robert N. Noyce when Fairchild Semiconductor named him as its inventor of the integrated circuit on its patents. For a less-than-flattering conception of his role in the process of designing the first integrated circuit at Fairchild Semiconductor, see Leslie Berlin, *The Man Behind the Microchip: Robert Noyce and the Invention of Silicon Valley* (New York: Oxford University Press, 2005).

5. We have intentionally employed all of McLuhan's major works in this description, essentially using McLuhan's notions in a fashion that we think best and concisely reflects his thinking. In this regard, we have found the following texts by McLuhan most useful: McLuhan (1951); Carpenter & McLuhan (1960); McLuhan (1962); McLuhan (1964); McLuhan & Fiore (1967); McLuhan (1968); McLuhan (1970); McLuhan & Watson (1970); and McLuhan (1943/2006).

6. For an overview of the information age and its development, see Chesebro & Bonsall (1989), Chapter 1, "The Information Society," pp. 13–37.

4

The Future—Predicting the Future of the Digital Revolution

In this chapter, we explore technological innovations and technological forecasting. Specifically, we examine what is involved if we want to identify reliable predictions, and what standards we should apply if we are to isolate forecasts that are actually more likely than not to occur.

However, predictions about the future have covered a wide range of possibilities and employed a diverse set of methods from the claims offered by local gypsies as they look into a crystal ball to what has now emerged as the science of the future, *futurism*, a mode of inquiry first recognized, isolated, and identified in Italy by Filippo Tommaso Marinetti in his 1909 *Manifesto of Futurism* (1909/1983).

Today, discussions of and predictions about the future—and even technological innovations—have explored a wide range of topics with a tremendous range of methods (see, e.g., the 10 methods surveyed by Martino, 1972).

One of the earliest of these sets of predictions was offered by Hal Hellman (1969), who was particularly interested in the likely outcomes of media engineering. In sharp contrast, in a volume that has the style of a political conspiracy, Katherine Albrecht and Liz McIntyre (2005, back cover) identified how major corporations and the government plan to trace "your every move" with radio frequency identification systems that will "in just a few short years," use this "explosive new technology" to tell "marketers, criminals, and government snoops everything about you."

One year later, the Pew Internet & American Life Project (Anderson & Rainie, 2006, p. iii) surveyed 742 Internet users via email about seven specific predictions projected to occur by the year 2020. Pew found that a majority of their panel of Internet users agreed on four of the seven predictions and the difference between those agreeing and those disagreeing was extremely close (e.g., 56% to 43% or 52% to 44%). While we might examine each of the predictions one-by-one, what is outstanding at this point is that there are legitimate differences among experts on what they anticipate will be.

A few years later, in 2008, Jonathan Zittrain, a professor of Internet Governance and Regulation at Oxford University and cofounder of the Harvard Law School's Berkman Center for Internet & Society, forcefully argued for and proposed controlling the future of the Internet in his volume *The Future of the Internet and How to Stop It.*

Beyond this proposal, in 2010, in its popular survey of the American adult population, the Pew Research Center for the People & the Press (July 14) found that 58% of Americans expect another world war in the next 40 years and 53% believe that a major terrorist attack on the U.S. with a nuclear weapon will occur by the year 2050. Of all of the age groups they examined, the highest percent—some 68%—of those believing that another world war will occur are among those aged 18 to 29.

Additionally, in another of its popular surveys (Pew Research Center, 2010), some 41% of adult Americans believe that Jesus Christ will return to Earth. Fully 58% of white evangelical Christians say Christ will return to Earth, by far the highest percentage of any religious group.

And, in terms of science, technology, and environments, most Americans see dramatic scientific and technological advancements on the horizon but also foresee "a grim environmental future" with "rising world temperatures, more polluted oceans and severe water shortages in the U.S." as "definite or probable over the next 40 years" (Pew Research Center, June 22, 2010, p. 1).

Such an informal survey of readily available technological forecasting is, at best, diverse and scattered, if outright contradictory. Indeed, it is literally impossible for all of these predictions to be right. As Peter Edidin (2005) put it, a

> common response to technological innovation has been to predict where it will lead, which is also an assertion of control of time. But…the crystal balls are almost always cracked. With some startling exceptions, prognosticators are usually dead wrong.

In even more dramatic terms, Douglas Rushkoff (2013) sought to explain "why futurists suck."

In terms of this chapter, these observations suggest that we must proceed with tremendous care. We need to examine not only what serious prognosticators have said about the likely future of technology, but also how and why they make the future claims that they make. In other words, we need to establish some standards for predictions that are based upon known multiyear trends and also ask if there is high agreement among experts in terms of what the projected trends will mean to people and their lives. Toward this end, we undertake three objectives in this chapter. These objectives are outlined in the preview of this chapter.

Questions Previewing This Chapter

1. *What would be an appropriate survey of 21st-century predictions about innovations in technology?*
2. *What techniques provide valid and reliable predictions about technology innovations?*
3. *Are there valid and reliable predictions about some 21st-century technology innovations?*

Ultimately, then, we need to consider the major predictive schemes that have already been developed to predict the future of technology innovations. We need to offer a review of the strengths and limits of these predictive schemes. With this review in mind, we propose that a prediction about any innovation in a technology must be based on a reasonably stable trend (trend extrapolation) as well as high agreement among experts (Delphi procedure) about the meaning and significance of this trend for people. Once these standards are established, we apply these methods and isolate what we think are reasonable and extremely likely predictions about the future of innovations in technology. Accordingly, we end this chapter by identifying valid and reliable predictions about the future of technology innovations during the next several decades.

Survey of 21st-Century Predictions About Innovations in Technology

A rich and varied set of predictions now constitute a "data base" about what the future of technology innovations will be. We examine 12 of these previously completed projections of the future of technology innovations. This survey serves several ends. First, it gives you background on the kinds of predictions that have been made. It also provides you with a sense of how they are justified as valid and reliable predictions. Finally, this survey provides a foundation for us to suggest what the strengths and weaknesses of these predictive schemes are. We use this as a way to introduce and justify the attention we will later be devoting to trend extrapolation and the Delphi procedure as methods for generating valid and reliable predictions about innovations in technology.

While 10 of the 12 predictions we examine here are, indeed, formulated to reflect a 21st-century perspective, we begin with two that were crafted in 1990. We select these two 1990 predictions because of the power and impact of these predictive schemes on 21st-century futurism.

Alvin Toffler's 1990 Powershift

In 1990, Alvin Toffler—who had published *Future Shock* in 1970—published *Powershift: Knowledge, Wealth, and Violence at the Edge of the 21st Century*. In a nutshell, this volume is a case study in power and empowerment in communication. The volume is guided by at least three major questions:

1. What is the traditional conception of power?
2. How are power and communication related?
3. Has the relationship between power and communication been changing?

Toffler begins with an exploration of the "traditional view" of power and communication. In this regard, power is the use of force and/or coercion while communication is the use of persuasion. Brembeck and Howell (1952, p. 24) aptly reflected this traditional view of persuasion when they defined it as "the conscious attempt to modify thought and action by manipulating the motives of men toward predetermined ends." With this definition of *persuasion* in mind, Brembeck and Howell (1952, p. 26) argued that the "basic process of persuasion reveals four general steps or stages, which may occur in the following order:

1. Gain and maintain attention;
2. arouse desires useful to the persuader's purpose;
3. demonstrate how these desires can be satisfied by acceptance of the persuader's proposition; and
4. propose the specific response desired.

Detaining the implications and meaning of this process, Bremeck and Howell (1976) argued that:

We define persuasion as *communication intended to influence choice*. The word *communication* denotes that this phenomenon is symbolic and interactive, the word *intended* suggests that the persuasive attempt has a predetermined goal, the word *influence* suggests that behavioral change of some sort is sought, and the word *choice* reflects the view that the receiver has options available to him. (p. 19)

Even more precisely, Brembeck and Howell (1976, p. 10) suggested power and communication are discrete concepts for power involving the use of physical violence, and communication enables choice: "Since we are dealing in the use of all verbal and nonverbal symbols, we exclude all nonsymbolic modes of influence from the province of persuasion, that is, the forms of physical violence."

Sometimes we want to make distinctions that help us order the world the way we want it. However, sometimes these distinctions are not used by others, and we cannot expect that others are making the distinctions we are. In terms of the distinction between the concepts of *power* and *communication*, is it best to think of these two concepts as separate and discrete or do they overlap in practical situations? The following statements can be used to argue for the use of power and communication as discrete and separate. Do you agree with each of these statements?

1. Force and coercion are **not** conscious attempts to modify thought and action by manipulating the motives of people toward predetermined ends.
2. Persuasion is **not** a form of force and coercion.
3. Force and coercion are only nonsymbolic (e.g., physical restraint, physical beatings, etc.).
4. Force and coercion are forms of physical violence, but they **do not** involve the use of verbal and nonverbal symbols.

Figure 4.1. The attempt to distinguish power and communication.

Toffler initiated four propositions that challenge the traditional view that power and communication are separate and discrete. He believed they are intimately involved, and they have been so for centuries. And, as the future unfolds, power will be increasingly defined by and in terms of communication, especially when communication is understood, believed, and defined to be knowledge. These four propositions include:

1. **The nature of power is changing**. As Toffler (1990, p. 3) put it, "We live at a moment when the entire structure of power that held the world

together is disintegrating. A radically different structure of power is taking form. And this is happening at every level of human society." And, as Toffler (1990, p. 7) articulated this position even more directly: "Power, which to a large extent defines us as individuals and as nations, is itself being redefined."

2. **A transformation in the nature of power is occurring** in the following stages:
   A. First Stage: Power = Brute Force
   B. Second Stage: Power = Wealth
   C. Third Stage: Power = Information.

3. **Knowledge is now the source of brute force and wealth.** As Toffler (1990, p. 9) argued the point, "There is, however, a much larger sense in which changes in knowledge are causing or contributing to enormous power shifts. The most important economic development in our lifetime has been the rise of a new system for creating wealth, based no longer on muscle but on mind."

4. **Knowledge will become the most important factor in power.** This kind of claim is far more complex, and Toffler (1990) began to make distinctions that are critical to this overall prediction:

The use of violence as a source of power will not soon disappear....

Similarly, the control of immense wealth, whether by private individuals or public officials, will continue to confer enormous power on them....

For it is now indisputable that knowledge, the source of the highest-quality power of all, is gaining importance with every fleeing nanosecond. (p. 470)

In our view, Toffler made a prediction that can be viewed as truly global, and we think the evidence for his prediction must also reflect such a scope. Likewise, in our view, Toffler needs to make some arguments from trends, namely that reasonable measures of both violence and wealth have declined as a source of power while information has dramatically increased as the means to resolve major conflicts over time. Such trends seem critical and central to the argument that Toffler is making. The question becomes whether or not *Powershift* adequately provides this evidence, especially in increasing ways over time. We encourage you to check out Toffler's *Powershift* to make such a determination.

Toffler's predictions about the changing nature of power from violence and wealth to information require deliberation. Two questions emerge:

1. Different types of personalities exist. Some people are trained and conditioned to respect and use physical violence more than others. Some people respect wealth and what it can do more than others. Some have been trained and conditioned to respect the insights that information reveals more than others. Our training and conditioning frequently determine what we perceive and know. Given these variations in personality types, will not the meaning of power and what changes people's behavior vary from one group to another?

2. If we examine a period of time, such as a century, and particularly the 21st century, do we see a decline in the use of violence and wealth to control outcomes of major conflicts and situations in which conflicts were resolved or avoided by appeals to information? Can these conflict-resolution patterns and use of power be compared and contrasted over time? Can equivalent measures of these three uses of power be operationalized and compared?

Figure 4.2. Personality bias and measuring trends.

John Naisbitt and Patricia Aburdene's 1990 Megatrends 2000: Ten New Directions for the 1990's

Originally predicted in a volume in 1982, Naisbitt and Aburdene suggested in their 1990 volume that these 10 shifts continued through the 1980s and into and through the 1990s:

1. Industrial Society > to Information Society.
2. Forced Technology > to High $\dfrac{\text{Tech}}{\text{High}}$ Touch.
3. National Economy > to World Economy.
4. Short Term > to Long Term.
5. Centralization > to Decentralization.
6. Institutional Help > to Self – Help
7. Representative Democracy > to Participatory Democracy
8. Hierarchies > to Networking
9. North > to South
10. Either/Or > to Multiple Options

In terms of technological innovations and specific developments in computers, Naisbitt and Aburdene (1990) offered four specific predictions regarding digital communication technologies:

1. **Biotechnology Vocabulary**: Biology may be our best metaphor for discussing concepts such as information-intensive, micro, inner-directed, adaptive, and holistic. As Naisbitt and Aburdene (1990) argued,

    In the information era we borrow from the vocabulary of biology.... Information feedback systems, biological and electronic, are mutually reinforcing. Computers are being used to help unlock the secrets of life: biology instructs new information software and systems. (pp. 241–242)

2. **Miniaturization**: Naisbitt and Aburdene (1990) maintained that "Computers are shrinking. In global financial transactions, electronic impulses have replaced paper" (p. 27).

3. **Globalization and the Global Economy**: In the view of Naisbitt and Aburdene, Telecommunications made the global economy possible in the first place; now that same technology is accelerating its development. Financial services, the most evolved sector of the global economy, have more to do with electronics than with finance or services, which have been with us a long time. What is new is high-tech telecommunications. (p. 94)

4. **Increased Importance of the Individual**: Naisbitt and Aburdene aptly argued: The very nature of an information economy shifts the focus away from the state to the individual. Unlike a widespread Orwellian-instructed view that computers would tighten the control of the state over individuals, we have learned that computers strengthen the power of individuals and weaken the power of the state.... At once, as we globalized our economies, individuals are becoming more powerful and more important than they were in the industrial era. (p. 95)

A host of issues have emerged within the context of these four predictions about the transformation of human life as a result of the computerization of everyday life. Some of these limitations are cast as questions in Figure 4.3, "Critical issues and Naisbitt and Aburdene's digital communication technology predictions."

1. Does the use of the biological metaphor constantly suggest that humans and computers might be more intimately linked? Are computer implants a viable and desirable objective for human beings, perhaps first as medical repairs for missing arms and legs but later for brain implants to enhance cognitive abilities?
2. What are the limitations of miniaturization especially in terms of transparency and honesty with others?
3. Are cultural, geographic, and language hampering the long-term development of a global economy?
4. Is the quest for more effective advertising and the use of technical devices such as cookies dramatically reducing the individual's sense of privacy? Is the individual's sense of the self and how he or she is linked to others increasingly created by computers?

Figure 4.3. Critical issues and Naisbitt and Aburdene's digital communication technology predictions.

Overall, in terms of our very specific concerns for long-term trends and predictions that most experts understand and can agree with, given the long-term (since 1982) and global framework (especially given their newer works on the United States and China), we find few futurists who have done as much as Naisbitt and Aburdene for as long as they have and at the scope they have in terms of topics and geography.

James W. Cortada's 2002 Making the Information Society

In 2002, James Cortada provided a detailed set of predictions about digital communication technologies in *Making the Information Society: Experience, Consequences, and Possibilities*. In his Chapter 10, about "The Future of Information in America" (pp. 369–411), Cortada offered seven predictions that are particularly rich in terms of the motivation for people. Moreover, with the insights we have all gained from computer uses and developments in the 2010s, some of these predictions now seem obvious while other predictions identify the year 2047 as a key year for the realization of major information and personal goals.

1. Computer chips will continue to shrink in size and cost, while expanding their capacity to hold ever-growing amounts of data (p. 382).

2. The rate of change in computer chips—smaller size and cost—will increase data capability—which will exert a host of changes in the human environment:

   A. By 2047, all information about physical objects, including humans, buildings, processes, and organizations will be online.

   B. By 2047, cyberspace will provide the basis for information, entertainment, and education.[1]

   C. By 2047, new levels of personal service, health care, and automation will exist.

   D. Between the "early years of the 21st century and mid-century, speed will grow by 9 orders of magnitude, mostly based on existing technological and known laws of physics" (p. 383).

3. In the immediate future we will see important advances that will increase our use of computer-based information because of how humans will "talk" to computers.

   A. While typing and reading responses on a screen is the most frequent mode of computer-human interaction today, as memory and speed increases in computers, we will be able to converse with machines and have them respond verbally.

   B. New interface systems with computers may be based on nonverbal reactions (gestures, gazes, facial expressions, and voice inflection) (p. 384).

4. Collaboration and communication among people will continue to expand (p. 385).

5. Computerization is becoming global, but pockets of massive non-use exist.

   A. While national differences will exist, reliance on information will increase, and information use will increasingly be more and more handled by individuals (p. 395).

   B. The United States will continue to lead the information revolution in terms of developing new information tools and rapidly exploiting them (p. 395).

6. Americans will continue to be motivated by the need for and need to develop more information, because:

   A. Information helps to get jobs done.

   B. Information provides a fast and efficient path to ends.

   C. Information can be used individually, and it enables Americans to do things for themselves (p. 397).

7. As Americans enter the 21st century, information itself will become more:
   A. list and ranking oriented.
   B. numeric.
   C. precise, relevant and therefore small in size, especially as the Internet becomes more popular and the norm for information.
   D. diverse in terms of sources and formats.
   E. immediately available just when needed (pp. 400–404).

Cortada provided an intriguing range of predictions. One of the reasons for surveying such a set of predictions is to ask what kind of focus or bias such predictions have and what are their limits and advantages. Figure 4.4 provides a more guided approach to such critical questions.

Cortada is able draw a host of his conclusions and predictions based solely upon numeric increases and decreases. Such a focus has often been identified as a focus on the *information society* or *information age.* The concepts have been amazingly useful for many calculations and determinations.

What do we know from a focus solely on numeric and ranking data?

When do we need to know what the content of numeric and ranking data are?

If you look back through Cortada's seven predictions, which are most useful? Those that are predominantly formal (numeric and ranking data)? Those that combine formal and content analyses?

Figure 4.4. Information age—form and content.

Ben Shneiderman's (2003) Leonardo's Laptop: Human Needs and the New Computing Technologies

In *Leonardo's Laptop*, Shneiderman (2003) adopted an orientation that some would find unusual. Shneiderman is well known for his volume, *Designing the User Interface: Strategies for Effective Human-Computer Interaction.* In 2009, he published—with his coauthors—the fifth edition of this 624-page book. This volume is profoundly applied, with a strong engineering, design, and user orientation. *Leonardo's Laptop* provides a sharp contrast to *Designing the User Interface.*

In *Leonardo's Laptop*, Shneiderman (2003, p. 3) sought to link the high-tech world "more closely to the needs of the people," a link that "still requires some new forms of thinking." To achieve this end, the creative genius and assimilative thinking of Leonardo da Vinci (1452–1519) is employed as a metaphor for what must be done if an appropriate "integration" of engineering and human values is to be achieved in contemporary computer-human interactions. Leonardo's ability to integrate disciplines might "guide us in repairing the split in our modern world." In Shneiderman's view, "Leonardo-like thinking could help users and technology developers to envision the next generation of information and communication technologies. The balance of the volume is devoted directly to this task of asking readers to think in different ways about computer technologies.

The 11 chapters of Shneiderman's (2003) volume are particularly intriguing and stimulating. Each one deserves consideration. For the sake of time and space, the concept of *universal usability* functions as a convenient and extremely appropriate illustration of the focus of this book.

In Chapter 3, "The Quest for Universal Usability," Shneiderman (2003) initially argued that:

> An important step toward the new computing is to promote the compelling goal of universal access to information and communication services. Enthusiastic networking innovators, business leaders, and government policymakers see opportunities and benefits from widespread usage. But even if they succeed in bringing low costs through economies of scale, new computing professionals will still have much work to do. They will have to deal with the difficult question, How can information and communication technologies be made usable for every user? (p. 36)

With roughly 20% of the American population now avoiding or ignoring computer technologies, it does become relevant to ask how computer technologies should be designed for a broad audience of unskilled users. In this regard, Shneiderman (2003) noted that, "Users of information and communication technologies also have a vital role to play in pushing for what they want and need." As Shneiderman argued in greater detail:

> Customer-oriented pressure will accelerate efforts to make new computing technologies usable and useful. Older technologies such as postal services, telephones, and television are universally usable, but computing technologies still are too hard for too many people to use. Low-cost hardware, software, and networking will benefit many users, but interface and information design improvements are necessary to achieve higher levels of success. (p. 36)

Within this context, Shneiderman defined "universal usability" as "more than 90 percent of all households being successful users of information and communication technologies at least once a week" (p. 36).

Shneiderman (2003) compiled arguments that warrant universal usability. For example, he argued that with universal usability the growth of the Internet will be a happy by-product, that e-businesses will expand, and that government services at all levels (e.g., national digital libraries, tax rules and filing, Social Security benefits, etc.) will improve when people are given access to them. Shneiderman noted that universal usability will require that the "gap between what users know and what they need to know" must be bridged. He believes that such accommodations are likely to "broaden the spectrum of usage situations, forcing technology developers to consider a wider range of designs which often leads, he believes, to innovations that benefit all users" (p. 48).

Ben Shneiderman is a professor in the Department of Computer Science at the University of Maryland as well as the founding director of the Human-Computer Interaction Laboratory at the University of Maryland. Additionally, given his publications as well as his academic appointments, there is no question that Shneiderman can adopt a computer science, if not strictly engineering, approach when considering the future of digital innovations and communication. Yet, he clearly abandoned these approaches in *Leonardo's Laptop*.

In *Leonardo's Laptop*, Shneiderman made a moral argument that humans throughout the world need and require. He did not make the argument based on trends and the agreement among experts about what those trends can and do mean to people. In this sense, he leaves the foundations that many futurists have adopted. His argument is based on what is moral and what can be done to achieve that moral foundation.

We do recommend and employ trend extrapolation and the Delphi procedure as ways to predict the future of digital technology with validity and reliability. Yet, we would ask if Shneiderman's approach in *Leonardo's Laptop* precludes other approaches for justifying what future can exist and how it can be used? In our view, Shneiderman is actually addressing the question of what the future **should be** in terms of global digital technologies, not what he thinks the future **will be**. In other words, Shneiderman is seeking to direct, forge, and control the future (securing the opinions of others to support a cause) more than he is seeking to predict what actually will happen and when it will actually happen.

Figure 4.5. Shneiderman's unique posture among futurists.

Pew Internet & American Life Project's (2005), The Future of the Internet *(September 2008),* The Future of the Internet II *(December 2008),* The Future of the Internet III, *and* The Future of the Internet IV *(2010)*

Pew Research Center for the People & the Press has used the Pew Internet & American Life Project as a vehicle to publish—as of June 30, 2013—some 15 studies on the future of the Internet or various key feature of the Internet. In all of these undertakings, essentially the same method is employed. A panel is created of seasoned and expert Internet users for each report; a questionnaire is emailed to these users and they are asked to numerically rank the degree to which they believe a predicted event or outcome is likely to occur. The results of this survey are usually reported in terms of those "agreeing" or "disagreeing" with the prediction.

The quality of such surveys depends on who is involved in the survey and how they are related to the object of study. In the Pew studies, the number of participants and their relationship to the Internet has varied tremendously. In every case, participants were also encouraged to provide written responses with their decisions if they wished, and in many of the studies these written responses are provided as part of the study. In terms of actual numbers, at least 500 participants were involved in each study. In the first study in 2005, Pew's panel included 1,286 respondents, of whom "about half are internet pioneers and were online before 1993" (p. 1). In the second futurism study in September, 2008, 742 Internet "stakeholders" participated, with biographies of 250 of these participants provided in the report. The sample is non-random, composed of those who opted in after being invited to participate (pp. 84–104). The third futurism study, in December 2008, involved 1,196 participants, 578 of whom were "leading Internet activists, builders, commentators" and 618 of whom were "stakeholders" (p. 3). The final general study of the future, in 2010, employed 895 "technology stakeholders" and "critics" (p. 3).

In terms of the Internet, these first four surveys were intentionally extremely broad (surveys from 2010 to the present have generally focused on specific topics such as "social relations," "Millennials," "apps and the web," "money," and "gamification"). In terms of particulars, we can use the 2005 survey as our extended example. In this survey, 1,286 participants were asked to indicate if they agreed or disagreed with each of 14 predictions, if they "challenged the prediction," or if they "did not respond." The first prediction explored in this survey dealt with "Network Intrastructure," and it specifically asked the respondents if they agreed or disagreed with the following prediction: "At least one devastating attack will occur in the next 10 years on the networked information infrastructure or the country's power grid." In terms of this prediction, 66% of the respondents agreed, 11% disagreed, 7% would challenge the prediction, and 16% had no response. Overall, the majority of respondents agreed on seven of the 14 predictions, with less than 50% agreeing on the other seven predictions.

While these Pew studies may not directly provide an extended discussion of the trends that have given rise to each prediction, the foundation for such trends is generally available in the other 10 to 20 studies carried out by Pew every year since 2000. And, the "expert" status of the participants on each panel normally minimizes such background and cognitive issues. In addition, if participants do not understand the prediction or feel they cannot agree or disagree, they can mark the "no response" categories. A high percentage of those checking the "no response" category would, of course, indicate that there may be a conceptual problem with the prediction itself.

In terms of the standards we introduce later in this chapter (e.g., trend extrapolation and the Delphi procedure), Pew Internet & American Life Project future studies, at least methodologically, adopt the kind of research standards we would recommend. Accordingly, we find these studies particularly valuable when we seek to predict the future of digital communication technologies.

We are extremely clear about our biases in terms of how to study predictions. Just as the Pew Internet & American Life Project has done, we recommend that predictions about the future be examined in terms of trend extrapolation (validity) and reliability (high agreement among observers).

Yet, we need to be cautious. Valid and reliable predictions may not be descriptions of what the future actually turns out to be. For example, in Pew's 2005 future study, some 66% of its respondents agreed that "At least one devastating attack will occur in the next 10 years on the networked information infrastructure or the country's power grid." Only 11% disagreed with the prediction. As we write this analysis some eight years after this prediction, while we are fully aware of the intensity and fear created by the September 11, 2001 attack on the World Trade Center in Manhattan, there is no indication that such an attack will occur in the next two years. In other words, while our assessment of a prediction may be valid and reliable, it may not tell us what the future actually will be.

We do think we need to be as scientific as we can be when anticipating our future, but also are fully aware that we may lack the power to be able to know with total confidence what that future will actually be.

Figure 4.6. Valid and reliable predictions and forecasts that actually occur.

Steven Johnson's 2005 Everything Bad Is Good for You: How Today's Popular Culture Is Actually Making Us Smarter

For Johnson (2005), his object of study is popular culture, and he is convinced that the content of the popular culture is affecting cognitive development and changing how we process information. We cast Johnson's work here as a form of futurism, for Johnson has ultimately argued that the popular culture is conditioning and changing how "young people" think. The content of popular culture, Johnson argued, is changing the cognitive development and thinking process of "young people," and is being affected by the introduction of new digital communication processes and devices such as laptop communication. In order words, Johnson is making a prediction: Young people will change and be different than what they would have been because of the power of the popular culture, a cultural conditioning system that is now being shaped and affected by new digital communication systems and devices.

As Johnson (2005, p. xiii) developed his argument, he articulated his thesis early on in *Everything Bad Is Good for You*: "The popular media steadily, but almost imperceptibly, is making our minds sharper, as we soak in entertainment usually dismissed as so much low brow fluff." More specifically, he noted that he is examining a "kind of thinking" that has become an "everyday component of mass entertainment" (p. 9). In Johnson's view, his explorations are "more systemic than symbolic, more about causal relationships than metaphor. It is closer, in a sense, to physics than to poetry" (p. 10). Johnson ultimately believes that the popular "culture is getting more intellectually demanding, not less" (p. 9) and that these cultural messages are a "new force altering the mental development of young people today," a force that is "largely a force for good," that is "enhancing our cognitive faculties, not dumbing them down" (p. 12).

For Johnson (2005), this "mental development" is controlled by two cognitive techniques fostered by digital communication techniques.

1. ***Probing***. In Johnson's view, "young people" are conditioned to employ ***probing***. Johnson described probing as "trial and error, by stumbling across things, followed by hunches," a "nuanced form of exploration" (p. 43). Johnson turned to the analysis by game scholar James Paul Gee, saying that Gee broke down probing as "looking around, clicking on something, or engaging in a certain action," which is then used to form a hypothesis, reprobing for effects, and then rethinking the original hypothesis (p. 45).

The first stage of probing is followed by a second cognitive process in which one dramatically narrows the cognitive process to one specific action.

2. *Telescoping*. Johnson noted that the probing process is also a form of telescoping in which one follows a pathway or pattern but constantly keeps narrowing choices thereby allowing one to focus on specific actions (p. 48). In Johnson's view, telescoping is a mental process of managing the "immediate problems while still maintaining a long-distance view" (p. 54).

From a broader perspective, Johnson is suggesting that our cognitive processes are undergoing a transformation. He is arguing that specific human thought processing is changing. This process is changing how we understand our environment or how we understand our reality. As we think differently, we will accordingly change our behaviors and values.

1. **Limits with the Johnson Analysis.** A cognitive approach often requires a written or perhaps even a visual conception of how the brain is functioning and how alternative stimuli are "rewiring" the links within the brain. Unfortunately, Johnson does not provide such a description.

2. **Digital Technologies Do Change How the Brain Functions**. We do think that people may actually be thinking differently after extensive use of digital technologies. In their November 2008 study (see also Small & Vorgan, 2009), Small and Vorgan pointedly argued that "The current explosion of digital technology not only is changing the way we live and communicate but also is rapidly and profoundly altering our brains" (p. 44). Small and Vorgan were extremely clear: "Because of the current technological revolution, our brains are *evolving* right now—at a speed like never before" (p. 44).

3. **Children Spend More and More Time With Digital Technologies**. Small and Vorgan (2008) initially argued that children are spending more and more time with digital communication devices. They summarized previous research that demonstrates that children are watching videos or DVDs with a total exposure of one hour and 20 minutes a day, that five- and six-year-old children spend an additional 50 minutes in front of the computer; the 2005 Kaiser Family Foundation study reported that those eight-to-18-years-of-age spend an average of eight-and-a-half hours with digital and video sensory stimulation a day.

4. **Digital Technologies Do Change Brain Activity and Human Behavior, Not Always in Positive Ways**. Small and Vorgan (2008) actually

studied two groups of people. One group was composed of computer-savvy subjects, and Small and Vorgan found that the left-front part of their brains was extremely active. A second group of "Internet-naïve subjects showed minimal, if any activation in this region" (p. 46). But, after just one hour a day for five days of searching and using the Internet, the left-front section of this second group's brains became active. Digital communication technologiesultimately create a different kind of thinking that involves "continuous partial attention" or a "process of keeping tables on everything while never truly focusing on anything" (Small & Vorgan, 2008, p. 47). In other words, extensive use of digital technologies enables people to react more quickly and accurately to visual stimuli and rapidly shift through large amounts of information. Additionally, while digital technologies enable us to track "our pals" online, these contacts may constitute a form of "false intimacy" compared to what is gained during face-to-face contact (p. 47). Moreover, the constant online contact of digital technologies may be ego-enhancing, feeding a sense of self-worth and ultimately becoming irresistible (p .47). With time, of course, excessive use of digital technologies can create "digital fog" and "techno-brain burn-out" (pp. 47–48).

Figure 4.7. Approaching change and future from a cognitive perspective.

Paul Messaris and Lee Humphreys's (2006) Digital Media: Transformations in Human Communication

In the "Introduction" (pp. xv–xx) to their 2006 book, *Digital Media: Transformations in Human Communication*, Messaris and Humphreys argued that:

> If we look at the future instead of the past, we can make some reasonable guesses about where digital media are headed—or, at least, where today's developers and users would *like* them to be headed—and it seems quite likely that it's going to take them a very substantial amount of time to get there. Measured by the yardstick of those future goals, the digital media of today still seem quite "young": they have a long way to go, and they are changing rapidly.

Specifically, "four broad trends are discernible." Messaris and Humphreys (2006, pp. xvi–xvii) argued that these four trends involve: (1) **Complete Simulation**—the ability to reproduce our real-world sense impressions so completely as to fool us into thinking that we are witnessing raw reality; (2) **Complete Interactivity**—"we would one day be able to interact with virtual worlds more naturally, using a broader repertoire of actions and communication modalities, such as eye movements or voice commands; (3) **Total Connectivity**—the ability to access all information, in all places, at all times; (4) **Complete Transparency**—"perhaps best thought of as a situation in which computers adapt to humans instead of the other way around."

1. While we don't have a complete and comprehensive statement from Messaris and Humphreys (2006) about their formal methods for studying the future, they do suggest that human goals may directly affect how the future unfolds. Perhaps identified by some as philosophically idealistic, many are confident that humans create their own destiny and their own future. We must admit that we are extremely sympathetic with that philosophy.
2. Likewise, while a "yardstick of future goals" should be employed to determine what "young people" want, it is unclear how those goals would be identified and measured. The attitude system and measures employed by Pew Internet & American Life Project (see Table 4.6 above) might be one way of determining what specific goals are.
3. We find the "four broad trends" identified by Messaris and Humphreys (2006) to be extremely insightful about predictions we are likely to be dealing with in the years to come.

Figure 4.8. Do we control our future?

Alvin and Heidi Toffler's (2007) Revolutionary Wealth: How It Will Be Created and How It Will Change Our Lives

As we have already suggested when we examined other Toffler books, the Tofflers (2007) presume that the industrial-based economy is giving way to a knowledge-based economy in which money is no longer the sole determinant of wealth. In the Tofflers' view, people have become "prosumers" who consume their own production. Indeed, the Tofflers think that people are now taking on "third jobs," work that they do without pay, such as parenting, painting their own houses, blogging, and improving their own diets. Prosumerism of this kind blurs the distinction between creators and consumer of goods and services. Corporations have benefitted from prosumerism and are adapting by eliminating labor costs as reflected in the creation of ATMs instead of bank tellers and automated or "personalized" checkout lines instead of human cashiers.

Within this context, the Tofflers (2007) proposed a time-space-lifestyle and economic knowledge construct to suggest elements of the industrial society that are being outstripped. The new elements of this construct are creating a "revolutionary wealth." Along with this new wealth, new taxes are being created. When someone fills out a form, for example, a "time tax" on the person filling out the form could be said to have been created. In this regard, for example, our cell phones (which we pay for in terms of the devices themselves and then the service to maintain them) are also now used to determine and track identity. Ultimately, however, the Tofflers maintain that revolutionary wealth can be used to eliminate poverty with a specific four-part antipoverty plan that ultimately would stabilize the world.

> Schemes about the future can be amazingly elaborate, complicated, and ultimately a synthesis of a host of variables. Such schemes are normally highly personal integrations. While intriguing and potentially useful for any number of specific reasons, the scheme as a whole remains extremely personal. The Tofflers, for example, are convinced that barter is on the increase and that capitalism and money are becoming extinct. Yet, there are elements surrounding these assumptions that can be insightful to a reader even though the larger claims about barter, capitalism, and money are misconceptions. In order words, the personal may be a motive for predicting the future and it may be an incentive to predict, but it also requires the integration of specific trends and shared agreements from others if the predictions are to be credible and actually likely to happen. Yet, the Tofflers—for over 30 years—have encouraged creative thinking about the future. We hope they continue their efforts.

Figure 4.9. The personalized view of the future.

Damien Broderick's (2008) Year Million

This volume seeks to identify what human life will be like in one million years. The back cover of this book raises its own skeptical view of the book itself: "In fourteen original essays, leading scientists and science writers cast their minds forward to 1,000,000 C.E., exploring an almost inconceivable distant future." A similar kind of skepticism exists in the "Introduction": "A million years—it's a haunting number, quite terrifying if you can put your imagination to work trying to grasp what it means, what it implies." In this regard, an analogy is used, and its use suggests, not just a level of skepticism, but a foundation for complete rejection of the scheme offered here. Accordingly, the editor of the volume, Damien Broderick (p. ix) maintained that a historical analogy might be appropriate: "Casting our minds a million years into the past, we find in the wide world no trace anywhere of familiar, or comforting intelligence." Indeed, a historical view of life one million years ago would suggest that human beings did not exist at all in any form we would understand now to be human. And, the editor seems to recognize that possibility about the future as well: "A million years hence, if any of our lineage survive, they will be very different from how we are today, far more alien to us than we are to early *Homo*" (p. x). As the editor concluded: "Is there any way to get a sense of those pitiless gulps of time, a million years echoing away to either side of us? Not really" (p. x). Nonetheless, the editor concluded that "We can try, though, by analogy" (p. x).

However, even the most creative of analogies do not provide a clue about the constraints that should govern a prediction about the year 1,000,000 CE. While we could try to summarize the entire volume, actually any one of the chapters reveals what we need to know about the volume. Therefore, we have randomly selected the conclusion of Rudy Rucker's chapter, "The Great Awakening" (pp. 228–249) to illustrate the effort undertaken in this volume.

Rucker presumed that, by the year one million, telepathy would be possible: "Of course, we won't stop at mere telepathy!" (p. 247). As Rucker put it, "By the Year Million, we'll have teleportation, telekinesis, and the ability to turn our thoughts into objects" (p. 247). As Rucker explained, "Teleporting can be done by making yourself uncertain about which of two possible locations you're actually in—and then believing yourself to be 'there' instead of 'here'" (p. 247). Rucker continued by suggesting that "we can work out this uncertainty-based method of teleportation as a three-step process" involving self-visualization of where you are and want to be, a mental concept of how you can "weave them together," and then a "quantum collapse of your wave function" (p. 247). Having teleported the self, Rucker suggested that one can teleport other objects by briefly "merging" with the object. In terms of support for such a process, Rucker drew on "a science-fictional idea in a Robert Sheckley story" (p. 248). What would be the ultimate fate of the human being? Rucker concluded, "By the Year Million, we'll be pushing on past the realm of the finite and into the transfinite realms beyond the worlds of this local universe" (p. 249).

1. Every method employed to verify a prediction sooner or later faces limitations that prevent a researcher from confirming a conclusion. For example, while researchers may be advocates, if not true believers, in trend extrapolation, they will note that as the trend is extended over years and perhaps even decades, more and more variables can and will affect the trend. As the variables are recognized, the trend slowly reveals more and more variance. With time, the researcher simply has no idea what is a "best line fit for" dealing with the variables affecting the trend. Time, in this case, reflects the complexity of not knowing what is controlling and affecting all human conditions.
2. The Million Year predictions encounter limitations, but long before the one million year mark is reached. There are simply too many variables—many of them unknown and inconceivable—that can occur before the one million year mark is reached.
3. The study of the future should not become science fiction. This distinction must be recognized if the identities of futurism and science fiction are retained. Admittedly, science fiction may be a motivation for studying the future, but futurism avoids speculation as the sole or primary foundation for a prediction of the future.

Figure 4.10. Year Million: Any examination of the human condition encounters limitations.

Pat LaPointe's (January 4, 2011) "Predictions for Social Media Metrics: 2011"

In the introduction to this essay, LaPointe advised, "Never hesitate to jump on the New Year prognostication band-wagon, I'm giving you a few predictions for some significant new elements and important evolutions in social media metrics in 2011" (p. 1). This advice provides the introduction for four predictions: (1) "**Rapid maturation**. Social media measurement will mature rapidly now that there is real money being spent by marketers in the social realm"; (2) "**Government Drives Better**. Social media tracking…will improve significantly, mostly due to government regulations"; (3) "**Online and offline will merge**. By year end, social media will no longer be equated with just online marketing tactics"; and (4) "**Predictive value is emerging**. The holy grail of social media is the "so what"—what will all the 'buzz' do for sales and profits?"

As we read LaPointe's statement, several basic questions about the nature of predictions emerge.

1. Are all of the terms within the prediction clear? What does a concept such as *mature rapidly* mean, and how can it be operationalized so that it is discrete from other kinds of change measures such as *little* or *modest change*?
2. Does the prediction identify a specific change and identify what kind of change will occur? Accordingly, while it is a prediction to say that "Things will continue as they have," the statement is an exaggeration, for everything changes over time. In everyday parlance (the vernacular), a prediction is a forecast of things that have not yet happened. As Angeles (1981, p. 222) argued in the *Dictionary of Philosophy*, a prediction is a "Declaration that something will happen before it happens." In this sense, a prediction identifies a change in the present state of affairs. Statements that note that a situation will continue as it has are simply not predictions.
3. Does the prediction make a difference? Certainly, context determines if a change is significant. The person making the prediction needs to explain how the change makes a difference.

Figure 4.11. Standards for assessing future predictions.

Lev Grossman's (February 21, 2011) "2045: The Year Man Becomes Immortal," Time

The cover of the February 11, 2011, issue of *Time* noted, "If you believe humans and machines will become one, welcome to the Singularity movement." In this *Time* essay, Grossman explored the multiple meanings embedded in Ray Kurzweil's (2005) *The Singularity Is Near: When Humans Transcend Biology.* While clearly written and organized, some translation of Kurzweil's book is understandable. His volume is 652 pages with a precise and careful organization especially useful for someone interested in the issues involved with long-term futuristic predictions. He explored issues in stages, implicitly using trend extrapolation materials to support his claims. Kurzweil also broke the pace and coherence of his own discussions by considering the counterstatements and reactions of others to his thoughts. Moreover, Kurzweil was careful to consider the "limitations" of his claims at all points in his volume. In all, while the work of a scholarly intellectual, *Time* magazine's summary actually gave Kurzweil and the Singularity the popular venue it deserved. Indeed, Kurzweil's thesis is powerful, if not shocking: he predicted that by the year 2045 human beings and machines will merge.[2]

Specifically, Kurzweil (2005) argued that we are headed for a period of Singularity. What is Singularity? Kurzweil argued that Singularity is:

> a future period during which the pace of technological change will be so rapid, its impact so deep, that human life will be irreversibly transformed. Although neither utopian nor dystopian, this epoch will transform the concepts that we rely on to give meaning to our lives, from our business models to the cycle of human life, including death itself. (p. 7)

As Kurzweil later noted, "The key idea underlying the impending Singularity is that the pace of change of our human-created technology is accelerating and its powers are expanding at an exponential pace" (pp. 7–8). In terms of its specific functions and outcomes, Kurzweil maintained that:

> The Singularity will allow us to transcend these limitations of our biological bodies and brains. We will gain power over our fates. Our mortality will be in our own hands. We will be able to live as long as we want (a subtly different statement from saying we will live forever). We will fully understand human thinking and will vastly extend and expand its reach. By the end of this century, the nonbiological portion of our intelligence will be trillions of trillions of times more powerful than unaided human intelligence.

> The Singularity will represent the culmination of the merger of our biological thinking and existence with our technology, resulting in a world that is still human but that transcends our biological roots. There will be no distinction, post-Singularity, between human and machine or between physical and virtual reality. If you wonder what that will remain unequivocally human in such a world, it's simply this quality: ours is the species that inherently seeks to extend its physical and mental reach beyond current limitations. (p. 9)

In terms of evolution, Kurzweil (2005) essentially offered a new conception of human history, a history composed of six epochs. And, he has specifically reported that the "Singularity will begin with Epoch Five and will spread from Earth to the rest of the universe in Epoch Six" (p. 14). Figure 4.12, "Kurzweil's Six Epochs," provides a convenient conception of these six epochs (pp. 14–33).

Overall, Kurzweil offers a new concept of human history and a future grounded in an intelligence transformed by its merger with computerized machinery. While grounded in trend extrapolations and exploring the kind and degree of agreement that exists with other specialists on a host of these questions, we are unsure of the validity and reliability of the future Kurzweil provides. Such a conclusion suggests that trend extrapolation and the Delphi procedure have limitations. We certainly acknowledge that under certain circumstances, traditional futuristic methods may be unable to respond to all of the future contexts that some can conceive of as a possible future.

Epoch 1—"A few hundred thousand years after the Big Bang, atoms began to form, as electrons became trapped in orbits around nuclei consisting of protons and neutrons. The electrical structure of atoms made them "sticky." Chemistry was born a few million years later as atoms came together to create relatively stable structures called molecules. Of all of the elements, carbon proved to be the most versatile; it's able to form bonds in four directions (versus one to three for most other elements), giving rise to complicated, information-rich, three-dimensional structures" (Kurzweil, 2005, pp. 14–15).

Epoch 2— "In the second epoch, starting several billion years ago, carbon-based compounds became more and more intricate until complex aggregations of molecules formed self-replicating mechanisms, and life originated" (p. 16).

Epoch 3—Among a host of functions, "in the third epoch, DNA-guided evolution produced organisms that could detect information with their own sensory organs and process and store information in their own brains and nervous systems" (p. 16).

Epoch 4—This stage traces "the evolution of human-created technology. This started out with simple mechanisms and developed into elaborate automata (automated mechanical machines). . . . To compare the rate of progress of the biological evolution of intelligence to that of technological evolution, consider that the most advanced mammals have added about one cubic inch of brain matter every hundred thousand years, where we are roughly doubling computational capacity of computers every year" (p. 16).

Epoch 5, The Merger of Human Technology with Human Intelligence—"Looking ahead several decades, the Singularity will begin with the fifth epoch. It will result from the merger of the vast knowledge embedded in our own brains with the vastly greater capacity, speed, and knowledge-sharing ability of our technology. The fifth epoch will enable our human-machine civilization to transcend the human brain's limitations of a mere hundred trillion extremely slow connections" (p. 20). The result, suggested Kurzweil, will be profound: "The Singularity will enable us to overcome age-old human problems and vastly amplify human creativity" (p. 21).

Epoch 6, The Universe Wakes Up— Kurzweil surveyed all of those factors that might affect, control, and restrict the roles and functions of human intelligence in the universe. In Kurzweil's view, human intelligence will not be limited because of human biology (see Epoch 5). Nor will factors such as the mass of the universe, the existence of antigravity, Einstein's cosmological constant, or restrictions related to the and functions of human intelligence

(p. 361). Indeed, for Kurzweil, the only limit to human intelligence, once it is linked to computerization, would be a black hole (p. 362).

But, in Kurzweil's view, based on recent reconceptions offered by Stephen Hawking, "the ultimate computers that we can create would be black holes. Therefore, a universe that is well designed to create black holes would be one that is well designed to optimize its intelligence (p. 364). In this context, Kurzweil's (p. 364) fundamental notion is relevant: "intelligence is more powerful than physics." By the time this notion rules, we will—of course—have a totally new conception and understanding of the universe and the roles and functions of human intelligence in this universe (p. 367).

Figure 4.12. Kurzweil's six epochs.

Since *Time*'s February 21, 2011 focus on Kurzweil, futurism and future predictions have continued to occupy an important role in the popular culture. For example, analyses such as "The 22nd Century at First Light: Envisioning Life in the Year 2100" (*The Futurist*, September–October, 2012) and "The Future of Science 50, 100, 150 Years from Now" (*Scientific American*, January, 2013) are readily available in public bookstores and at local news stands. We see no end of these kind of publications and analyses in the foreseeable future, and such explorations may actually be a healthy and useful stimulus for human creativity.

Trend Extrapolation and the Delphi Procedure as Methods for Future Predicting[3]

Trend Extrapolation

Trends have always been with us, and they are part of the popular vernacular. For example, *Webster's New Collegiate Dictionary* (1981, p. 1236) offers several conceptions, which include a "general drift, tendency or bent of a set of statistical data as related to time" and a "linear calculation that shows predicted performance based on historical data." In terms of technological trends, Martino reported that "new levels of capability are reached as a result of advances which transcend the limitations of previous technical approaches" and "usually arise in some area of technology which is considerably different from the area in which progress was halted by a barrier, or by the concerned technology having reached saturation." For example, progress with computers from the 1930s and 1940s required an innovation and the introduction of miniaturization in computer design as well as the development of the vacuum tube in the 1950s. While the base technologies themselves changed from the 1930s to the 1950s, the increases in computer speed and computer performance as well as the decreases in computer storage costs were all measured by the same standards even though computer technologies had shifted to miniaturization and vacuum tubes during this period.

The trends used for technological prediction have varied tremendously. In his volume, *Trends: How to Prepare for and Profit From the Changes of the 21st Century*, published in 1997, Celente made forecasts that apparently began four years later, in the year 2000, although the date for each of his predictions is often extremely difficult to determine. For example, one of Celente's "trendposts" is that online education will dramatically affect higher education. Noting distance-learning opportunities offered by Brown University and the University of Pennsylvania in 1996, Celente (1997) argued that:

> TRENDPOST Interactive, on-line learning will revolutionize education. Demand for "distance learning" software, hardware, and service will generate a mighty new component within the telecommunications and publishing industries. Especially in its early stages of development, distance learning will provide rich opportunities for small entrepreneurs, scholars, artists, educators, and inventors, as well as for established communications giants. (p. 249)

For others, trends have been far longer and more clearly identifiable. Beckwith (1967) used 500-year trends for "scientific predictions of major social trends" in areas such as government, population, work and wages, production control, finance, agriculture, industry, commerce, houses and cities, communication, education, health care, family life, crime, religion and philosophy, and science. Based upon these 500-year trends, Beckwith (1967, pp. 308–336) extended the direction and curve of his trends (a best line fit) and suggested what life will be like in the year AD 2500 and the years AD 2500 to 3000.

Collections of such trends are readily available. Some historical volumes can be very useful, such as Beckwith's (1967) volume or McHale's (1973) *World Facts and Trends*. Moreover, *The New York Times* has been publishing a variety of trends (mostly since the early 1970s) on a host of social issues (e.g., spanking children as good or bad, men are more suited to politics than women, marijuana should be legal, percentage of people who believe in life after death, percentage of people who believe sex before marriage is "always wrong," percentage of people who believe homosexuals should be allowed to teach in colleges and universities, percentage of those who have seen an X-rated movie in the last year, percentage of people saying that they are "not too happy," "pretty happy," or "very happy," etc.). Of course, for some 15 years, Pew Research Center's Pew Internet & American Life Project has traced trends in media access, usage, reactions, and so forth. Accordingly, while unusual and unexpected questions might emerge, generally speaking, trend data on virtually any question, especially questions dealing with communication technologies, is readily available now.

Delphi Procedure

Definition

The Delphi procedure is a method for forecasting future events. The procedure frequently possesses the following seven characteristics:

First, it employs experts to make predictions about the future. The word *experts* in this context should be used carefully. Sometimes children (ages 5 to 17) playing video games might be considered a group of "experts" if the question is how and why children react the way they do to video games.

Second, it seeks to determine if agreement exists about the nature and date of future among experts or if it is possible to secure such agreement among experts. Hence, it is appropriate to talk about a panel of experts to underscore the agreement that exists, or can be secured, among experts.

Third, the procedure frequently enables experts to interact about their predictions or provide feedback to each other about their predictions. Indeed, formal stages may be created in which the Delphi experts are contacted periodically to determine if their opinions or degrees of agreement have changed.

Fourth, if the Delphi procedure is carried out in stages, the panel of experts may be allowed to interact and reflect upon their findings at the completion of each stage.

Fifth, the interaction among experts is very likely to include influence and persuasion. The experts implicitly and explicitly seek to influence others about the validity and reliability of any specific predictions at each stage. The attempts to influence and persuade can be a central feature of how agreement is reached by the experts on the panel.

Sixth, in addition, the procedure normally seeks to determine and measure the degree of likelihood that a future event will exist as a prediction as well as when it is predicted to occur.

Seventh, and finally, the procedure normally seeks to determine and measure the degree of confidence that the panel of experts has about its predictions.

History

The Delphi procedure gains part of its identity from its history as a research method. The technology-forecasting studies that eventually led to the development of the Delphi method started in 1944. By 1946, Project RAND (an acronym for Research and Development) was created to study the "broad subject of intercontinental warfare other then surface." Project RAND sought to determine what a discipline could do if scientific laws did not exist, and the effort was to determine if the testimony of experts could somehow be employed. The problem, then, was how to use this testimony and, specifically, how to combine the testimony of a number of experts into a single useful statement. The Delphi method therefore recognized human judgments as legitimate and useful inputs to generate forecasts.

Yet, the Delphi procedure also possesses limitations. Single experts sometimes suffer biases; and, group meetings suffer from "follow the leader" tendencies and reluctance to abandon previously stated opinions (Gatewood & Gatewood, 1983; Fowles, 1978).

In order to overcome these shortcomings of the basic notion of the Delphi method, theoretical assumptions and methodological procedures were developed in the 1950s and 1960s at the RAND Corporation. Forecasts about various aspects of the future are often derived through the collation of expert judgment. Dalkey and Helmer developed the method for the collection of judgment for such studies (Gordon & Hayward, 1968).

Fowles (1978) asserted that the word *Delphi* refers to the hallowed site of the most revered oracle in ancient Greece. Forecasts and advice from gods were sought through intermediaries from this oracle. However, Dalkey (1968) stated that the name "Delphi" was never a term with which either Helmer or Dalkey (the founders of the method) were particularly happy. Dalkey (1968) acknowledged that it was rather unfortunate that the set of procedures developed at the RAND Corporation, and designed to improve methods of forecasting, came to be known as "Delphi." He argued that the term implies "something oracular, something smacking a little of the occult," whereas, as a matter of fact, precisely the opposite is involved; it is primarily concerned with making the best you can of a less-than-perfect kind of information.

One of the very first applications of the Delphi method carried out at the RAND Corporation is illustrated in the publication by Gordon and Helmer

(1964). Its aim was to assess the direction of long-range trends, with special emphasis on science and technology, and their probable effects on society. The study covered six topics: scientific breakthroughs, population control, automation, space progress, war prevention, weapon systems (Gordon & Helmer, 1964). The first Delphi applications were in the area of technological forecasting and aimed to forecast likely inventions, new technologies, and the social and economic impact of technological change (Adler & Ziglio, 1996). In terms of technology forecasting, Levary and Han (1995) stated the objective of the Delphi method is to combine expert opinions concerning the likelihood of realizing the proposed technology as well as to determine when expert opinions concerning the expected development time will occur. When the Delphi method was first applied to long-range forecasting, potential future events were considered one at a time as though they were to take place in isolation from one another. Later on, the notion of cross impacts was introduced to overcome the shortcomings of this simplistic approach (Helmer, 1977).

The outcome of a Delphi sequence is nothing but opinion. The results of the sequence are only as valid as the opinions of the experts who make up the panel (Martino, 1978). The panel viewpoint is summarized statistically rather than in terms of a majority vote.

Advantages of the Delphi Procedure

Martino argued that there are three advantages to using a Delphi procedure for technological forecasting:

- The total information available through the pooling of expert information in a Delphi procedure is greater than the information possessed by any single member.
- A group of people are more likely to consider as many, if not more, factors external to the field of the forecast, and therefore avoid mistakes when making forecasts.
- Groups tend to be more willing to take risks than individuals.[4]

Limits of the Delphi Procedure[5]

The Delphi method has been supported as a procedure, but it has also been critically examined. The most extensive critique of the Delphi method was made by Sackman (1974) who criticized the method as being unscientific, and Armstrong (1978), who wrote critically of its accuracy. Martino (1978) underlined the fact that Delphi is a method of last resort in dealing with extremely complex problems for which there are no adequate models. Helmer (1977) stated that sometimes reliance on intuitive judgment is not just a temporary expedient but in fact a mandatory requirement. Makridakis and Wheelright (1978) summarized the general complaints against the Delphi method in terms of (a) a low-level reliability of judgments among experts and therefore dependency of forecasts on the particular judges selected; (b) the sensitivity of results to ambiguity in the questionnaire that is used for data collection in each round; and (c) the difficulty in assessing the degree of expertise incorporated into the forecast. Martino (1978, pp. 31–52) listed major concerns about the Delphi method:

- Discounting the future: Future (and past) happenings are not as important as the current ones, therefore one may have a tendency to discount the future events.
- The simplification urge: Experts tend to judge the future of events in isolation from other developments. A holistic view of future events where change has had a pervasive influence cannot be visualized easily. At this point cross-impact analysis is of some help.
- Illusory expertise: Some of the experts may be poor forecasters. The expert tends to be a specialist and thus views the forecast in a setting that is not necessarily the most appropriate one.
- Sloppy execution: There are many ways to do a poor job. Execution of the Delphi process may lose the required attention easily.
- Format bias: It should be recognized that the format of the questionnaire may be unsuitable to some potential societal participants.
- Manipulation of Delphi: The responses can be altered by the monitors in the hope of moving the next round responses in a desired direction.

Using Trend Extrapolation and the Delphi Procedure as a Critical Framework for Assessing and Creating Predictions

The Pew Internet & American Life Project has employed both trend extrapolation and the Delphi procedure for over 10 years when offering future predictions. An early use of both of these methods occurred in 2008 in Pew's third overall view of the future of the Internet. This study produced eight predictions that Pew forecast for the year 2020 as summarized below (Anderson & Rainie, 2008, pp. 5–6) in Figure 4.13, "Eight Predictions by Pew Internet & American Life about the Future of the Internet in the Year 2020 in Rank Order by Degree of Agreement Among Delphi Participants."

Most Likely to Occur

1. **Next-generation research will be used to improve the current Internet; it won't replace it.** 78% of experts and 81% of total respondents agree; 6% of experts and 9% of all respondents disagree; and 16% of experts and 11% of total respondents did not respond.
2. **The mobile phone is the primary connection tool for most people in the world.** 77% of experts and 81% of total respondents agree; 27% of experts and 19% of total respondents disagree; and 0% of experts and 0% of total respondent did not respond.
3. **Talk and touch are common technology interfaces.** 64% of experts and 67% of total respondents agree; 21% of experts and 19% of all respondents disagree; and 16% of experts and 14% of total respondents did not respond.
4. **Few lines divide professional time from personal time, and that's OK.** 56% of experts and 57% of total respondents agree; 29% of experts and 29% of all respondents disagree; and 15% of experts and 14% of total respondents did not respond.
5. **Many lives are touched by the use of augmented reality or spent interacting in artificial spaces.** 55% of experts and 56% of total respondents agree; 30% of experts and 31% of all respondents disagree; and 15% of experts and 13% of total respondents did not respond.

Unlikely to Occur

6. **Transparency heightens individual integrity and forgiveness.** 45% of experts and 44% of total respondents agree; 44% of experts and 45% of all respondents disagree; and 11% of experts and 10% of total respondents did not respond.
7. **Social tolerance has advanced significantly due in great part to the Internet.** 32% of experts and 33% of total respondents agree; 56% of experts and 55% of all respondents disagree; and 13% of experts and 13% of total respondents did not respond.
8. **Content control through copyright-protection technology dominates.** 31% of experts and 31% of total respondents agree; 60% of experts and 61% of all respondents disagree; and, 9% of experts and 8% of total respondents did not respond.

Figure 4.13. Eight predictions by Pew Internet & American Life about the future of the Internet in the year 2020 in rank order by degree of agreement by Delphi participants.

As Figure 4.13 suggests, if agreement among experts is taken to be a critical foundation for assessing the reliability of a future trend, then not all of these predictions are equally reliable. Indeed, at least 50% of the experts agree on the first five predictions identified on Table 4.13. In terms of predictions #6 through #8, the experts are either evenly divided or a majority disagree with the stated prediction. At least, these last three predictions might be best understood as "unknown" or "unsure." We cannot really say these three predictions will not occur; we can only say that we don't know or are unsure about the likely outcomes of predictions 6, 7, and 8.

In this sense, we can use the Delphi procedure as a way of distinguishing what we are confident about and what we are unsure of in terms of future predictions.

This same kind of critical posture might be used to assess the strength and quality of trends. Figure 4.14 provides some guidelines and questions for you in judging the quality of trend data.

1. Is the trend long enough in terms of time to make an extrapolation? Some have used 500 year trends while others have used 4-to-5-year trends. How long or large should a trend be before it is treated as valid and reliable? Is the extrapolation restricted by the time frame governing the trend?
2. Has the same unit of observation and measurement been used for each year of the trend? For example, what variations in the characteristics of subjects can and should be tolerated from year to year?
3. As data from a given year are compiled, will there be variations among the data? Is a "best line fit" that averages the variations in data appropriate? Is it valid in terms of what the predictions seek to explore?
    What other standards should be employed to evaluate the quality and usability of a trend extrapolation?

Figure 4.14. Standards and questions for evaluating the quality of trends.

Conclusion and Future Possibilities

In this chapter, we have not only examined what previous prognosticators have suggested about the future of Internet and digital communication technologies, but we have also examined how and why they make the future claims that they make as well as how valid and reliable such predictions are. In other words, we have sought to establish some standards for evaluating predictions that are based upon known multiyear trends that also have high agreement about what these multiyear trends mean to people and their lives.

In all, no matter what we might hope, we cannot ignore the future. Every corporation we know takes special care to predict what is happening to their product or service in terms of its various publics. Such "planning" is part of what occurs through the business world today. Effective planning, we would hold, must therefore consider what specific methods should be used for judging the quality of its predictions. You should be able—with confidence—to identify predictions that are likely to occur and when they will occur and to contrast this with predictions that are unlikely to ever occur or will occur at a time period when action cannot be initiated to control outcomes. In our view, thinking about the future—its likely scenarios—is always an exciting endeavor. We hope you share this view.

Key Terms

Delphi Extrapolation
Prediction
Reliability
Trend
Validity
Universal Usability

Chapter Exercise

Current Issues: Divide the class into four to five groups, and assign a major publication (e.g., *The Wall Street Journal*, *The New York Times*, *The Washington Post*, or *The Futurist*) to each group. Each group should conduct a content analysis of their publication to determine the degree to which each publication offers forecasts or predictions about the future. Some of these forecasts or prediction may deal only with anticipated daily events (e.g., weather reports and traffic problems) while others may suggest that major events of far more critical significance will occur (e.g., when Congress will act on specific legislative proposals) to major sociocultural transformations (e.g., technological breakthroughs). What does each publication use as a reason to make people believe that a change will occur? What kind of evidence does each publication use to establish a forecast or prediction as credible? When all of the groups have reported, determine what forecasts or predictions are most likely to be made? Which forecasts and predictions seem most believable? Which forecasts and predictions actually occurred?

Debate It: Split the class into two groups to debate the resolution: *Resolved: That predictions become self-fulfilling* (i.e., regardless of circumstances, many events occur only because people begin to think about them and make plans about certain anticipated events).

Ethical Consideration: While some people ignore forecasts and predictions as non-real, others believe they have a self-contained power and control the lives of people. For example, for some people, forecasts and predictions found in the Bible have been perceived as a matter of their personal beliefs, as real, and as extremely likely to occur. What kind of ethical issues are involved when people believe that certain forecasts and predictions will occur? Does the anticipated future become a "reality" that determines what people are and should be doing? To what degree must nonbelievers adhere to the restrictions of this anticipated reality?

Everyday Impact: Daily schedules and "calendar events" are treated by most people as future events that will occur. And, making plans for the future is part of everyday life, including educational, vocational, and interpersonal goals that ultimately become anticipated lifestyle events that have the power and value of actually lived events. Keep a one-day log of any day in your life, and after a week has passed, go back through the activities recorded in the log determining if a specific action responded to immediate circumstances or anticipated circumstances. Determine the degree to which your life activities are present- or future-oriented. Have you maintained the proper balance between present and future events?

Future Focus: Review all of the forecasts and predictions made in this chapter. Select the one forecast or prediction that you think is most likely to occur. Provide reasons for your position. Compare your forecast or prediction to those of others in the class.

Endnotes

1. At this point, Cortada provides an outstanding reference and recommendation for future reading: Bell and Gray (1999).

2. For details regarding Kurzweil's notions of the human body, see Kurzweil and Gross-man (2004).
3. For an example of how the authors of this volume have used these methods, see Chesebro (2009).
4. Martino (1972), pp. 18–19. For a discussion of "The Accuracy of Delphi," "The Precision of Delphi Estimates," and "The Reliability of Delphi," see Martino, pp. 31–52.
5. All of these "limitation" references are provided in Martino (1972, pp. 31–52).

5

Can or Should a Purpose Be Attributed to Digital Technologies Such as the Internet?

The Internet is a global system of computer networks, sharing a common or standardized protocol that serves billions of users throughout the world. Content placed on the Internet can potentially be believed by millions of people, and, accordingly, it can control certain policies and actions in major areas of the world. But, given how decentralized the Internet is by language, culture, and technologies, it seems virtually impossible to control the Internet with just one international entity, governing body, or government. Indeed, for some, the Internet should not be controlled by any single entity, even if it could be. In this context, the Internet is viewed as a resource for all unique individuals, and issues of individual access and expression on this system fall within the domain of privacy. For others, the Internet—the information contained on it, how and who accesses it, and how information is used—can directly affect the security and sovereignty of individual nation-states and governments. Accordingly, individual nation-states and governments can feel that they must regulate and control how the Internet is used.

In this chapter, we explore how the Internet is and should be controlled. The question of control has become especially difficult because of the rapid and complicated growth of the Internet. As we noted in Chapter 1, for example, some 14% of adults in the United Sates used the Internet in June 1995, some

36% used it in March 1998, and some 53% U.S. adults accessed the Internet by February 2001.

The degree and kinds of control imposed on the Internet determine its essential purpose. By analogy, if we had this argument about television in the United States, some might claim that television—especially for the nation's young—must be judged by both its educational and entertainment values. What we mean by *educational* and *entertainment* would be extremely difficult to determine and it would be even more difficult to suggest who should be responsible for determining and applying these judgments. And, of course, it would have to be decided what penalties the responsible group should apply if they detected violations. These issues grow exponentially when we think about the Internet. Many websites are international, and it is unclear how international bodies can even control such sites. Indeed, we can also examine how international groups, such as the Internet Corporation for Assigning Names and Numbers (ICANN), have functioned during the last 10 years to try to determine what international agencies might function as Internet control systems. In all, then, questions of if, how, and what should be regulated become tremendously difficult in an international arena.

In this chapter, we trace the developments of the first attempts to attribute a purpose to the Internet, the success of recent control mechanisms, and issues embedded in recent conflicts regarding control of the Internet.

Questions Previewing This Chapter

1. *What were some of the early attempts to attribute a purpose to the Internet?*
2. *How successful have recent mechanisms been at controlling the Internet?*
3. *What issues are involved in recent contemporary Internet conflicts?*

Early Attempts to Attribute a Purpose to the Internet

Our point of departure here is with the "inventory" of the World Wide Web. We turn first to Tim Berners-Lee to suggest what the motivating purpose for the Internet might have been.

Original Purpose

Described as the "inventory of the World Wide Web," in his 1999 book *Weaving the Web: The Original Design and Ultimate Destiny of the World Wide Web*,

Tim Berners-Lee (1999) argued that the Internet is "much more than a tool for research or communication; it is a new way of thinking and a means to greater freedom and social growth than ever before possible."[1] Specifically, Berners-Lee presumed that all information is interconnected and that the Internet provides the means for such a total immersion, a "vision encompassing the decentralized, organic growth of ideas, technology, and society" (p. 1). Rather than regulate and restrict options for individuals, Berners-Lee suggested that, "The vision I have for the Web is about anything being potentially connected with anything. It is a vision that provides us with new freedom, and allows us to grow faster than we ever could when we were fettered by the hierarchical classification systems into which we bound ourselves" (pp. 1–2).

Given the view expressed by several governments recently (specifically that access to information should be restricting based on national security issues), in sharp contrast, Berners-Lee (1999) originally argued that the controlling purpose of the Internet was to give everyone permanent access: "An essential goal for the telecommunications industry (and regulatory authorities) should be connecting everyone with *permanent* access" (p. 159). By 2003, Berners-Lee's focus was even clearer. He argued that "universality" was the "essential" purpose of the Internet:

> The Web was designed to be a universal space of information, so when you make a bookmark or a hypertext link, you should be able to make that link to absolutely any piece of information that can be accessed using networks. This universality is essential to the Web: it loses its power if there are certain types of things to which you can't link. (pp. xii–xiii)

A Purpose is Partially Realized

In 2001, Bernardo Huberman began a very different set of explorations in which he sought to demonstrate that the Internet actually was realizing its objective of providing an active communication mechanism for all individuals. He began with the premise that "in spite of its haphazard growth, the Web hides powerful underlying regularities: from the way its link structure is organized to the patterns that one finds in its use by millions of people" (2001, p. vii). Specifically, Huberman tracked and measured the size of websites, who initiated these websites, and described the pattern of interactions among websites. While larger firms were certainly very active on the Internet, the average website was not a product of a major corporation. Using a rich variety of tables and graphs, Huberman was able to conclude that, "In terms of its web pages and links, the Internet is strongly individualistic" (p. 24). Indeed, in Huberman's view, "it should be possible to predict who could be friends with whom by analyzing text, links, and mailing lists" (p. 40).

The Harsh Realities of the Politics of the Internet Emerge

The harsh political realities began to vividly emerge at the outset of the 21st century. This "harsh reality" is slow to emerge, and it is clearly not built systematically upon earlier actions and conceptions. And, in our view, it ultimately has radical political overtones, but overtones that are not so readily identified in earlier interactions. Accordingly, the emergence of these "harsh realities" proceeds in rather awkward steps, because those involved simply do not provide the kind of clear links and interrelationships we might prefer.

From our perspective, we begin here with the publication of Michael Hart's (2000) volume *The American Internet Advantage*. The book's title is particularly instructive, suggesting that the United States' domination of the Internet is somehow possible, desirable, if not a very "natural" kind of situation. And, in his volume, Hart specifically argued that the United States has dominated the Internet in terms of its development, improvement, and use. In Hart's view, these dimensions of the American culture merged and interacted to create the Internet, which included American science efforts (major American universities have provided high-tech science and mathematics for decades), the United States government (in a major client/customer relationship with high-tech businesses), and American businesses (especially in terms of venture-capital required for startup companies in Internet areas). In all, in Hart's view, the combination of industry (wealth), university (science), and government (politics) links in the United States have created U.S. dominance of the Internet.

Others have objected in very different ways to the notion that a single government can and should control and dominate the Internet. Indeed, a radical counter-statement to the kind of analysis offered by Hart emerges almost immediately. For example, in the 2000 and 2001 volume *The Cluetrain Manifesto: The End of Business as Usual*, Levin, Locke, Searls, and Weinberger—in the vein of Martin Luther when he challenged the power of the Catholic Church with his theses—argued for "95 Theses" as "enabling conversations among human beings that were simply not possible in the era of mass media." In the preamble to these 95 Theses, Levin, Locke, Searls, and Weinberger argued that:

People of Earth…

A POWERFUL GLOBAL CONVERSATION HAS BEGUN.

Through the Internet, people are discovering and inventing new ways to share relevant knowledge with blinding speed. As a direct result, markets are getting smaller—and getting smarter faster than most companies.

These markets are conversations. Their members communicate in language that is as natural, open, honest, direct, funny, and often shocking. Whether explaining or complaining, joking or serious, the human voice is unmistakably genuine. It can't be faked.

Most corporations, on the other hand, only know how to talk in the soothing, humorless monotone of the mission statement, marketing brochure, and your-call-is-important-to us busy signal. Same old tone, same old lies. No wonder networked markets have no respect for companies unable or unwilling to speak as they do.

But learning to speak in a human voice is not some trick, nor will corporations convince us they are human with lip service about "listening to customers." They will only sound human when they empower real human beings to speak on their behalf. . . .

The result will be a new kind of conversation. And it will be the most exciting conversation business has ever engaged in.

From this foundation, the 95 Theses are proclaimed in single declarative sentences. While we cannot reprint them all here, we can provide you with a sample:

95 Theses

1. Markets are conversations....
6. The Internet is enabling conversations among human beings that were simply not possible in the era of mass media....
7. Hyperlinks subvert hierarchy... .
11. People in networked markets have figured out that they get better information and support from one another than from vendors. So much for corporate rhetoric about adding value to commoditized products.
12. There are no secrets. The networked market knows more than companies do about their own products. And whether the news is good or bad, they tell everyone....
36. Companies must ask themselves where their corporate cultures end.
37. If their cultures end before the community begins, they will have no market....
49. Org charts worked in an older economy where plans could be fully understood from atop steep management pyramids and detailed work orders could be handed down from on high....
74. We are immune to advertising. Just forget it.
89. We have real power and we know it. If you don't quite see the light, some other outfit will come along that's more attentive, more interesting, and more fun to play with....

95. We are waking up and linking to each other. We are watching. But we are not waiting.

Adopting the more traditional role and style of the critic, which many academics in particular find more understandable, in *Dark Fiber: Tracking Critical Internet Culture*, Geert Lovink (2002) argued for the creation of what he called *net criticism*. In Lovink's view, the Internet is being closed off by corporations and governments intent on creating a business and information environment free of dissent. As he put it, he wants to "wrestle the Internet from corporate and state control." Calling himself a "radical media pragmatist" (p. 12; see also pp. 218–225), Lovink envisioned an Internet that goes beyond engineering and brings the humanities, user groups, social movements, nongovernmental organizations, artists, and cultural critics into the core of Internet development. The people, Lovink believes, want a "digital revolution" (pp. 254–272). And, to guide this revolution, he argued for "net criticism" (pp. 11, 15–19; see also pp. 160–174). Lovink said that the overall development pattern for the transformation of the Internet should occur in four stages:

1. Discovery—A realization such as that found in Rheingold's (1992) *Virtual Communities* (e.g., appreciation statements for web page developments and Lovink's pp. 5–8);
2. Colonization (encouraging each website to formulate its own identity and functions);
3. Socialization (hyperlinking among all sites) and
4. Control (based upon an ongoing conversation about process of socialization among all websites). (pp. 330–345)

While *The Cluetrain Manifesto* and Lovink's *Dark Fiber* may function as dramatic and vivid arguments for Internet individual freedom, these efforts have truly been overshadowed by the attention given to hackers. As a concept, the *hacker* has truly become a confounded issue. When first introduced within a computer context in 1985, Steven Levy associated the term with the brilliant and eccentric nerds from the late 1950s through the early 1980s. And, when he updated his volume *Hackers: Heroes of the Computer Revolution* in 2010, he considered Bill Gates, Mark Zuckerberg, Richard Stallman, and Steve Wozniak as "hackers" who were the heroes of the computer revolution. Yet, Levy's conception of *hackers* was not the only image available for this group.

A year before Levy's implicit conception of the *hacker*, in a very different domain, psychologist Sherry Turkle (1984) offered a very different view of the hacker. Specifically, Turkle (pp. 199–219) filled out their image with seven particular attributes.

First, Turkle (1984) suggested that hackers' self-conception was as nerds, loners, and losers, and provided this description:

> In the MIT culture it is computer science that occupies the role of the "out group," the ostracized of the ostracized. Computer science becomes a projective screen for the insecurities and self-hate of others in the community. And many of the computer-science students accept this reflection of themselves as archetypical nerds, loners, and losers. (pp. 199–200)

Second, Turkle (1984) maintained that hackers are known by many names, such as *computer person, computer wizard, computer wheels, computer freaks,* or *computer addicts.* "Whatever the label, they are people for whom computers have become more than a job or an object of study, they have become a way of life" (p. 200). Turkle continued:

> Engineers rationalize, indeed sometimes apologize for, the overintensity of their relationships with machines....What is different for many hackers is that the means-end relationship is dropped. The fascination is with the machine itself. Contact with the tool is its own reward. Many hackers are young men for whom at a very early age mastery became highly charged, emotional, colored by a particular desire for perfection, and focused on triumph over things. Their pleasure is in manipulating and mastering their chosen object, in proving themselves with it. (p. 201)

Third, Turkle (1984) implied that being a hacker is a form of addiction, arguing,

> Hackers have been the centerpiece of numerous articles in the popular press expressing grave concern about the dangers of "computer addition." The nature of this concern varies. There are fears that young people will fall victim to a new kind of addiction with drug-like effects: withdrawal from society, narrowing of focus and life purpose, inability to function without a fix. Others fear the spread, via the computer, of characteristics of the "hacker mind." And hackers are almost universally represented as having a very undesirable frame of mind: they prefer machines to sex, they don't care about being productive. (p. 206)

Fourth, Turkle (1984) suggested that hackers suffered from a form of compulsive mastery, stating,

The computer supports growth and personal development. It also supports entrapment. Computers are not the only thing that can serve this role; people got "stuck" long before computers ever came on the scene. But computers do have some special qualities that make them particularly liable to become traps. (p. 208)

This compulsive mastery, Turkle (1984) argued, is confounded by:

- Parents support computer use.
- Lucrative adult work.
- The discipline allows for "obsession for 'perfect mastery.'" As Turkle detailed this: "MIT hackers call this 'sport death'—pushing mind and body beyond their limits, punishing the body until it can barely support mind and then demanding more of the mind than you believe it can possibly deliver" (p. 210).
- "There are few women hackers. This is a male world. Though hackers would deny that theirs is a macho culture, the preoccupation with winning and of subjecting oneself to increasingly violent tests makes their world peculiarly male in spirit, peculiarly unfriendly to women. There is, too, a flight from relationship with people to relationship with the machine—a defensive maneuver more common to men than to women" (p. 210).

Fifth, Turkle (1984) maintained that the hacker faces profound loneliness, but this loneliness is also a form of safety for the hacker. As Turkle argued her point:

It is a culture of people who leave each other a great deal of psychological space. It is a culture of people who have grown up thinking of themselves as different, apart, and who have a commitment to what one hacker described as "an ethic of total toleration for anything that in the real world would be considered strange." Dress, personal appearance, personal hygiene, when you sleep and when you wake, what you eat, where you live, whom you frequent—there are no rules. But there is company.

The people who want to impose rules, the inhabitants of the "real world," are devalued, as is the "straight" computer-science community. They are in a means-end relationship with the computer. They want it to "run" their data, facilitate their experiments. The "straights" control the resources and pay the salaries, but they do not share a true allegiance to the machine....The hackers are always trying to "improve" the system. This can make the system less reliable as a tool for getting things done, because it is always changing. The hackers also make the system more complex, more "elegant" according to their aesthetic, which often makes it more difficult for other people to use. But hackers have to keep changing and improving the system. They have built a cult of prowess that defines itself in terms of winning over ever more complex systems. (pp. 213–214)

Sixth, Turkle (1984) argued that hackers avoid relationships to avoid being "burned" in these relationships. As she articulated her position:

> The hacker culture appears to be made up of people who need to avoid complicated social situations, who for one reason or another got frightened off or hurt too badly by the risks and complexities of relationships. . . . Hacking is a way out of isolation without what many hackers refer to as "complicated" relationships with other people. "Complicated" can be synonymous with "risky." Hackers talk a lot about getting burned. And, if you need to feel in total control, "getting burned" is one of the worst things that can possibly happen to you." (pp. 216–217)

Seventh and finally, Turkle (1984) offered what has probably become the most socially demeaning of all attributes of the hackers in the contemporary popular culture. She argued that, as a class, hackers are anti-sensual:

> Hackers are not alone in denying the sensual. But it is fair to say that the hacker culture crystallizes something problematic in engineers' relationships with what Burt called the "flesh things."

> This sensuality goes beyond the overtly sensual. There is sensuality in music, literature, art, and there are sensual relationships with the world of things: the musician caresses his or her instrument; its shape, its tonality, its touch, can be pleasing and exciting. But the prototypical hacker's taste in each of these realms tends not toward a sensual caress but toward an intellectual contact. (p. 219)

In additional to Turkle's characterization of the hacker, hackers were shortly associated with all sorts of corporate and government computer break-ins that released sensitive information to competing corporations, especially international competitors, and military information that could threaten domestic security.

At best, then, two different conceptions of the *hacker* were proposed in the mid-1980s. One view cast the hackers as the creative heroes of the computer revolution. The other view was created by Turkle. In our view, Turkle's view has emerged and dominated in the popular culture. But, the hacker as security threat also began to emerge in the late 1980s.

Accordingly, when hackers emerged as an organized movement in the year 2004, many people immediately dismissed their claims because of the conception of hackers Turkle had provided as well as their belief that hackers were threats to computer security. If nothing else, then, the hacker's initial counterstatement required that the most fundamental of conceptions be provided.

Indeed, in 2004, *A Hacker Manifesto* was released by McKenzie Wark. As his point of departure, Wark (pp. 001–002) maintained that all social classes now fear the "relentless abstraction of the world" except for the "hacker class," for hackers seek to systematically destroy abstraction. This foundation provides a base for five overall principles and nine specific principles. Figure 5.1 provides an overview of the "overall" and "specific" principles in *A Hacker Manifesto*.

Five Overall Principles:

- New basic rights are created in a digital technology era.
- A new class has been created by the digital technology era.
- A new purpose now exists because of the digital technology era.
- Every individual, who creates with digital technology, possesses a new identity, an identity as a *hacker*.
- *A Hacker Manifesto* is not unlike the proclamations of a radical confrontation for the information age.

Nine Specific Principles in *A Hacker Manifesto*:

1. Reality exists on two levels:
   A. One level—a physical reality.
   B. Second level—what people "make of" a physical reality.
2. "Making" something that goes beyond a physical reality is a process of abstraction.
3. Hackers are the perfect example of those who abstract—the abstracting is taking raw data and creating:
   A. new concepts
   B. new perceptions
   C. new sensations (002)
4. Hackers are inherently a creative class, a class of people producing new abstractions as programmers, artists, writers, scientists, and musicians. (013)
5. Hackers are the producers of knowledge, knowledge in the arts, sciences, philosophy, and culture. (004)
6. The producers of knowledge seek to move into a "new world, long imagined—a world free from necessity." (011)
7. The "ruling class" seeks to control "innovation and turn it to its own ends, depriving the hacker of control of her or his creation." (012)
8. "A vectoralist class wages an intensive struggle to dispossess the hackers of their intellectual property." (021)
9. Hackers—creators of new concepts, perceptions, and sensations within digital technologies—should "come together with workers and farmers—with all the producers of the world—to liberate productive and inventive resources from the myth of scarcity." (023)

Figure 5.1. Principles in *A Hacker Manifesto*.

Embedded in the principles offered by Wark (2004) is a dramatically different conception of the hacker than the view offered by Turkle (1984). For Wark, the hacker seeks to secure the basic rights required by a politically responsible citizen. As a politically responsible citizen, the hacker's task is to construct new concepts and perceptions from the raw data or reality created by the digital revolution. These new concepts and perceptions constitute the new knowledge of the digital culture. And, ultimately, as a group, hackers will engage in whatever struggles are necessary to make these new concepts and perceptions available to others. In this context, if a corporation or government intentionally seeks to block new concepts and perceptions from others, as a matter of morality it becomes the duty of the hacker to secure and redistribute this information.

In all, while it evolves over time in some rather haphazard ways, the foundation for differences exists among the various agents and agencies linked to the development of the digital technologies. Before truly profound conflicts come into existence, it is appropriate to review the mechanisms that control confrontations and conflicts among these potentially warring factions of the digital age.

The Success of Recent Control Mechanisms

We should immediately offer two major qualifiers here: (1) It is not apparent to us that there is any clear set of criteria to employ when discussing and dealing with "mechanisms" (individuals, agencies, and/or governments) to control the Internet; and (2) It is unclear to us that anyone, any agency, or even any government has ever successfully controlled the Internet for any significant period of time when major conflicts about the control exist among various nation-states. Given these qualifiers, our task here is actually to illustrate how some control mechanisms have been initiated and why we think these mechanisms are virtually incapable of regulating how the Internet functions.

We want to illustrate recent efforts to control the Internet by examining the purpose, functions, and operations of the **Internet Corporation for Assigned Names and Numbers** or **ICANN** as we prefer to call it here. In our view, this analysis of ICANN suggests—as much as it is possible—how a single government can—under the right circumstances—restrict and control certain aspects of Internet policies and procedures. Following this discussion, we turn to the United States' effort to create an Internet "net neutrality" policy. First we deal with the United States' experiences with ICANN.

ICANN

Before the establishment of ICANN, the United States government was solely responsible for and controlled the name and number domain system used to identity websites on the Internet. Before the creation of ICANN, this function was under the supervision of both the Clinton and George W. Bush presidential administrations.

ICANN was created by the United States as a nonprofit organization designed to oversee the system used to identity websites by name and number, a task previously undertaken by other organizations such as the Internet Assigned Numbers Authority (IANA), which is now under the supervision of ICANN. This identification process necessarily involves the coordination and assignment of address blocks to regional registries, creating and maintaining the registration of all of the various websites, or what ICANN calls "domain name space," and the technical management of root registries.

Accordingly, it is ICANN that makes sure every website has its own, or unique, name. This means that when you type in the name of a website, only that website pops up on your screen. It is also ICANN that regulates what the extension of a website address can be (e.g., *com*, *edu*, *org*, and so forth). These extensions let you know what kind of website you are contacting. Likewise, if you have ever wondered why "xxx" or "por" is not used to designate pornography websites, it is ICANN that has decided not to use such extensions.

Moreover, ICANN was designed and created to more generally, as its primary operational principles, preserve the stability of the Internet, to represent the global Internet community in the various organizational units of ICANN, and to employ a "bottom-up" consensus-based set of procedures and processes ("Memorandum of understand," 1998).

While the mission of ICANN may be to represent the global Internet community, in June 2005, the Department of Commerce of the United States government announced that it would not relinquish control of the 13 root servers underlying the Internet. In November 2005, in Tunis, world government participants at the United Nations World Summit on the Information Society began renegotiations on this control issue. On July 26, 2006, the United States renewed its contract with ICANN, and in this document, the United States Department of Commerce is given final, unilateral oversight of all ICANN functions. In July 2008, the United States Department of Commerce restated its control of ICANN and also stressed that it has "no plans to transition management of the authoritative root zone file to ICANN" (Mueller, 2002, provides the most comprehensive history of these events).

The control and administration of ICANN can be viewed as one of the many questions regarding how the Internet should be configured, its role around the world, and how governments and other regulators should structure policy about the Internet. In November 2007, Pew Internet & American Life Project surveyed those attending the Second Internet Governance Forum in Rio de Janeiro, Brazil. This meeting was aimed at

> creating a global conversation about the future of the Internet and perhaps recommen-
> dations to the United Nations and the World Summit on the Information Society about
> policies that might be developed to promote widespread public access to the Internet and
> how the Internet might be configured. (Book, Anderson, & Rainie, 2008, p. 1)

Of all of the issues covered in this survey, perhaps the most powerful was Pew's findings regarding the planks of an Internet Bill of Rights. Some 76% of respondents supported freedom of information as "a core ethic of online life" and 75% agreed that such "a policy ensuring freedom of expression on the Internet should be adopted." In addition, 77% of survey respondents believed that "the private sector should take the lead in solving access problems" and that "only open and neutral Internet access can close the digital divide" (Book, Anderson, & Rainie, 2008, pp. 1–2).

In terms of the United States and its apparent control over the Internet's ICANN, two findings are particularly relevant:

1. 47% of survey respondents agreed with the statement that, "The Internet has no center of gravity—no one concentration of central control." Thirty-six percent said the Internet does have a concentrated center of power. Of these respondents, "65% said the center of the Internet's influence or concentrated power is in the United States, and 22% of those who said there is a power center cited the countries of the Northern Hemisphere" (Book, Anderson, & Rainie, 2008, p. 4).
2. "When queried about the work of the Internal Corporation for Assigned Names and Numbers (ICANN), 45% of respondents agreed with the statement that ICANN 'is not effective' and should be placed in a more neutral, global control structure" (Book, Anderson, & Rainie, 2008, p. 5).

At the end of September 2009, Rhoads (2009; see also Fost, 2009, and AP, 2009) reported that:

The U.S. government said Wednesday it had ended its 11-year contract with the nonprofit body that oversees key aspects of the Internet's architecture, after demands from other countries for more say in how the Web works.

The move addresses mounting criticism in recent years that no country should have sole control over important underpinnings of the Internet, such as determining domain name suffixes like "com."

The criticism has intensified as Internet usage has soared around the world and become critical to economics and governments. Some countries, including China, have suggested they would build their own version of the Internet if the matter wasn't resolved.

"This reflects the globalization of the Internet," said Rod Beckstrom, chief executive of the body, called the Internet Corporation for Assigned Names and Numbers, or ICANN. "By America relaxing some control and inviting other countries to have an active hand, that increases the possibility that the global Internet will remain unified," Mr. Beckstrom said in an interview.

At the outset of this section, we offered two major qualifiers: (1) It is not apparent to us that there is any clear set of criteria to employ when discussing and dealing with "mechanisms" (individuals, agencies, and/or governments) to control the Internet; and (2) It is unclear to us that anyone, any agency, or even any government has ever successfully controlled the Internet for any significant period of time when major conflicts about the control exist among various nation-states. Given this point of departure, we don't see that the United States made any major international policy mistakes in assuming responsibility for its governance role when dealing with ICANN, nor do we believe it was a mistake for the United States to decrease its control and involvement with the governance of ICANN. In our view, some matters can initially require that a single nation exert independent action, and it is also true that a matter can evolve into what is truly an international matter requiring international action. While the timing of each of these realizations might be a matter of dispute, timing itself is often something that can be handled in any number of ways. In all, more than anything else, the United States' actions with ICANN reveal the dynamic nature of international relations and the need to remain ever alert to the need for change.

The United States' Net Neutrality Policy

Net neutrality—or as it is sometimes called *Internet neutrality* or *network neutrality*—is predominantly an ideal or principle that holds that all Internet data, Internet providers, and governments should be treated equally in terms of origin, site, type of attachment equipment, users, content, platforms, and modes of communication. Who or how such equality would be granted to all Internet providers and/or governments has never been clear (e.g., see BBC News, 2012). Accordingly, controversy exists regarding whether or not net neutrality can be the foundation for a law or laws (e.g., *The Wall Street Journal*, September 18, 2009).

Within this context, it is appropriate to note that the Federal Communications Commission (FCC) provided "guidelines" for net neutrality in 2009 (Schatz & Johnson, September 22, 2009). These guidelines are provided in Figure 5.2.

- Consumers are entitled to access any legal Internet content
- Consumers are entitled to use any Internet applications or services
- Consumers are entitled to connect to any devices that won't harm the network
- The same rules apply to cable/DSL and wireless Internet
- Internet providers can't block or slow competitors' online services

Figure 5.2. FCC guidelines for net neutrality.

Seeking to implement its 2009 guidelines for net neutrality, in 2010, the FCC proposed a 10-year plan that would establish a high-speed Internet as the United States' dominant communication network. The proposal argued that the Internet is displacing the telephone and broadcast television industries. Additionally, the proposal carried provisions for a subsidy for rural parts of the country without Internet access, the auctioning off of broadcast spectrum space for wireless devices, and the development of a new universal set-top box that connects the Internet and provides cable services (Stelter & Worthham, March 13, 2010).

Within the context of implementing its net neutrality principles legally, the FCC, in court in early April 2010, argued that Comcast was violating FCC regulations. The FCC challenged Comcast's position that some bandwidth should be redistributed to larger services because smaller sites were hogging scarce Internet bandwidth, which was slowing all Web traffic (see Wyatt, April 6, 2010; and Liberman, April 12, 2010). The U.S. Court of Appeals said that the FCC overstepped its authority when FCC regulators barred Comcast from interfering with other Internet companies. In a unanimous decision, the Court ruled that the FCC does not have an explicit right to regulate broadband service or to develop rules to regulate cable TV or phone services.

The FCC net neutrality plan was perceived by some as extremely controversial. For the week of August 9–13, 2010, for example, the issue of FCC's net neutrality was the number one choice of bloggers. Some 19% of blog reactions criticized Google for seemingly softening its support of net neutrality (Pew Research Center, August 19, 2010). And on December 23, 2010, the FCC published a 194-page report and summary of various positions of key FCC members falling under such subheads as "Preserving the Free and Open Internet," "The Need for Open Internet Protections," "Open Internet Rules," and "The Commission's Authority to Adopt Open Internet Rules," "Enforcement," and so forth. In January 2011, *The Daily Online Examiner* released a survey that suggested that only 21% of Americans supported FCC regulation of the Internet (Davis, January 6, 2011). Likewise, while it was not opposed to net neutrality in principle, The Electronic Frontier Foundation questioned if the FCC could find a way to impose neutrality rules without also paving the way for censorship in the future (Davis, February 4, 2011; for context, see also Davis, November 2, 2009). And, in May 2010, Wyatt noted that the FCC review of its own regulations is now "hotly debated."

While the movement for net neutrality has not been resolved, the political intensitydevoted to the topic may not continue with the same intense energy and attention it received during the last half of the 2000s. Indeed, in February 2011, Secretary of State Hillary Rodham Clinton argued that the Internet has been used both to topple repressive regimes and crush dissent. While admitting that net neutrality has been "hotly disputed by some technology experts and human rights activists," Secretary Clinton more moderately noted that the Internet can be a force that leads to democratic change, but that it is also not a "magic bullet" that brings down all repressive regimes (in Landler, February 15, 2011).

We remain curious how the net neutral policy will be handled in the years to come.

Recent Contemporary Internet Conflicts

The United States' responses to the Internet have been active, sometimes contradictory, and, it would seem, always controversial. All of these responses are part of the process defining what the purpose of the Internet should be. One such response has been Richard A. Clarke's book *Cyber War: The Next Threat to National Security and What to Do About It* (2010), a volume that provided a chilling warning about America's vulnerability in what this former presidential adviser and counterterrorist expert described as "a terrifying new interactional conflict—Cyber War." In addition, Imperva released a 2012 report in which it described a 25-day cyber attack carried out in three phases. In Imperva's view, this attack constituted a "new age of hacktivism" (p. 3) carried out by "real people with real techniques, who are attempted to "steal data," using "skilled hackers" and "laypeople" (p. 3), in a campaign designed to initiate a "distributed denial of service" (DDoS) that culminated in "hundreds of thousands" of attacks (almost 50,000 on Day 24 and almost 60,000 on Day 25) (p. 12). At the same time, in 2012, Andy Greenberg published his volume defending hackers: *This Machine Kills Secrets: How WikiLeakers, Cypherpunks, and Hacktivists Aim to Free the World's Information.* These hackers, Greenberg maintained, are part of a larger social history that began in 1969 with the whistleblower Dr. Daniel Ellsberg, who sought to expose the RAND corporation's ("a military think tank") role in "a secret history of America's involvement in Vietnam" (p. 12). With equal concern, Ackerman (2013) reported that "cyberattacks target finance hubs" and are a "'rapidly evolving' threat to market infrastructure from computer crime." And, more recently, Savage (2013) reported that Supreme Court Chief Justice John G. Roberts, Jr. has selected "conservative," Republican, and executive branch people to serve as judges to the Foreign Intelligence Surveillance Court, a "secret" court in charge of "deciding how intrusive the government can be in the name of national security" (p. A1).

In all, in dramatic and explicitly political terminologies, a conception of the image, position, and definition of the Internet is currently being resolved and determined. We are in the midst of the process that will shape and decide what the purpose of the Internet will be. We do want to focus on an event that is very likely to additionally shape the understood purpose of the Internet. As we are writing this chapter right now, there are really four major Internet cases or examples that are being considered by the United States government. Theoretically, any one of these four examples could be used to illustrate how the Internet's purpose is emerging. The outcome of each of these cases is unknown right now, and we truly are dealing here with a case for which we do not now know the outcome. But, such a case is intriguing. We will be giving you background on one of these cases and provide you with a sense of the issues involved. Ultimately, however, you will be the only one who knows how the case actually develops; you will be the one who decides what you think the outcome of this case means in terms of how the United States defines the purpose of the Internet; and, when all is said and done, you will be the one who decides if this case is really exerting a significant influence on how the purpose of the Internet is defined in the United States.

First, let's identify the four cases we think are ongoing and likely to define, at least in part, what the United States' purpose for the Internet is.

One potential case study would be General James E. Cartwright. General Cartwright has been the focus of an investigation about disclosure of classified information about American-led cyber-attacks on Iran's nuclear program (LaFraniere, July 21, 2013, p. 18).

A second potential case is that of State Department adviser, Stephen Jin-Woo Kim. Mr. Kim's case is compounded by a co-investigation of Fox News reporter James Rosen (LaFraniere, July 21, 2013, p. 18).

A third potential case is that of Army Private Bradley Manning who began downloading classified information soon after he was deployed to Iraq in 2009. Ultimately, he released 700,000 documents to WikiLeaks for publication on the Internet. He was arrested in May 2010, and he pled guilty in March 2013 to several criminal counts. He is currently facing 22 charges including aiding the enemy which carries a maximum sentence of life in prison without parole, and is currently in the middle of his court martial (Aylward, July 3, 2013; Bumiller, June 8 2010). Deliberations during his court martial have been intriguing (see Savage, July 27, 2013, for the details regarding the culmination of this case). The prosecutor portrayed Private Manning as an "anarchist" and "traitor." David E. Coombs, the defense lawyer, offered a rival narrative, arguing that his client was a "humanist," "whistle-blower," and a "young, naïve, but good-intentioned soldier who had human life and his humanist beliefs center to this decision—whose sole focus was 'maybe I just can make a difference, maybe make a change.'" So, while this case is about to end, we will shortly find out if Private Manning is legally a "traitor" versus a "whistle-blower." While the final sentencing phase has not yet been held as this is written, he was found "not guilty" of "aiding the enemy" or treason, but was convicted on six counts of violating the Espionage Act of 1917 as well as other charges, for which he faces a theoretical maximum sentence of 136 years in prison "although legal experts said the actual term was likely to be much shorter" (Savage, July 30, 2013).

A fourth potential case is that of Edward Joseph Snowden. Mr. Snowden is a former American technical contractor for the United States National Security Agency (NSA) and former employee of the Central Intelligence Agency (CIA) who leaked details of several top-secret U.S. and British government mass surveillance programs (see: Shane & Sanger, 2013, p. A1) to London's *The Guardian* in the spring 2013, which were published by *The Guardian* in June and July 2013. Shane and Somaiya (2013) reported that these leaks were among the most significant NSA security breaches in the history of the United States.

We have selected the case of Mr. Snowden for our focus here, because he is not currently apprehended nor is he being tried in the legal system of the United States right now. At this point in time, his case is still "open" and additional developments are expected that we think will influence how the case is finally handled. Additionally, if possible, we want to avoid some of the precise and unique legal issues imposed by the legal system of the United States military. In all, from our point in time, what happens to Mr. Snowden now and in the immediate future can only be tracked directly by you as Mr. Snowden moves through the situation he is facing.

On June 21, 2013, three charges of violating the Espionage Act and stealing government property were revealed against Mr. Snowden. The charges carry a maximum prison sentence of 10 years each for a total of 30 years.

As a point of departure, it is appropriate to note how Mr. Snowden perceived his situation when the news of his indictments was made public and also how President Obama reacted. Mr. Snowden felt that the leaks he provided could promote a public debate about civil liberties. And, while deploring the leaks, President Obama also endorsed the same goal of a vigorous public discussion of the "trade-off" between national security and personal privacy (Shane & Weisman, 2013, p. A1). In very general terms, then, both Mr. Snowden and President Obama "agree" on the basic principle or value at stake in terms of the Internet.

Yet, the "agreement" reached by Mr. Snowden and President Obama is open to at least three perspectives.

First, it would be possible to say that a purpose of the Internet is to provide a free and open forum for diverse individuals to express themselves with as much transparency as possible, and in the process, necessarily preserve individual privacy to guarantee that individual expression is as fully articulated at all times as possible.

Second, it would be possible to say that a purpose of the Internet is to preserve and protect the sovereignty of each nation-state.

Or, third, it would be possible to say that the purpose of the Internet is to move back and forth between individual civil liberties on the one hand, and national security on the other hand. In our view, in terms of its purpose, the Internet must—at key moments—respond to both civil liberties and national security simultaneously, and there will potentially be massive disagreements about how this balance should be achieved in any particular set of circumstances.

At the same time, the question of government efficiency and effectiveness was also an issue. Mr. Snowden had fled to Hong Kong and then to Russia. The United States government appeared to have a difficult time securing Mr. Snowden's return to the United States. Moreover, the significance of Mr. Snowden's leaks was not yet clear to many Americans. Accordingly, in June 2013, General Keith B. Alexander (director of the NSA) told Congress that "dozens of terrorist threats had been halted by the agency's huge database of logs of nearly every domestic phone call made by Americans" (Sanger, 2013). Yet, in July 2013, Dennis C. Blair (director of national intelligence) also reported that in the four years from 2005 through 2008, 153 cases involving leaking national security information had been referred to the Justice Department and that none of them had led to an indictment.

In part, the efficiency and effectiveness of the United States government turned on how it established and processed its national security and personal privacy issues. The United States gave responsibility for these national security and personal privacy issues to the NSA. And, the NSA's ability to gather phone data on millions of Americans hinged on a "secret court ruling that redefined a single word: 'relevant'" (Valentino-DeVries & Gorman, 2013, p. A1). In 2006, President Bush reauthorized the PATRIOT Act, which had originally been passed to deal with the international terrorist issues following the destruction of the World Trade towers on September 11, 2001. The 2006 reauthorization allowed the courts to approve surveillance of any information that the court determined might possibly be significant or have a demonstrated bearing on the investigation at hand. With the approval and appointments made by the Chief Justice of the Supreme Court, a special 11-member Foreign Intelligence Surveillance Court (known as FISA) was appointed.

The FISA court has been viewed by some as unique in several ways.

First, the decisions and actions of the FISA court have been viewed as secret. *The Wall Street Journal* identified the FISA rulings as "secret" (Valentino-DeVries & Gorman, July 8, 2013, p. A1) while *The New York Times* maintained that FISA "vastly broadened the powers" of the NSA "in secret" (Lichtblau, July 7, 2013, p. 1). In greater detail, Lichtblau (July 7, 2013, p. 1) reported that, "In more than a dozen classified rulings, the nation's surveillance court has created a secret body of law giving the National Security Agency the power to amass vast collections of data on Americans while pursing not only terrorism suspects, but also people possibly involved in nuclear proliferation, espionage and cyber attacks." As Valentino-DeVries and Gorman (July 8, 2013, p. A1) put it, "In classified order starting in the mid-2000s, the court accepted that 'relevant' could be broadened to permit an entire database of records on millions of people, in contrast to a more conservative interpretation widely applied in criminal cases, in which only some of those records would likely be allowed."

Second, FISA is one-sided. As Lichtblau (July 7, 2013) put it,

> Unlike the Supreme Court, the FISA court hears only from one side of the case—the government's—and its findings are almost never made public. A Court of Review is empaneled to hear appeals, but that is known to have happened only a handful of times in the court's history, and no case has ever been taken to the Supreme Court. In fact, it is not clear in all circumstances whether Internet and phone companies that are turning over reams of data even have the right to appear before the FISA court. (p. 4)

Third, informing the public is not a clear objective of the FISA court. As Lichtblau (July 7, 2013) maintained,

> Reggie B. Walton, the FISA court's presiding judge, wrote in March that he recognized the "potential benefit of better informing the public" about the court's decisions. But, he said, there are "serious obstacles" to doing so because of the potential for misunderstanding caused by omitting classified details. (p. 4)

The consequences of all these events related to the Snowden affair are by no means over or exhausted. While we are in no way exhaustive here, at least four of these consequences are now vivid.

First, the Snowden affair has had a negative impact on the United States' image in international affairs. Newspaper headlines make the case extremely clear: "Europeans Voice Anger Over Reports of Spying by U.S. on Its Allies" (Castle & Schmitt, July 1, 2013); "Outrage in Europe Grows Over Spying Disclosures" (Erlanger, July 2, 2013); and "Brazil Voices 'Deep Concern' Over Gathering of Data by U.S." (Rohter, July 8, 2013). The long-term consequences of the United States' surveillance on its international relationships are simply unknown at this point.

Second, as the United States' surveillance system has emerged, it has become clear that other nations have also engaged in such programs (Erlanger, July 5, 2013). If they are legally precluded from conducting such surveillance, they may buy or barter the results of such searches from other nations (Castle, July 5, 2013).

Third, the United States' efforts to bring Mr. Snowden back to the United States for trial have, in no way, silenced Mr. Snowden. On July 12, 2013, speaking from Moscow's Sheremetyevo Airport to his audience and a massive group of reporters, Mr. Snowden noted that one month earlier, as a United States employee, he "had the capability without any warrant to search for, seize, and read your communications. Anyone's communication at any time. That is the power to change people's fates." In commenting on his decision to release United States security information, Mr. Snowden said, "it was the right thing to do and I have no regrets." Specifically, Mr. Snowden argued, "I took what I knew to the public;

so what affects all of us can be discussed by all of us in the light of day, and I asked the world for justice."

Fourth, the Snowden affair has generated a political backlash. As Risen (July 18, 2013) put it, "The Obama administration faced a growing Congressional backlash against the National Security Agency's domestic surveillance operations on Wednesday, as lawmakers from both parties called for the massive collection of private data on millions of Americans to be scaled back." A week later, the United States' House of Representatives voted 205 to 217 on a resolution to "rein in N.S.A." (Weisman, July 25, 2013, p. A1). While the resolution was defeated, Weisman reported that "the 205-to-217 vote was far closer than expected and came after a brief but impassioned debate over citizens' right to privacy and steps the government must take to protect national security."

Since the end of July 2013, the NSA has continued to generate issues both inside and outside of Washington, D.C., in at least four ways.

First, the scope of NSA spying has seemed enormous to many. James W. Clapper, Jr., director of national intelligence, has testified that the "scale of eavesdropping by the N.S.A., with 35,000 workers and a $10.8 billion a year, sets it apart," specifically noting that the scope of United States' spying "probably dwarfs everyone on the planet" (in Shane, November 3, 2013, p. 1). In this context, Gorman and Valentino-DeVries (August 21, 2013, p. A1) noted that NSA "officials" have reported that its computer system "has the capacity to reach roughly 75% of all U.S. Internet traffic." More directly, on October 21, 2013, *USA Today* reported that, "For seven years, the National Security Agency (NSA) has been collecting detailed phone data on hundreds of millions of Americans not suspected of anything" (p. 8A).

Second, reason exists to believe that many of the NSA's eavesdropping activities were illegal or at least violated legal privacy rules. For example, the FISC which is responsible for supervising NSA activities, examined the 2009 activities of the NSA Shane (September 11, 2013) reported that

> The agency [NSA] told the court [FISC] that all of the numbers on the alert list had met the legal standard of suspicion, but that was false. In fact, only 10 percent of 17,800 phone numbers on the alert list in 2009 had met that test. (p. A1)

FISC Judge Walton specifically wrote that

> The government has compounded its noncompliance with the court's orders by repeatedly submitting inaccurate descriptions of the alert list process. It has finally come to light that the F.I.S.C.'s authorization of this vast collection program has been premised on a flawed depiction of how the N.S.A. uses [the phone call data].

In terms of the NSA's defense, a "senior American intelligence official" admitted "the sting of the court's reprimand but said the problems came in a complex, highly technical program and were unintentional" (Shane, September 11, 2013, p. A14). A similar report occurred regarding "violated privacy rights" in 2011, which "found that there has been 2,776 such 'incidents' in a one-year period" (Savage, August 17, 2013, p. A12; see also Savage & Shane, August 22, 2013). In light of the decision by the FISC, Gorman, Barrett, and Valentino-DeVries concluded that, "The National Security Agency violated the Constitution for three years by collecting tens of thousands of purely domestic communications without sufficient privacy protections, according to a secret national-security court [FISC] ruling."

Third, beyond eavesdropping on activities of individuals directly, NSA has apparently also collected personal and telephone data from websites such as Facebook (Bilton, August 27, 2013), Google (Rushe, August 14, 2013), and Yahoo (Savage, Miller, & Perlroth, October 31, 2013). These data are often sought by and shared with other federal agencies (Lichtblau & Schmidt, August 4, 2013, p. 1).

Fourth and finally, the NSA's scope of eavesdropping activities have apparently functioned internationally, within the internal affairs of other nation-states, in ways that have surprised, if not shocked, some nation-states. For example, in October 2013, it was confirmed that the NSA had been tracking and collecting records, between December 2012 and early January 2013, of some 70 million French citizens as well as 60 million Mexican citizens (see, e.g., Entous & Gorman, October 30, 2013, pp. A1, A12; Smale, 2013). In addition, Obama's administrative review of NSA indicated that NSA has been monitoring some "35 world leaders," who were "outraged" by the report (Gorman & Entous, October 28, 2013, p. A1).

In terms of the question we raised at the outset of this chapter regarding the purpose of the Internet, we are convinced that there are competing forces at work, and the issues are nothing less than the conflicts between government security and individual privacy rights. Attempts to resolve these issues have already emerged. The United States Senate plans on a "total review" of the U.S. spying program in an effort to resolve the competing issues (Gorman, October 29, 2013). President Obama has announced plans to overhaul the U.S. spying program (Gorman, Lee, & Hook, August 9, 2013) and to "halt eavesdropping on leaders of American allies" (Landler & Sanger, October 29, 2013, p. A1), although Savage and Shear (August 10, 2013, p. A1) reported that, "Mr. Obama showed no inclination to curtail surveillance efforts" and would only "move to that degree of openness and safeguards that would make the public 'comfortable.'" On the international front, Germany is considering asking Snowden to testify about the spying situation that Germany has encountered (Smake, November 7, 2013). Finally, complicating this set of issues, in an ongoing legal dispute, the U.S. Department of Justice plans to raise the matter of "evidence being used against a defendant from a warrantless wiretap" in order to "set up a Supreme Court test of whether such eavesdropping is constitutional" (Savage, October 27, 2013). Indeed, we should be watching how Congress, the President, the American legal system, and other international nation-states respond to the NSA in the months and years to come.

In all, we do not believe that the issues of government security and individual privacy are at all resolved. For some, individual privacy has already been lost. For example, after reviewing the scope of NSA activities, Russia computer-security expert Eugene Kaspersky concluded that, "There is no more privacy" (in Sonne, September 3, 2013). For others, in the years to come, there will be a constant "balancing act" between government security and individual privacy rights (Mazzetti & Shane, August 10, 2013). So, when all is said and done, the Snowden affair is really only just beginning. We invite you to track this set of circumstances as they evolve and develop. As you watch this affair, what do you think it tells us about the purpose of the Internet? Can the Internet balance civil liberties and national security or must it shift from one to the other of these goals depending on circumstances?

Conclusion and Future Possibilities

In this chapter, we have asked whether or not a purpose can and should be attributed to the Internet. At this point, we cannot provide you with a definitive answer. Too many circumstances are still in play, and how those circumstances are resolved will determine what we say the Internet is designed to do or will accomplish. As Crovitz (November 26, 2012) put it, "Who runs the Internet? For now, the answer remains no one, or at least no government." But, in the meantime, we expect that many people—still uncomfortable with how the current situation is defined and understood—may decide to redefine some basic concepts. *Hacking* may become a "form of protest" (Sengupta & Perlroth, March 5, 2012, p. B1) and *national security* may be more vividly redefined and understood by the image of the towers and death the United States was left with on September 11, 2001. As we work out—and rework—what we think the purpose of the Internet is, we do expect our terminologies and perceptions of events to continually undergo change.

Chapter Exercises

Current Issues: The digital age has made the concept of *privacy* more complicated and involved than at any other time in history. At one point, what was "private" was the "content of consciousness that is in accessible to anyone except the person having it" or "knowledge that a person has, such as his pains, pleasures, feelings, emotions, that cannot be known by others, cannot be made available to public knowledge, and cannot be directly verified by public knowledge" (Angeles, 1981, p. 226). In a digital age, especially in an age when people automatically leave tracers or markers (e.g., cookies) when they use a website, the content of the consciousness is able to or can be derived by others. Pains, pleasures, feelings, and emotions can be known by others, and while they may not be made public in the sense that everyone has access to the website history of others, a significant few may gain information about another's activities online. Moreover, without a user's awareness or knowledge, website observers can make judgments and categorizations of users based upon what users do and do not do, thereby, without informing the user in any way, the website observed creates an understanding or knowledge about a user that the user has no idea has even been made. While privacy once involved content only available to a single person's consciousness, what is *private* can now be known by others. Given the capabilities of website observers, how would you define *privacy* today? Can a definition be fashioned that provides an individual with the right to determine when his/her individual and personal thoughts and behaviors are known by others? If we assume that Internet access and use is now a common mode of accessing information, does privacy exist now?

Debate It: *Resolved: That the written permission of any effort to track, obtain, or use knowledge of a user's online behavior online must be secured from the user prior to any tracking, obtaining, and/or using the behavior.* Assuming that a sufficient penalty can be determined in specific cases, the issue is whether or not a control of a behavior can be determined by a user prior to the use of the behavior.

Ethical Consideration: While transparency of all individual behaviors may be a goal of websites, especially social media websites, for a specific individual, transparency of an individual's behavior should be a right provided by the individual. Certainly, when sufficient grounds exist and are demonstrate, others may have a right to know if individuals engage in certain behaviors (e.g., murder, rape, theft, etc.), but an entire cluster of behaviors do not fall within the framework of the "right to know," especially if the behaviors have no known negative effects on others. Decide where you stand on this issue. What should be the ethical guidelines in terms of these issues?

Future Focus: As tracing mechanisms improve in the years to come, website observers will be able to know more and more about every individual's behavior. Individuals will need to plan on appropriate privacy counter-activities. As experience with the National Security Agency has determined, given concerns such as "national security," it may be impossible to formulate known and public safeguards for these appropriate individual privacy counter-activities. Future activities may emphasize extensive legal regulation of this process. Should websites provide degrees of privacy security for individual use?

Endnote

1. Front and back book flaps of Tim Berners-Lee, *Weaving the Web: The Original Design and Ultimate Destiny of the World Wide Web* (Harper San Francisco, 1999).

6

Business of the Internet: Marketing

When considering the business of the Internet, there are a host of options to explore. Certainly, attempting to examine this entire subject in a single chapter or even single book would be a futile experience. It is much more recommendable, then, to examine a key participant or topic related to the business of the Internet, particularly those that have been especially influential or noteworthy.

We have chosen to examine Internet marketing. This area has been selected because of its influence on the business of the Internet as a whole, and also because of its prominent position in the public mind-set. For the members of the general public typically not attuned to such matters, this area, because of its prominence or familiarity, serves as a resource for comprehending the business of the Internet as a whole. While all-encompassing statements are usually flawed and ripe for dispute, it is nevertheless generally safe to maintain that every person who has ever been on the Internet has experienced some form of marketing. For many non-commerce sites, advertising is the primary if not sole form of revenue.

Traditional notions of advertising and marketing cannot fully be applied to online marketing, given the accuracy of targeting consumers (which means a decreased need for the number of advertisements) and given the methods through which the promotion of a company, product, or service is actually conducted by consumers in addition to or rather than marketers and producers themselves.

Accordingly, within this chapter, beyond exploring the various types of Internet marketing, we will examine how Internet marketing differs from previous forms of marketing and how the Internet has transformed marketing as a whole.

Questions Previewing This Chapter

1. *What are the types of Internet marketing?*
2. *How has the Internet transformed marketing?*

Internet Marketing

The term *marketing* is used, specifically, rather than advertising, which is commonly associated with the promotion of companies, products, and services through both the Internet and through other means, because marketing involves a host of endeavors, of which advertising is a subset. Within this section of the chapter, we will examine advertising, but will also examine marketing techniques and strategies more broadly.

Using very traditional notions of advertising, it is possible to isolate the revenue generated by advertisements on the Internet. In the United States, specifically, Internet advertisement spending continues to increase and remains second only to television (eMarketer, 2012). Worldwide, digital advertising also continues to grow and accounts for over 20% of all advertising dollars, with digital advertising spending expected to reach 163 billion U.S. dollars in 2016 (eMarketer, 2013,).

While impressive, these numbers do not accurately convey the capabilities and influence of Internet advertising or marketing. For instance, while it might appear that television dominates the advertisement spending in the United States, it could be argued that this amount is higher because television advertising has less accuracy and influence than Internet advertising. Consequently, less money has to be spent online, because Internet advertisements are more likely to be experienced by target audience members and are ultimately more influential than through other means. Further, it is also possible that these advertisements are subsequently shared among users of the Internet, which not only provides free advertising and would not be included in revenue numbers, but also makes these advertisements more meaningful and powerful.

As will be discussed, the Internet has profoundly transformed advertising specifically and marketing in general. New methods of marketing have been generated, traditional techniques have been enhanced, and it is now possible to evaluate consumer behaviors and activities along with marketing effectiveness in ways simply impossible without the Internet.

In what follows, we will examine the primary types of Internet marketing. In doing so, it is important to recognize that these are not "pure" forms of marketing. Rather, there is a great deal of overlap among some of these types of marketing, with multiple strategies evident in a single marketing attempt. For instance, a banner may have been placed through affiliate marketing, using behavioral and contextual marketing strategies. Likewise, as we will discuss, there remains some question as to whether native marketing is a unique form of marketing or a type of content marketing. Further, social networking site marketing may be both a marketing advertisement and marketing strategy. These types of Internet marketing are presented separately in order to focus on each one, although other forms may be mentioned within each separate section. Figure 8.1 lists and provides a brief summary of each type of Internet marketing.

Affiliate Marketing	Marketing conducted by third parties who are compensated based on effectiveness
Banner Marketing	Marketing conducted using embedded advertisements on web pages
Behavioral Target Marketing	Marketing conducted by tracking online activity and crafting advertisements most suitable to users based on that activity
Content Marketing	Marketing conducted by offering users sponsored content
Contextual Marketing	Marketing conducted by matching advertisements to subject matter
Email Marketing	Marketing conducted through email
Native Marketing	Marketing conducted by offering users sponsored content that appears to be part of or is unique to a particular site
Search Engine Marketing	Marketing conducted through search engines
Social Marketing	Marketing conducted through social networking sites and through online relational connections

Figure 8.1. Types of Internet marketing.

Banner Marketing

We will begin by introducing an Internet marketing original—banners. **Banner marketing** uses advertisements embedded within a website. These advertisements appear as boxes or rectangles at various points on the page or as new browser windows appearing in front of or behind the original page.

Although some marketing attempts took place on the early Internet, banners marked the true beginning of digital advertisements. Hotwired (now Hotwire) was the first site to sell banner advertisements, beginning in 1994. These original banners were sponsored by 14 different companies, with the first-ever advertisement coming from AT&T. The advertisement was somewhat visionary, reading, "Have you ever clicked your mouse right here? You will" (Singel, 2010).

Banner marketing may seem somewhat trite at this point in time. However, it was revolutionary when it first appeared. Because of the Internet, for the first time, marketers had a very good idea exactly how many people came into contact with an advertisement and how many people actually interacted with an advertisement. And, the AT&T advertisement was correct; people did click on these advertisements and continue to do so. In spite of fewer numbers of users clicking on banners now than in the past, billions of dollars of online advertising is spent on them.

There are a number of issues encompassing banner marketing, however, which call into question its present impact. Beyond the decreasing numbers of users who are clicking on such advertisements, these issues involve whether or not they are actually seen.

Programs are now available that can eliminate these advertisements completely, at least in their pop-up and pop-under versions. **Pop-up advertisements** are those appearing as a new browser window in front of an original page. Even if not closely viewed or clicked on, users must at least acknowledge their existence in order to remove them and continue browsing the original page. **Pop-under advertisements** appear as a new browser window behind the original page. Users are generally not aware of their existence until removing their primary browser screen. The thinking behind pop-under advertisements is that users may find them less annoying than pop-up advertisements, and those users may be more likely to attend to them if their primary task has been completed. Users tend to dislike both types of popping, however, as indicated by the wide availability and utilization of blocking software.

Another issue involves banners actually appearing on pages. They can, of course, be ignored by users. Beyond being ignored, though, they may not be seen by users. A study found that 54% of these types of advertisements were not seen by users. One reason for this is because of placement. If the banner is located on the bottom of the screen, for example, the user may not scroll down that far. Similarly, banners located at the top of the screen may not been seen because users have already scrolled down below it. A second reason is that these banners may not have loaded in their entirety before a user has moved on. This is especially common with slow connection speeds and was certainly the case when banners first began appearing in 1990s. A third and final reason involves fraudulent activity. Marketers may be paid in part by impressions and clicks. An *impression* occurs when an advertisement is displayed on a user's screen. A *click* occurs when someone clicks on the banner. Some software makes it appear as if a user is visiting a page on which the banner is displayed, but these users are not real. Nevertheless, it appears as if an impression has been made (Vranica, 2013).

Banner marketing has been around for quite some time and will likely continue to be used to a certain extent. However, the inability to place materials in locations more likely to be seen by users along with the likelihood that users tend to view these advertisement as distractions to be avoided and removed calls into question their effectiveness. Such issues have led to the rise of these next forms of Internet marketing, which users may see as a more natural and useful part of their Internet experience.

Content Marketing and Native Marketing

Content and native marketing will be discussed together, because there has yet to be a clear consensus on what native marketing entails and whether it is a unique form of marketing or simply a type of content marketing (see Vega, 2013; Wasserman, 2012). However, we will clear things up a bit here.

Content Marketing is conducted by offering Internet users sponsored content. This content may include research studies, how-to guides, or other materials that targeted users may find valuable. Offline and historical examples of content marketing include sponsored magazines from farm equipment manufacturers and airlines, for example, as well as sponsored cookbooks from brand-name foods. Content marketing works by increasing brand awareness and trust. It may also compel people to purchase products, as in the case of sponsored cookbooks (either physical or digital). Online content marketing is also used to bring Internet users to a particular site.

Native Marketing is conducted by offering users sponsored content that appears to be part of or is unique to a particular site. The term *native marketing* has been traced to a speech given by Fred Wilson at an OMMA (Online Media, Marketing and Advertising) Global conference in September 2011, where he talked about advertisements that were "unique and native" to sites (Wasserman, 2012). Since that time, native marketing has become a significant part of Internet marketing. In simplest terms, native marketing is sponsored content that appears as if it is part of a website. It is *native*, a natural inhabitant of the site.

Alternative View

It might be argued that native marketing is not unique to the Internet. Hardcopy newspapers and magazines occasionally include "articles" that appear similar to other articles yet usually have the word advertisement running across their headers and footers. Likewise, some television commercials for medical products airing during news broadcasts are produced and presented in a manner that makes them appear to be a medical segment of the broadcast rather than an advertisement. However, regardless of how authentic these methods might be for other technologies, they are not achieved to the same extent as those appearing online. Furthermore, as will be discussed, such Internet marketing strategies as contextual marketing and behavioral targeting can be legitimately matched to surrounding content and to the interests of users.

Examples of Internet native marketing include BrandVoice (Forbes), Partner Posts (BuzzFeed), Promoted Tweets (Twitter), and TrueView Videos (YouTube). These appear within the regular content of sites, provide material similar to regular content, and tend to blend in with regular, non-marketed material. In addition to materials included in newsfeeds and general content, music sites such as Spotify feature sponsored playlists, and review sites such as Yelp (Yelp Ads) include advertisements among search items. In the United States, the Federal Trade Commission has issued requirements that these materials must be recognizable as advertisements, but the actual clarity of this distinction for users is unclear (Yadron & Ovide, 2013).

As mentioned above, there are disagreements as to whether native marketing is simply content marketing or a unique marketing tool. Both views are probably correct. If we were to fully distinguish one from the other, though, native marketing could be considered that which appears to be part of a particular site, while content marketing does not necessary appear to be a natural part of a site. Further separating native and content marketing as two distinct forms of marketing, content marketing would likely refer to stand-alone products such as a complete online magazine or research paper, whereas native marketing would likely refer to smaller items embedded within stand-alone products or platforms (websites).

Email Marketing

As with banner marketing, **email marketing**, where marketing materials are delivered directly to the inboxes of Internet users, is a long-established form of Internet marketing. There are two types of marketing emails: opt-in email and unsolicited email. **Opt-in email** is sent to users who have granted permission, knowingly or unknowingly, to receive messages from a businesses or nonprofit agency. These messages typically include such materials as updates about organizations, discount coupons, and information about sales or campaigns. **Unsolicited email** is sent to users who have not granted permission to receive messages from a business or nonprofit agency but whose addresses have been discovered and targeted.

Email marketing is connected to marketing through other technologies, most notably through postal services and telephone. Within the United States, marketing materials account for over 60% of all mail received by households (United States Postal Service, 2011). Until the National Do Not Call Registry was implemented in 2004 to allow the blocking of phone numbers, most households received numerous unsolicited telemarketing calls each day.

However, email marketing has two distinct advantages over such marketing through other technologies. Most notably, it is quite a bit cheaper. Marketing through email does not require costs associated with the development of physical materials to send through the mail. It also does not require the cost of postage or the salary of numerous people making calls from a phone center. Email marketing also enables organizations to better determine the impact of their marketing by tracking whether users open the messages and whether they click on the materials offered by these messages (Silverpop, 2013).

Spam

Unfortunately for marketers, email marketing is also viewed with much the same derision as "junk mail" delivered through the postal service and unsolicited phone calls from telemarketers. Both opt-in email and unsolicited email are commonly referred to as "spam," so named from a Monty Python comedy skit about the ubiquitous nature of the eponymous meat. Just as the meat seems to have always been around, Internet spam seems to have been there since the Internet's beginning.

The first spam message actually occurred in 1971 when Peter Bos, an engineer at MIT, sent an anti-Vietnam message to fellow engineers using the Compatible Time-Sharing System (CTSS). The message read, "There is no way to peace. Peace is the way" (Brunton, 2013). Credit for the first e-mail spam goes to Gary Thuerk, who sent a message on ARPANET in 1978 inviting 400 of the approximately 2,600 users who had email accounts at the time to an open house sponsored by the Digital Equipment Corporation (Brunton, 2013; Worthen, 2008). The outcry from users was so significant that a decade passed before a man named Robert Noha adopted the pseudonym *Jay-Jay* to "flood" users with fund-raising attempts (Brunton, 2013; Morozof, 2013). The first commercial email spam, or at least the first notorious commercial spam, was sent on Usenet in 1994 by a law firm specializing in immigration and known as the Green Card Lottery Spam (Brunton, 2013).

Just as pop-up and pop-under advertisements caught the ire of Internet users, so too did spam. And, just as software was created to prevent pop-ups and pop-unders, email systems soon provided users with spam blockers. Unlike banner marketing strategies, however, spam caught the full attention of lawmakers. In the United States, the **Controlling the Assault of Non-Solicited Pornography and Marketing Act of 2003 (CAN-SPAM)** was passed in an attempt to regulate spam. This act requires marketing email to be labeled as such, use legitimate email address, and allow receivers to opt out of receiving additional messages. The provisions of this act are very similar to Australia's **Spam Act of 2003**. Taking things even further, the European Union's **E-Privacy Directive** and Canada's **Fighting Internet and Wireless Spam Act** (FISA) prohibit unauthorized email altogether.

In spite of consumer opposition and in spite of laws regulating the use of email as a marketing tool, spam is an effective means of marketing. This effectiveness is based on the advantages separating this form of advertising from similar methods through other technologies outlined above. It is also based on the number of people who open and click on links contained within these messages (Silverpop, 2013). And, this effectiveness is perceived based on the amount of spam still being sent. Spam represents an average of 72% of all email, which is the lowest it has been in five years but still a substantial amount (Gudkova, 2013).

Alternative View

In the United States, the CAN-SPAM Act of 2003, which regulated spam, and the Do-Not-Call Implementation Act of 2003, which resulted in the implementation of the National Do Not Call Registry the following year, were both met favorably by consumers. However, some people may question the government's intrusion in these areas and non-intrusion when it comes to junk mail sent by its own postal service, especially considering spam, telemarketing phone calls, and junk mail are generally placed within the same unsolicited and unwelcomed categories. Instead of the government attempting to reduce the amount of junk mail littering mail boxes, as it did with the amount of spam littering in-boxes, the United States Post Office is actually encouraging businesses to send more junk mail in order to combat its financial woes (Levitz, 2011).

Search Engine Marketing

As the name implies, **search engine marketing** uses search engine results as a marketing device. As with other forms of online marketing, this sort of marketing is exclusive to the Internet. There are two methods of search engine marketing: search engine optimization and paid inclusion.

Search Engine Optimization

The process of increasing the visibility of a website by implementing strategies to ensure a better rank or placement in search results is known as **search engine optimization (SEO)**. Generally, more people are likely to click on links placed near the top of search engine results, so the better a website's placement in search results the more traffic that website can expect. For better or worse, a website's success is totally predicated on its positive placement in search results (Cabage, 2013). Search engine optimization is occasionally referred to as organic, because the favorable placement of sites occurs naturally rather than through paid inclusion, which will be examined below.

To optimize search results, web producers utilize strategies that must adhere to regulations established by search engines. Such strategies encompass the inclusion of the use of popular and relevant key words, the creation of relevant links to and from the site, and the measurement and conveyance of the site's popularity. These permissible practices are referred to as **white hat SEO**.

Although certain requirements are in place to ensure natural results, some people attempt to circumvent search engine rules and to manipulate the results. Among these strategies has been including an abundance of key words in various forms, the placement of which on the page has nothing to do with the actual content. Similarly, key words have been included on the page written in the same color as the background, so that they are hidden from visitors to the site. Some sites include numerous links to other pages with nothing to do with the content, and sometimes these other pages reciprocate by including links to the original site. Fictitious sites have also been created and linked to in order to increase the number of links established by a site. These prohibited practices are referred to as **black hat SEO**.

Black hat SEO practices are prohibited by search engines and might result in the banishment of a site or a significant decrease in its ranking. However, they are common occurrences and are sometimes exposed in a dramatic and embarrassing fashion. Most notably, a firm hired to optimize search engine results for J. C. Penny was exposed by *The New York Times* for such practices (Segal, 2011). Although successful in placing the J. C. Penny website at the top of many search results, the firm, which was quickly fired, had done so using prohibited techniques.

Search engines actually attempt to prevent black hat SEO practices. Google, in particular, has updated its page rankings and results specifically to combat these prohibited techniques. Since the beginning of the present decade, in a likely nod to white hats and black hats, these updates have been named after white and black creatures. The Google Panda Update was released in 2011, and the Google Penguin Update was released in 2012 (Stetzer, 2013).

Paid Inclusion

The placement of relevant advertisements among organic search results is known as **paid inclusion**. Such search engines as Bing (Bing Ads) and Google (Adwords) generate much of their revenue through advertisements. As with some banner advertisements, the payment for these search engine advertisements is determined in part by the number of clicks by users. Accordingly, it would be in the best interests of search engines to make these advertisements appear as genuine as possible. Yet, within the United States, they are required to classify or mark advertisements as such according to Federal Trade Commission guidelines (Efrati, 2013). At the same time, it has been shown that users frequently have difficulty distinguishing between these advertisements and genuine search results (Wall, 2012).

Contextual Marketing

Marketers also attempt to target user interests by engaging in **contextual marketing**, through which advertisements appearing as banners, pop-ups, and pop-unders are selected based on webpage content. For instance, if a user opens a webpage featuring content about fishing, a banner advertising new fishing poles might appear on the page. If that user then moves to a cooking website, a pop-up advertisement about a sale on cookware might appear.

Marketers have targeted consumers through other technologies in similar ways. For instance, television commercials and magazine advertisements are selected based on audiences that are most likely to be interested in the respective content. So, as above, a fishing program on television will feature a commercial for fishing poles, and a cooking magazine may feature an advertisement about cookware.

However, media content is not always as narrow as a fishing program or a cooking magazine. By scanning key words and other material in a webpage, contextual advertising through the Internet enables advertisements to be displayed with more accuracy than those of other technologies and to immediately adapt to changes in content.

Behavioral Target Marketing

While contextual marketing utilizes website content to determine advertisements, **behavioral target marketing** draws from user activities. This type of Internet marketing is accomplished by tracking online activity and crafting advertisements most suitable to users based on that activity.

Behavioral target marketing relies heavily on cookies, which are pieces of data or files sent by websites and stored in a user's browser. Some cookies keep track of activity during a single browsing session, while other cookies keep track of activities from past sessions.

First-party cookies are those specific to a website being visited. These are used to remember user preferences, passwords, and other materials, benefiting users by making their experience more personalized and easier. Without cookies, for example, form data might not be remembered within a single session, and shopping carts on a commerce site might only be able to hold one item at a time. Making things easier for users ultimately benefits sites by increasing the likelihood that satisfied users will return. However, sites can also use this information to study how users interact with the content and modify this content according to their goals for user activity. Of course, commerce sites also want users to place more than one item at a time in their shopping carts.

Third-party cookies are those used in behavioral target marketing and are placed within user browsers by third-party marketers on sites that are visited. These types of cookies track such activities as which sites are visited, how long people spend on websites, and what content is used. This information is then used by companies gathering the information, or sold to other companies. Advertisements are then developed and presented based on analyses of this information.

Ubiquity of Tracking

The tracking of user activity is a common occurrence. The ubiquity of tracking began emerging into public consciousness following a *Wall Street Journal* study that found the 50 most popular websites in the United States, which accounted for 40% of all page views in the United States alone, installed 3,180 tracking files on a test computer. On average, individual sites were found to install 64 tracking files into user browsers (Angwin & McGinty, 2010).

Hoofnagle and Good (2012) examined the placement of cookies by the most popular 100, 1,000, and 25,000 websites respectively, as established by Quantcom. This exploration used two types of crawls. A *shallow crawl* represents a visit to just the homepage. A *deep crawl* represents visits to six links on a given site. Only a shallow crawl was conducted for the top 25,000 sites. The findings of this study are provided in Figure 8.2.

	Percentage of Sites With Cookies	Number of Cookies	Number of First-Party Cookies	Number of Third-Party Cookies
Top 100 Sites (Deep Crawl)	100	6,485	992	5,493
Top 1,000 Sites (Deep Crawl)	97.9	65,381	8,658	56,723
Top 25,000 Sites (Shallow Crawl)	87	476,492	111,069	365,423

Figure 8.2. Cookie data for most popular sites. Hoofnagle and Good, 2012.

Support and Opposition for Behavioral Targeting

From a marketer standpoint, behavioral targeting increases the likelihood that advertisements will reach the intended audiences, and it becomes possible to determine the effectiveness of an advertisement. It is argued that this targeting also benefits users by not bombarding them with advertisements for which there is no interest or relevance (Angwin, 2010; Angwin & McGinty, 2010). Some companies responsible for these tracking devices also claim that not using them might lead to additional paywalls, should Internet advertising become less effective, and that there could be an increase in the number of advertisements and the intrusiveness of these advertisements (Pellman, 2013). It is also argued that the information is anonymous since users are not identified by name (Angwin, 2010; Pellman, 2013).

Opponents of behavioral targeting, however, maintain that third-party cookies and other devices are a violation of user privacy (Angwin, 2010). As with pop-up and pop-under advertising, Internet browsers have created tools that enable users to prevent the tracking of their activities. Nevertheless, tracking devices are becoming more clandestine and some are able to reproduce themselves even when eliminated by users. Furthermore, there is great disagreement among marketers, consumer and privacy advocates, and the producers of Internet browsers about what counts as tracking and targeting. As Singer (2013) argued, "until the ad industry, technology firms, and advocacy groups come to an agreement … the don't-track-me browser settings represent little more than symbolic consumer flags that companies are free to ignore."

Affiliate Marketing

Affiliate marketing is that which is conducted by third parties who are compensated based on their effectiveness. These third parties, not to be confused with the cookie variety, might be individual users with a personal blog or a professional site dedicated specifically to affiliate marketing. Effectiveness is predominantly based on the number of users attracted to a site as a result of affiliate influence or on the amount of sales garnered by an affiliate.

An example of a company's use of affiliate marketing is Amazon Associates (www.affiliate-program.amazon.com), which enables members to promote and link to products sold on Amazon in return for a commission on sales resulting from customer referrals. Amazon provides potential affiliate marketers with a variety of tools to use in the process.

Affiliate marketers in general use a variety of Internet marketing strategies in order to reach consumers. These strategies include banner and email marketing techniques, as well as the development of actual websites. For instance, comparison shopping websites, coupon websites, and review websites are often developed by affiliate marketers in order to attract users interested in particular products.

Social Marketing

Marketers are increasingly recognizing the influence of relationships through **social marketing**, that which takes place through social networking sites and through online relational connections among users. Social marketing can be separated into two forms: (a) that which relies on social networking sites specifically and (b) that which relies on relational connections among users.

Social Networking Sites

Given the popularity of social networking sites discussed in Chapter 8, it is not surprising that marketers use these sites as marketing tools. Fifteen million businesses have pages on Facebook alone (Kostsier, 2013).

At the most basic level, marketing through social networking sites is done in order to promote and maintain consumer awareness. Promoting and maintaining consumer awareness is accomplished in part through social networking site updates. Once connected to a business, organization, or product, users will receive updates in newsfeeds, which will likely at least be recognized if not consumed by the user.

On more advanced levels, marketing through social networking sites is done in order to better understand consumer reaction and to engage in customer service. For instance, consumer comments, both positive and negative, provide information pertaining to marketing campaigns as well as products and services themselves.

It also enables companies to engage in **social care**, customer service through social networking sites. Social care lets consumers "choose when and where they voice their questions, issues, and complaints, blurring the line between marketing and customer service" (Nielsen, 2012a, p. 14). Nearly half of social networking site users engage in social care, and 30% prefer this form of contact over using the telephone (Nielsen, 2012a). The ability of the consumer to have such direct contact with a company can embolden the user's expectations, with 81% expecting a same day response to questions and complaints posted on newsfeeds, and 30% of Twitter users expecting a response within 30 minutes (Oracle, 2012). At the same time, however, such service and marketing seems to benefit companies. Customers who engage with companies through social networking sites are more loyal to these companies, and they spend up to 40% more money with the companies than customers who do not (Barry, Markey, Almquist, & Brahm, 2011).

Promoting Relational Connections

Separating social networking sites from the promotion of relational connections is somewhat artificial. The influence of these connections on consumers takes place or is the result of materials within social networking sites. However, this influence is certainly not confined to social networking sites. Further, while the previous section focused on contributions from marketers themselves, this section will primarily examine the contributions of fellow consumers, sometimes using those materials provided by marketers. In what follows, we will examine the influence of social networks and the trust placed in the opinions of others as compared with actual advertising. We will then discuss viral advertising, perhaps the most recognizable form of such marketing. Then, we will explore consumer online brand-related activities.

Trust and Influence in Social Networks. Social networking sites enable marketers to create relationships among business, organization, products, and consumers. However, the true value of social networking sites comes from relationships shared among consumers themselves. Much more powerful than an actor in a commercial extolling the virtues of a product is a friend extolling the virtues of a product. Beyond opinions of friends, the behaviors of friends influence people. If a friend purchases a product, there is perceived trust and value in that product. Something similar happens when a friend likes a product on a social networking site, forwards a commercial for a product, or makes a comment about a product. And this influence is not limited to immediate friends, but through patterns of interaction that extend to friend's friend's friends (Adams, 2012; Christakis & Fowler, 2009).

The trust placed in the views and opinions of members within a person's social network has been supported by various studies, which also found little trust in advertising. In a global survey, Nielsen (2012b) discovered that when evaluating information about a product or service, 92% of consumers say they trust recommendations from people they know. The next most trusted source is real people they do not know; 70% of consumers say they trust evaluations posted online by other consumers. When it comes to other media, less than half of consumers trust advertisements on television (47%), in magazines (47%), in newspapers (46%), on radio (42%), or shown before movies (41%). Compared with other types of Internet marketing, half of consumers say they trust emails that they have signed up to receive, while only 40% trust advertisements in search engine results, 33% percent trust banner advertisements, and 33% percent trust display advertisements on mobile devices.

Another survey, conducted by Forrester Research (Stokes, 2013) in the United States and Europe resulted in different numbers but the same overall sentiment. Seventy percent of United States consumers trusted brand or product recommendations from friends and families, while 61% of Europeans indicated such trust. In terms of other types of Internet marketing, sponsored advertisements on search engine results were trusted by 27% of United States consumers and 24% of European consumers. While not specifying whether emails were opt-in or unsolicited, the Forrester survey found emails from companies or brands to be trusted by only 18% of United States consumers and 11% of European consumers. Finally, banner advertisements were trusted by only 10% in the United States and 8% in Europe.

Viral Advertising. Especially given the distrust of traditional marketing material, marketers increasingly seek to take advantage of the trust and influence of social networks. Of course, marketers have long recognized the potential value of word-of-mouth marketing, the spreading of information about brands, products, and services by consumers (Arndt, 1967). However, the Internet has dramatically altered how, and by what extent, this phenomenon can be accomplished along with the ways in which it could be manipulated by marketers.

Multiple terms have been used to represent such forms of marketing, and as with many areas of research, have been used interchangeably even when examining different things (Golan & Zaidner, 2008; Petrescu & Korgaonkar, 2011; Porter & Golan, 2006). Although there is some overlap among these related terms, there are differences among them, as noted in Figure 8.3.

Word-of-Mouth Marketing (WOM)	The verbal spreading of information about brands, products, and services by consumers
eWOM/Word-of-Mouse Marketing	The digital spreading of information about brands, products, and services by consumers
Viral Marketing	Online and offline marketing activities designed to compel consumers to spread commercial messages among themselves
Buzz Marketing	Consumer communication resulting from viral marketing
Viral Advertising	Unpaid digital distribution of marketer-generated advertisements forwarded from consumer to consumer

Figure 8.3. Social marketing terms. Adapted from Petrescu and Korgaonkar, 2011, p. 211.

Here we will look specifically at viral advertising, since it is the most recognizable and common among these forms of marketing. Viral advertisements take a variety of forms, including pictures, songs, infographs, lists, quizzes, and videos. Of these, videos tend to be the most successful.

This form of advertising has a few advantages over traditional advertising, such as television commercials. Among these advantages, their length can be expanded without concerns for fitting within established parameters or paying for additional time. Also, these advertisements are more likely to be viewed without interruption or competition for attention, since they are watched according to a user's schedule. Further, if the advertisement has been forwarded or posted by someone within a person's social network, it comes with that person's implicit endorsement.

Of course, a minimum condition of viral marketing is that these advertisements are spread by consumers to other consumers. Not all attempts at creating a viral advertisement are successful. According to Nobel (2013), successful viral advertising requires attracting attention, retaining attention, sharing, and persuading. So, this content must be personally enjoyable or meaningful. However, that alone is not enough to cause people to forward content. Accordingly, people tend to forward material for some sort of personal gain, perhaps through identity construction or relational enhancement. Figure 8.4 describes common, successful types of viral video advertisements and why they may be forwarded. In addition to the identity-based reasons included as intentions, the forwarding of any type of video may also be done for relational maintenance or enhancement.

Type of Content	Example Videos	Description of Content and Intention of Forwarder
Humor	Kmart Ship My Pants and Big Gas Savings (www.youtube.com/user/Kmart)	Content that is humorous; forwarder may wish to be perceived as someone who enjoys and wants to be associated with such material
Controversy	SodaStream Banned Commercials (http://www.youtube.com/user/SodaStreamGuru)	Content that has gained notoriety for being overly risqué or provocative or for being banned; forwarder may wish to be perceived as someone who can find, and is not troubled by, such content
Celebrity	Google Chrome for Mom (Mom!) —Featuring Stewie Griffin (http://www.youtube.com/user/googlechrome)	Content that features a celebrity or fictional character; forwarder may wish to be perceived as a fan of content or as knowledgeable of receiver interests

Figure 8.4. *Continued*

Values*	Dove Real Beauty Sketches (realbeautysketches.dove.us)	Content that features values or morals; forwarder may wish to be perceived as someone holding these values or morals and as wanting to share them
Groups*	Ram Trucks God Made a Farmer (www.youtube.com/user/ram)	Content that features a specific group of people; forwarder may wish to be perceived as a member of the group or as someone who admires the group
Privileged Content*	Google How It Feels Though Glass (www.youtube.com/user/Google)	Content that is presented as available to only a few people or as providing an early look at a brand, product, or service; forwarder may wish to be perceived as someone possessing special information
Hidden Gems*	Samsung Extreme Sheep LED Art (www.youtube.com/user/BaaaStuds)	Content that is unique or unconventional; forwarder may wish to be perceived as someone who appreciates such material or as someone with the ability to uncover such material online

Figure 8.4. Types of viral video advertisements. * Adapted from Nobel, 2013.

Consumer Online Brand-Related Activities: COBRAs. Sharing videos and other marketing content is not the only consumer activity online. **Consumer Online Brand-Related Activities (COBRAs)** is an emerging term used to denote brand-related consumption, contribution, and creation by consumers online. These activities encompass the use and sharing of content by consumers as well as consumer interaction, expression, creation, and a host of other online endeavors. Such activities frequently occur on social networking sites, but they are certainly not limited to those sites. Further, as with the sharing of video and other marketing content specifically, consumers usually engage in these activities for some sort of personal or relational gain.

Muntinga, Moorman, and Smit (2011) established three categories of COBRAs and investigated the consumer motivation underlying each of the following categories: consumption, contribution, and creation.

Consuming brand-related content includes such activities as watching brand-related video, listening to brand-related audio, viewing brand-related pictures, playing brand-related games, reading brand-related threads on forums, reading brand-related social networking site updates, and reading brand-related product reviews.

Consumers engage in such activities to gain information, entertainment, and remuneration. Through gaining information, consumers are motivated to discover what is happening within brand-related communities, to increase their understanding of brands and content, to secure knowledge prior to making purchases, and to gather brand-related ideas from other consumers. As entertainment, brand-related consumption provides enjoyment, relaxation, and provides something through which to pass the time. Finally, through remuneration, consumers hope to achieve tangible gain from such consumption. Some products and brands provide incentives such as free merchandise or downloaded music for visiting their sites, for instance.

Contributing to brand-related content includes such activities as rating products, joining a brand-related social networking site page, and making comments on brand-related forums or on social networking sites.

Consumers contribute brand-related content for personal identity development, integration and social interactions, and entertainment. Brand-related contributions enable consumers to create a connection between themselves and a brand or product. For instance, if a product is marketed as being used by adventurous people, consumers may wish to convey themselves as adventurous by association. Consumers may wish to be seen as knowledgeable about a product or seen as someone who is capable of making contributions. Consumers are also able to enact membership into groups and connect with like-minded consumers through their contributions. Finally, consumers gain enjoyment and relaxation through these contributions.

Creating brand-related content includes such activities as producing a brand-related website, blog, or fanpage, uploading brand-related media content, and writing brand-related reviews.

Consumers create brand-related content for personal identity development, integration and social interaction, entertainment, and empowerment. This category includes the empowerment of consumers, who see themselves as "brand ambassadors—people who display their enthusiasm for a brand and, importantly, enjoy convincing others that the brand is worth using or purchasing. They [are enabled] to reach and influence not only other customers but also companies"

(Muntinga, Moorman, & Smit, 2011, pp. 33–34). In many ways, then, the motivations for the creation of brand-related content are much like those for contributing to brand-related content.

Conclusion and Predictions

Marketing is among the most recognizable and notable uses of the Internet for business. Through its capabilities and through the relational activities of its users, the Internet has transformed and continues to transform marketing in previously unimaginable and profound ways. Unique from any other method of marketing, the Internet has become the most effective marketing tool ever conceived and developed.

As we look toward the future, a few predictions concerning the business of the Internet emerge. First, *businesses will increasingly integrate with the Internet to the point that there will be little distinction among marketing, shopping online, and shopping in a physical location.* Indeed, shopping experiences taking place in a physical location will require digital technologies. In some cases, rather than placing physical items in a shopping cart, consumers will simply touch a screen. Mobile devices are already being used to scan items consumers wish to purchase (i.e., Tode, 2013). Ebay has experimented with window-display screens placed in various locations that enable people to purchase items to be delivered by courier within an hour (Etherington, 2013). Marketing will continue developing into a constant part of all shopping experiences.

Despite attempts at regulation, *advertisements and other marketing strategies will become fully incorporated into a person's life.* They will appear without a person knowing or even recognizing them as such, but they will nevertheless influence behaviors. Native marketing, for example, will become so commonplace, that it will not be given any thought and truly seen as part of a person's natural world.

Behavioral marketing will become an accepted and expected practice. In spite of concerns for privacy, people will likely accept this practice because they will be given something in return. On a basic level, they will be provided advertisements that are meaningful to them and suggestions for purchases that are entirely accurate and beneficial. On a cruder and potentially more problematic level, consumers may be promised discounts or other incentives to enable marketers to follow their every move.

As with other predictions, the passing of time will inform us as to their accuracy. However, there is little doubt that the Internet will continue transforming the nature of business and the global economy.

Key Terms

Affiliate Marketing

Banner Marketing

Behavioral Target Marketing

Black Hat Seo

Buzz Marketing

Controlling the Assault of Non-Solicited Pornography and Marketing Act of 2003 (CAN-SPAM)

Consumer Online Brand-Related Activities (COBRAs)

Content Marketing

Contextual Marketing

Digital Capitalism

Email Marketing

E-Privacy Directive

eWOM/Word-of-Mouse Marketing

Fighting Internet and Wireless Spam Act (FISA)

First-Party Cookies

Native Marketing

Opt-In Email

Paid Inclusion

Pop-Under Advertisements

Pop-Up Advertisements

Search Engine Marketing

Search Engine Optimization (Seo)

Social Care

Social Marketing

Spam Act of 2003

Third-Party Cookies

Unsolicited Email

Viral Advertising

Viral Marketing

White Hat Seo

Word-of-Mouth Marketing (Wom)

Chapter Exercises

Current Issues: Have students search for and report on the most current media reports regarding Internet-based marketing. What positive and negative images are being attached to marketing on the Internet at the present time? How might these images contribute to the overall image of Internet-based marketing specifically and to perceptions of the Internet in general?

Debate It: Split the class into two groups. Select one of the resolutions below. Have one group support the resolution and have the other group oppose the resolution in a debate.

Resolved: Behavioral target marketing should be allowed.

Resolved: Behavioral target marketing should not be allowed.

Ethical Consideration: Have students consider whether it is ethical for marketers to monitor Internet use. Would their answer change depending on whether a question is phrased "people's Internet use" or phrased "your Internet use?" Would their answer change depending on whether or not users are aware of these activities? Would their answer change depending on whether marketers could convince them that doing so would benefit user experience? Would their answer change depending on whether doing so would prevent being charged a fee for Internet use?

Everyday Impact: A person's understandings of business, marketing, and the Internet develop and continue to be modified through their own experiences, popular culture, and media reports. Have students consider how their perceptions of Internet-based marketing have developed and how they may have changed over time. Do they view such marketing more positively or more negatively than in the past? Has their awareness of Internet-based marketing increased over time? Have they become more cautious or less cautious about such marketing?

Future Focus: In 1994, Rust and Oliver predicted the death of traditional advertising and its replacement by a new paradigm by 2010. Specifically, "the reason for advertising's impending demise is the advent of new technologies that have resulted in the fragmentation media and markets, and the empowerment of consumers. In the place of traditional mass media advertising, a new communications environment is developing around an evolving network of new media, which is high capacity, interactive, and multimedia. The result is a new era of producer-consumer interaction" (p. 71). Traditional advertising may not have completely died by 2010. However, there has certainly been a rise in new marketing strategies, and the effectiveness of traditional advertising can be brought into question. Ask students what changes in advertising they anticipate sixteen years from this moment? Also, Rust and Oliver predicted not only the dominance of this form of advertising in 2010 but also that it would "last well into the middle of the century" (p. 71). Do they think the latter prediction will be correct?

7

Internet as a Knowledge and Information Generating System

The Internet is changing your brain. That statement is perhaps a bit overdramatic but happens to be completely true. The Internet is changing the way people think, literally changing how a user's brain functions. The Internet is also changing how knowledge is perceived, gained, and used in society. Although, based on other technologies, these changes are not very surprising.

Every technology developed by humans has changed the ways in which people think, has changed the nature of knowledge, and has changed society. Writing, for instance, shifted information from a reliance on acoustic perceptions, which provided the basis for orality, to a reliance on visual perceptions. It subsequently gave rise to linear and sequential processing of information as well as to abstract thinking. The eventual development of the printing press made knowledge more standardized and permanent, which in turn altered how knowledge could be stored, accessed, and evaluated. Print further influenced the development of such new institutions as universities. It also influenced shifts in social power, facilitating mass education and decreasing the social power and prominence of religious institutions (see Chesebro & Bertelsen, 1996, for complete analysis of media cultures).

While it is not surprising that Internet use is influencing knowledge, how knowledge is being influenced by Internet use is astonishing and is worthy of exploration. The full impact of the Internet will not be realized for decades or perhaps centuries. However, it is possible to begin characterizing its impact on knowledge and examining its impact on existing institutions. And, it is especially important that we begin these explorations now, because the ways in which a technology is used and perceived in its early stages will influence its continued development, its future use, and its institutional and societal impact.

Within this chapter, we explore the interconnection of knowledge and technologies, specifically examining knowledge in two media cultures as well as the rise of knowledge institutions. Next, we explore nine transformations and characteristics of knowledge established through the Internet.

Questions Previewing This Chapter
1. *How have technologies impacted the development of knowledge?*
2. *How is the Internet influencing knowledge?*

Technologies and the Historical Development of Knowledge

The development and nature of knowledge are interdependent with the dominant technology used by a society. What it means to know something, how information is evaluated, how information is stored, how information is controlled, how information is shared, and how additional information is generated all differ depending on technology. Recognizing the interdependence of knowledge and technologies assists in understanding the Internet's influence on knowledge. However, tracing the entire history of knowledge would cause us to drift off course. Accordingly, in what follows, we will briefly examine two different media systems, orality and literacy, in order to illustrate the interdependence of knowledge and technology. We will then examine the historical rise of institutions of knowledge.

Knowledge in Oral and Literate Cultures

In an oral culture, knowledge is based on lived experience and evaluated using personal and immediate experience. It is dependent on sound, rhythm, formula, nonverbal behavior, and images. Knowledge is controlled by leaders in small communal groups and shared and maintained by repetitive storytelling that necessitates the physical presence and participation of the community. Any additional accumulation of knowledge requires lived experience and is maintained only if included within the stories of the group.

Knowledge in literate cultures, as mentioned within the introduction, involves linear patterns and hierarchies. Rather than lived experience and factual reality, personal and philosophical interpretations are emphasized. Societal knowledge can be stored and transferred using a collection of symbols imprinted on a material object, which means that physical presence is not necessary to transfer knowledge. Those individuals who possess these material objects, who have access to these material objects, and who have the ability to understand the symbol system are able to control societal knowledge and to engage in societal governance. Additional knowledge can be accumulated by sharing existing information through the material objects, making personal interpretations of that information, and storing new information using symbols imprinted on additional material objects.

Through orality, literacy, as well as every ensuing technology, the nature of knowledge has been transformed. As the nature of knowledge transformed, so too did the societies in which that knowledge was embedded. In addition to these societal transformations, institutions of knowledge were developed to manage, protect, and control this knowledge as well as produce new knowledge.

The Rise and Fall of Institutions

These **institutions of knowledge** have resulted from the ways in which knowledge is perceived by a society and from the capabilities offered through the use of a given technology. McNeely and Wolverton (2008) noted that Western intellectual life has been dominated by the following institutions: (a) libraries, (b) monasteries, (c) universities, (d) the Republic of Letters, (e) disciplines, and (f) laboratories. These and such other institutions as museums and the news media have developed through the efforts of individuals, communities, religious institutions, and governments. Each institution has managed, protected, controlled, and generated knowledge with varying degrees of success, and power has been continually transferred among the groups who had the most control over knowledge.

Examining these shifts from an institutional perspective, McNeely and Wolverton (2008) maintained

> these institutions have safeguarded knowledge through the ages.... Each coalesced in reaction to sweeping historical changes that discredited it predecessor or exposed its limitations. And each parlayed dissatisfaction and disillusionment with existing ways of knowing into an all-encompassing new ideology.... In times of stability, these institutions carried the torch of learning. In times of upheaval, individuals and small communities reinvented knowledge in founding institutions. (p. xvi)

From our perspective, technologies have been the reason and driving force for institutional transformation. At the very least, they have enabled these transformations to take place. Consider the above institutional approach from a technological perspective. Knowledge has been safeguarded through the use of each existing technology, orality, literacy, electronic, and digital. When dominant within a society, each technology was the standard-bearer of knowledge. Each new technology reacted to and built upon previous technologies, exposing the preceding technology's limitations on the management, protection, control, and generation of knowledge. Technological upheaval led to the reinvention of knowledge and the foundations of new institutions of knowledge.

The Internet represents technological upheaval. With this technological upheaval comes an upheaval of knowledge. Weinberger (2011) described this upheaval as being "in a crisis of knowledge at the same time that we are in an epochal exaltation of knowledge" (p. xiii). Specifically,

> we fear for the institutions on which we have relied for trustworthy knowledge, but there's also a joy we can feel pulsing through our culture. It comes from a different place. It comes from the networking of knowledge. Knowledge now lives not just in libraries and museums and academic journals. It lives not just in the skulls of individuals. Our skulls and our institutions are simply not big enough to contain knowledge. Knowledge is now the property of the network, and the network embraces businesses, government, media, museums, curated collections, and minds in communication. (p. xiii)

The nature of knowledge is changing along with institutions of knowledge and ways in which knowledge is managed, protected, controlled, and generated. In the next section of the chapter, we will examine transformations of knowledge resulting from the Internet.

The Internet and Knowledge

The Internet is transforming knowledge in profound and previously unimagined ways. And, unlike previous technologies, we are truly able to study the impact of the Internet from its very beginnings. In what follows, we will isolate specific characteristics and transformations of knowledge resulting from the Internet. These characteristics and transformations of knowledge include the following: (a) massive information, (b) good and bad information, (c) constant information, (d) connected information, (e) multiple sources of information,

(f) personalized information, (g) public information, (h) transforming expertise, and (i) transforming aptitude and thinking.

Massive Information

Humans have access to and utilize more information now than at any point in history, and the amount of information available to be created and used is increasing at phenomenal rates. Examining the use of 20 sources of information in the United States, research conducted at the Global Information Industry Center found that people in the United States consumed information for 1.3 trillion hours in 2008, nearly 12 hours per day on average (Bohn & Short, 2009).

Such massive amounts of information are available through the Internet that it is difficult to fully comprehend. Further, it is impossible to accurately determine or know the expanse of the Internet. And, like most things, the answer depends on who is being asked and what is being counted. Still, we can certainly speculate about how massive it may actually be and how much information may be available.

One Big Encyclopedia

To begin our exploration of the size of the Internet and how much information is available, we will examine one single site dedicate to providing information: Wikipedia. Using information provided by the site, there exist 27 million articles comprised of over 11 billion words in approximately 286 languages. The English version of Wikipedia alone includes over four million articles with nearly two-and-a-half billion words, making it 50 times larger than *Encyclopedia Britannica* (which has now ceased publishing a print version). Reading 16 hours per day, it would take a person over seven years to read the entire thing (Wikipedia, 2013a, 2013b).

Thirty Trillion Pages

The numbers for Wikipedia alone are staggering, but we must keep in mind that Wikipedia is just one drop in the Internet bucket. Google maintains that it searches 30 trillion pages one billion times per month (http://www.google.com/insidesearch). The information provided on each page would likely cover more space than is available on a standard 8.5 x 11-inch piece of paper. However, for the sake of comparison, we will assume that the information provided on each page can fit neatly on that space. Thirty trillion standard pieces of paper would take up nearly 700,000 square miles. Putting this amount of space in perspective, the state of Texas could be covered 2.5 times; the entire country of England could be covered 14 times. Laying these sheets of paper side-by-side, top-to-bottom, they would circle the Earth around 210,000 times.

Big Data

Of course, digital information is measured in bits rather than in pages or physical space. A **bit** is one unit of data, coded as either a 1 or a 2. Table 7.1 provides standard measurements of data.

Table 7.1. Digital Units of Data.

Bit	One basic unit of data coded as either 1 or 0
Nibble	4 Bits
Byte	8 Bits
Kilobyte	1,000 Bytes *
Megabyte	1,000,000 Bytes
Gigabyte	1,000,000,000 Bytes
Terabyte	1,000,000,000,000 Bytes
Petabyte	1,000,000,000,000,000 Bytes
Exabyte	1,000,000,000,000,000,000 Bytes
Zettabyte	1,000,000,000,000,000,000,000 Bytes
Yottabyte	1,000,000,000,000,000,000,000,000 Bytes

Note. * One kilobyte specifically equals 1,024 bytes.

We can examine both the numbers associated with all digital content as well as numbers associated solely with Internet content. Examining all digital usages means including such content as telephone conversations, televisions, and pictures taken on a camera, which may not seem like an accurate way to determine the size of the Internet. However, a great deal of such content is assisted by or transferred through the Internet, and this phenomenon is becoming more and more common.

Research conducted through the International Data Center placed the amount of data created, replicated, and used in 2011 at over 1.8 zettabytes, "nearly as many bits of information in the digital universe as stars in our physical universe" (Gantz & Reinsel, 2011, p. 1). This number is expected to reach 4 zettabytes by 2020, with nearly 40% of that information stored or processed in cloud computing providers (Gantz & Reinsel, 2012).

Examining the Internet specifically, Cisco (2012) has projected that global Internet use will reach 1.3 zettabytes by 2016. According to this forecast, "the gigabyte equivalent of all movies ever made will [be conveyed] every 3 minutes … [and] "networks will deliver 12.5 petabytes every five minutes" (p. 1).

This massive amount of information, and the subsequent development of computers and software to manage and process these levels of information, enables the ability to use **big data**. The term *big data* generally refers not just to the data itself but also what can be done with that data. Mayer-Schonberger and Cukier (2013) explained,

> big data refers to the things one can do at a large scale that cannot be done at a smaller one, to extract new insights or create new forms of value, in ways that change markets, organizations, the relationship between citizens and governments, and more. (p. 6)

According to these authors, big data will result in major shifts in the ways in which researchers understand data: (a) using large sets of data rather than smaller sampled sets of data, (b) accepting messiness rather than statistical exactness, and (c) focusing on correlation rather than causation.

We are already seeing the utilization and potential benefits of big data. Amazon.com and Netflix.com use massive amounts of data to personalize displays and generate suggestions for customers. One-third of Amazon's sales and three-fourths of Netflix's orders come from these recommendations. Google Flu Trends uses billions of Internet searches to track locations of flu outbreaks. Through the use of this big data, such stores as Wal-Mart and Target are able to predict when people are most likely to purchase Pop Tarts, and even pregnancy, based on weather patterns and previous sales. With a focus on correlation rather than causation, mentioned above, it is of little concern why these trends exist but rather that they can be discovered and used (Mayer-Schonberger & Cukier, 2013).

Good and Bad Information

The Internet provides access to the greatest thoughts ever developed by humankind. It also provides access to the worst thoughts ever developed by humankind. Included among the information available through the Internet are histories of the world's cultures, significant discoveries, great artistic endeavors, keen critical evaluations, and even proper methods of removing a splinter. Depending on a person's needs at a given moment, this information is particularly valued and welcomed.

Also included among the information available through the Internet are claims that drinking Mountain Dew is a medical cure, videos of caterwauling teenagers whose friends have provided assurances of singing ability, hate speech, and what to do if abducted by aliens. Again, depending on a person's needs at a given moment, this information might be valued and welcomed, but probably not.

Compared to previous technologies, it is much easier to distribute information through the Internet. Worthless and questionable ideas have always existed, but they were previously confined to a limited number of people rather than being released and available to millions of people around the world in a matter of seconds. Previous technologies have also exerted more of a gate-keeping influence on information. Not everyone is able to write a book, produce a television program, or develop a radio show. Further, editors, distributers, and owners are generally in place to control the information. Aside from nation-states with strict censorship, the Internet enables any idea to be distributed with few obstacles and limited control.

Alternative View

There are a few alternative views to be taken here. First, we seem quick to dismiss methods for surviving alien abduction, but if aliens attack, this information will be invaluable. The Earth revolving around the Sun was once considered an outlandish idea. A number of great and valuable ideas have withered away without consideration because of limitations and control of previous technologies.

Second, just because it is more challenging to spread one's ideas through other technologies it does not mean that the ideas available through books, television, and radio are automatically correct and valuable.

Third, a great deal of information available on the Internet is accurate and subject to editorial and other controls. Online forums are often monitored, for instance. Information appearing on major news websites undergoes the same editorial scrutiny as that given information distributed through other technologies. In terms of accuracy, an early study of Wikipedia found the average number of errors per article to be four, comparable to the three errors found in the average Encyclopedia Britannica *article (Giles, 2005).*

Finally, questionable information does not necessarily rise to the forefront when using the Internet. Just because something is available does not mean that it will be used or even discovered. A great deal of less-than-valuable or inaccurate information can be found only if purposely searching the fringes of the Internet.

Constant Information

The Internet also enables people to have constant access to information. Rainie and Wellman (2012) noted that "with high-speed broadband and always accessible mobile devices, information flows through people's lives more quickly than at any time since most people lived in small villages" (p. 228). The Internet enables people to access information at any time and essentially any location.

Through the Internet, a person can access the answer to nearly any question that can be asked in a matter of seconds. The average search takes nine seconds to type, and Google Instant, which provides results as a person types, cuts that time by two to five seconds (About Google Instant, 2013). At its fastest average, then, an answer can be accessed in four seconds—about the amount of time it took you to read the first clause of this sentence.

Connected Information

Somewhat unique compared with other sources of information, that which is available on the Internet frequently provides links to materials in support of assertions and ideas. These links give users opportunities to verify information presented and to better understand how ideas were formulated. When reading a newspaper or watching television news, it is generally necessary to trust that the information provided is accurate. When consuming news through the Internet, the value and trustworthiness of this information is perhaps enhanced through the ability to verify what is presented. For readers of academic information, such support through references and footnotes is generally expected and necessary. However, the Internet increases the ease through which this material is accessed and likely increases its potential use.

The connected nature of information available through the Internet, subsequently, may increase the likelihood that users will go beyond what is offered in front of them. Doing so has the potential to increase their understanding of the topic area and has the potential to increase their awareness and understanding of matters related to that topic area.

Accordingly, the connected nature of knowledge on the Internet generates new connections and ideas within personal knowledge as well as the generation of new connections and ideas within collective public knowledge. For instance, following links to other material, a user can personally uncover previously unknown connections among procedures, people, events, places, and other matters.

Collective public knowledge is generated through the cumulative nature of the Internet and the interconnectivity of not only information but also users. Retaining all information ever posted to it, knowledge generated through the Internet is additive and developed by people from all walks of life with various degrees of expertise and understanding. As Weinberger (2011) noted, new knowledge and expertise has resulted from "people who don't know one another, spread out across a global network, who have answered a question, gathered data, refined results, contributed to a web of blogs, or even built an encyclopedia" (p. 54). Accordingly, the Internet, "as a place that connects lots of people who are different from one another, is not only finding expertise but also generating it" (p. 57).

Multiple Sources of Information

The Internet also increases the number of sources used to gather information. News consumers, for instance, are no longer restricted only to local newspapers, radio, and television or interactions with members of one's immediate social network. Rather, information can be gathered through numerous sources available through the Internet.

Studies indicate that news consumers do use a variety of sources to gather information. Ninety-two percent of American news consumers indicate that they use multiple sources to gather news on an average day. Forty-six percent of these news consumers use four to six separate types of sources. While using the Internet specifically, news consumers indicate that they typically use two to five news sources for information (Purcell, Rainie, Mitchell, Rosenstiel, & Olmstead, 2010). It is possible that the availability and use of multiple news sources online enables news consumers to be more critical. Certainly, this availability and use provides news consumers with opportunities to compare perspectives and verify information.

Beyond news consumers, Internet users conducting research or wishing to find the answer to a particular question also have a variety of sources at their disposal. Of course, as mentioned previously, some information available through the Internet is more accurate than other information. As such, someone using the Internet to find information must not only be able to find the answers but also be able to determine which of the answers provided is most accurate, something we will discuss in more detail below.

Personalized Information

While a host of news and other sources of information are available through the Internet, the consumption of this information is increasingly personalized.

On the one hand, this means that people are less likely to focus only on major sources of information. Information consumed can be focused on issues of individual importance (Rainie & Wellman, 2012). A decreased reliance on these major sources of information also means a potential redistribution of power among sources of information and the possibility to challenge generally accepted ideas and perspectives.

On the other hand, it is possible that this personalization of information is causing views and understandings to become more limited and insular (Weinberger, 2011). For instance, through personalizing a newsfeed, a person can receive only that information that he or she might find interesting. Consequently, a great deal of other information remains undiscovered by this news consumer, limiting his or her viewpoints, understandings, and knowledge.

Of course, as mentioned above, news consumers generally use multiple sources of information. However, news consumers frequently use sources of information that support their existing views (Stroud, 2008, 2010). If those sources of information share similar information and perspectives, then, their preferred and narrow use may actually cause people to become more extreme in their views. For instance, a liberal news consumer visiting only liberal sites may develop more partisan views as a result, likewise, a conservative news consumer visiting only conservative sites. Sunstein (2009) has warned that such activities, and specifically the filtering out of information a person does not wish to receive, will lead to increased polarization and have political consequences.

Alternative View

Countering these concerns, research conducted by Garrett (2009) suggests that consumers of Internet news can "use the control afforded by online information sources to increase their exposure to opinions consistent with their own views without sacrificing contact with other opinions" (p. 676).

Public Information

The Internet has made much of the world's knowledge available to everyone. This access has made information more available to all social groups than was the case with any previous technology. Specialized information and secret information are now more readily available to nonspecialists. And, private information is now more accessible.

Access to Information Among Social Groups

The Internet is making information more accessible to everyone in societies and not just a limited few. Information has historically often been available to a select few people in societies. The ability to read and availability of books and other written sources, for instance, were once limited to members of privileged classes (Manguel, 1996). Struggles encompassing a lack of access among some groups and censorship by nation states continue to exist. Compared to all previous technologies, however, the Internet has made information more accessible than at any previous point in history.

Access to Specialized Information

The Internet has also made specialized information that is usually limited to specific professions available to people outside of those areas. Consider the vast amount of medical information now available to people outside the medical field through the Internet. Information about symptoms, treatments, and other matters related to illness, along with current research and debate previously available only through instruction or access to medical books and journals are now readily accessible. This access to specialized information has enabled people to be more savvy consumers of goods and services. In the case of travel agencies, for instance, access to information is making some professions and services progressively more obsolete.

Access to Secret Information

The Internet is also making it more difficult to keep information private, while making it increasingly easy to spread this information. For businesses, internal company memos and email correspondence along with employee information, research and development plans, and other information is frequently leaked or discovered. For schools, student information is often inadvertently provided through emails or is found to be still available on the thought-to-be-erased hard drives of publically sold outdated computers. For governments, secret information is often the target of other governments, social activists, and such sites as Wikileaks (Greenberg, 2012).

Transforming Expertise

The Internet has also transformed expertise in terms of the availability of expert information and access to experts. The Internet has further influenced sources of expertise and how information and advice are evaluated and assessed.

Availability of Specialized Information

One component of expertise is knowledge about a particular area. An expert about a historical event might be expected to have a wealth of information pertaining to that event committed to memory. Through the Internet, a person can access much of that same information in a matter of seconds. While deeper levels of understanding may not be initially achieved, the basic foundation of expertise is readily available.

Access to Experts

Access to experts in particular areas is readily available through the Internet. A quick search will provide lists of contact information for people with sought-after information and experience. Some sites specialize in offering access to experts. Multiple sites offer real-time interaction with people to help guide decisions or to provide additional information about an issue. Other sites specifically exist to provide expert advice and assistance about a variety of matters.

Sources of Expertise

Specialized information is increasingly available in part because of access to the experts mentioned above and in part because it can come from a number of different sources. For instance, someone having difficulty with their cell phone could contact the manufacturer or service provider for assistance. However, that person could just as easily (and probably more easily) visit an Internet forum through which other cell phone owners could provide the needed answers. So, while more accessible than before, traditional experts may be less sought out for answers.

Evaluating and Assessing Expertise

These transformations of expertise have led to changes in its evaluation and assessment. Expert advice now seems less valuable than in the past, given the availability of this information and the great number of sources available. Further, given the popularity of such sites as Tripadvisor and Yelp, people may be finding information and advice provided by nonexperts more valuable. This nonexpert information and advice is also available from our social network through comments and evaluations on social networking sites. It is further available through such sites as Netflix that provide recommendations based on the actions of people similar to a given visitor to these sites.

Transforming Aptitude and Thinking

The critical evaluation of Internet sources and content underscores transformations in aptitude and thinking resulting from the Internet. Such transformations include developing the ability to find information and the ability to evaluate information, as well as encompassing changes in how people interact with and manage information.

Searching for Information

As discussed previously, the Internet provides users with unprecedented access to massive amounts of information available almost instantaneously. Users must, however, develop skills to locate information among the trillions of pages comprising the Internet. As with changes in expertise, more important than knowing information is the ability to access information.

To some extent, the search for information on the Internet is done for users, through such means as personal newsfeeds. However, there often exists a need for users to locate specific information. In these cases, users must develop skills related to determining which websites might be the best sources to locate this information. Personal information about a friend or acquaintance might be located using social networking sites, tomorrow's weather might be located on a number of weather sites, product information might be discovered on sales or product review sites, health information might be located on a health website.

Search Engines

Yet, many users do not go directly to a specific site. Instead, they are likely to begin their search for information by using a search engine. For instance, nearly 80% of people looking for health information on the Internet begin their search using a search engine. Only 13% go directly to a specific medical site (Fox & Duggan, 2013).

Search engines remain the dominant tool when searching for information on the Internet. And, it appears as if most users are confident in their ability to use these sites. Ninety-three percent of users report being confident in their ability to use search engines, with 91% reporting being able generally find what they are searching for. Moreover, these searches tend to be positive experiences. Eighty-six percent of users report learning something new or important that assisted them or enhanced their knowledge (Purcell, Brenner, & Rainie, 2012).

Users of search engines are not only confident in their abilities but also tend to be confident in search engines themselves. Sixty-six percent of users believe search engines are fair and unbiased. Seventy-three percent of users found much of the information discovered when using search engines to be accurate and trustworthy (Purcell, Brenner, & Rainie, 2012). At the same time, users must also develop skills related to the evaluation of the information discovered.

Evaluating of Internet Sources and Messages

The Internet requires users to develop new strategies for determining evaluating messages. Rieh and Hilligoss (2008) discovered that evaluation occurs incrementally. Predictions are initially made about the information resource prior to seeking the information. Sources that have proved valuable in the past or which have been recommended by trusted others will often be utilized. Information is then evaluated while being consumed, and evaluated again when utilized.

As with the evaluation of all sources and information, rigor when evaluating Internet sources depends on the importance of the information being sought. Users tend to be concerned about the credibility of a source when gathering information about important issues such as academics, health, and finances. Source credibility is also an important concern when the information will involve other people. Source credibility is less significant when gathering information deemed less important such as that involving entertainment and when gathering information that will only impact oneself (Rieh & Hilligoss, 2008).

Also, as with the evaluation of all sources and information, users rely on cues and shortcuts. Sundar (2008) noted that unique cues related to modality, agency, interactivity, and navigability can be utilized by users when evaluating Internet sources. Such cues could involve realism, social presence, responsiveness, telepresence, flow, browsing capabilities, and other features.

So, while similarities exist when evaluating credibility of Internet sources and when evaluating other sources, there are certainly differences in how these evaluations are made and what features are used to make these evaluations. Table 7.2 presents the findings of an early study conducted by Fogg and colleagues (2003), with credibility concerns about a healthcare website and a financial website ranked in order of importance for users. Table 7.3 presents the findings of a more recent study conducted by Dochterman and Stamp (2010) in which user credibility concerns about websites in general were examined.

Table 7.2. Evaluation of Website Credibility.

Criteria of Credibility	Characteristic/Example of Criteria
Design Look	Layout, typography, images, color scheme
Information Design/Structure	Site organization and navigation
Information Focus	Brevity and breadth of information
Underlying Motive	Sales and prosocial motivations
Usefulness of Information	Value of information to serve user purpose
Accuracy of Information	Trustworthiness and correctness of information
Name Recognition and Reputation	Recognition and reputation of site operator
Advertising	Pop-up ads diminished credibility of site
Bias of Information	Opinionated information
Tone of the Writing	Use of slang or marketing language
Identity of the Site Sponsor	Information about the sponsor provided
Functionality of the Site	Broken links and unhelpful search features
Customer Service	Openness to customers and accountability
Past Experience With Site	Previous positive and negative experiences
Information Clarity	Concise information and wordy information
Performance on a Test	Site performance when used for specific task
Readability	Font
Affiliations	Seals of approval from other sources

Note. From Fogg et al. (2003).

Table 7.3. Determination of Web Credibility.

Criteria of Credibility	Characteristic/Example of Criteria
Authority	Affiliations of sites and contributors
Page Layout	Ease of navigation and overall appearance
Site Motive	Self-serving and prosocial motivations
URL	Entire address, but especially domain suffix
Cross-Checkability	Links to verify information
User Motive	Academic and entertainment motives
Content	Language level and typos
Date	Timeliness and updates
Professionalism	Proficiency of web design
Site Familiarity	Previous visits to site
Process	Site credibility initially, during use, and later
Personal Beliefs	Personal views of topic and Internet in general

Note. From Dochterman & Stamp (2010).

Managing of Information

The Internet is also transforming how people manage and handle information, and in doing so, is changing our brains. Of course, any technology use or ability impacts our brains, which are surprisingly malleable. For instance, brain scans have also revealed changes in neural activity and development when reading (Dehaene, 2009). Changes in brain development have been discovered when new London cab drivers are trained to learn the city streets (Woollett & Maguire, 2011).

Such changes in brain activity and development have also been observed when using the Internet (Small, Moody, Siddarth, & Bookheimer, 2009). Indeed, the brains of new Internet users can be seen rewiring themselves after a mere five hours of Internet use (Small & Vorgan, 2007). Although we won't examine these neurological changes in depth here, we will examine how the ways in which people think are changing as a result of the Internet.

As discussed earlier, users can be less concerned about recalling basic facts than ever before. Memorizing the capitals of nation states, for instance, seems unnecessary when this information can be discovered in a matter of seconds using the Internet. Contact information for friends no longer must be memorized, since it is readily available on the Internet, and digital correspondence can be sent by simply clicking on a person's name or typing the first few letters of his or her name. There is no longer a need to remember locations or directions, since a person can be guided essentially anywhere though GoogleMaps or similar services. As also discussed earlier, many choices are even made for us now, such as evidenced by the success of Amazon purchase recommendations and Netflix viewing recommendations. Brooks (2007) referred to this phenomenon as the **outsourced brain**, maintaining that thinking and decision-making are being externalized and networked.

Such transformations of the brain and thinking have been met with concerns. Morozov (2013) pondered whether the new smart is making us dumb. Previously inanimate objects such as trash cans, cars, and even spoons are increasingly digital and connected to the Internet, providing us with information about our activities and guiding our behaviors. A great deal of thinking is being done for us.

At the same time, it can also be argued that the Internet is forcing people to think more than ever before, simply in different ways, for better and for worse. Internet users are hyper-informed. Information online appears infinite, updating and breeding without reprieve. Promising bits of information saturate the

Internet. The current human brain has a limited capacity to process, organize, and store all this meaningful data.

This limited capacity of our brains can lead to cognitive overload. As Carr (2011) explained, **cognitive load** is all the information flowing into working memory at given moment in time.

> When the load exceeds our mind's ability to store and process the information … we're unable to retain the information or to draw connection with the information already stored in our long-term memory. We can't translate the new information into schemas. Our ability to learn suffers, and our understanding remains shallow. (p. 125)

Further advancing cognitive overload, information online can oftentimes lack input and meaningful organization. The Internet user is regularly forced to approach all pieces of data equally without logical discrimination while simultaneously attempting to filter the relevant, the irrelevant, the significant, and the insignificant (Keen, 2007). Sifting through these mounds excessively will often lead to what Small and Vorgan (2008) termed **techno-brain-burnout**. After several hours online, the habitual scanning and filtering can cause notable stress that impairs cognition, leads to anxiety, and alters "the neural circuitry in the hippocampus, amygdale and prefrontal cortex—the brain regions that control mood and thought" (p. 48).

Referenced above, Carr (2008, 2011) most notably examined the potential changes and consequences of the Internet on our brains and subsequent ability to think, through an original article published in *The Atlantic* asking "Is Google Making Us Stupid?" and in a subsequent book entitled *The Shallows*. Indicated by the title of his book, Carr (2011) is concerned that the Internet promotes shallow thinking.

> When we go online, we enter an environment that promotes cursory reading, hurried and distracted thinking, and superficial learning. It's possible to think deeply while surfing the Net, just as it's possible to think shallowly while reading a book, but that's not the type of thinking the technology encourages and rewards. (pp. 115–116)

As a result, it is possible that such abilities of higher levels of thinking as the analysis and synthesis of information may be diminished. Also, as a result, it is possible that "we are evolving from being cultivators of personal knowledge to being hunters and gatherers in the electronic data forest" (p. 138).

Alternative View

At the same time, Carr (2011) also noted research indicating that some cognitive skills such as the processing of visual cues, fast-paced problem-solving, and working memory capacity may be improved as a result of the Internet. Since these skills benefit our search for information, it might be argued that we are adapting to this technology and that the ability to gather and sort through information is necessarily improving.

In spite of these concerns, the majority of Internet experts and critics generally tend to be hopeful about the future. The Pew Internet & American Life Project surveyed 895 Internet experts and critics about the future of the Internet (Anderson & Rainie, 2010). Three-fourths of those surveyed agreed with the following statement: "By 2020, people's use of the Internet has enhanced human intelligence; as people are allowed unprecedented access to more information they become smarter and make better choices. Nicholas Carr was wrong: Google does not make us stupid" (p. 2). Further, 65% agreed with this following statement: "By 2020 it will be clear that the Internet has enhanced and improved reading, writing and the rendering of knowledge" (p. 2). The passing of time will fully reveal whether the Internet has benefited or hindered thinking. The answer will likely include both benefits and hindrances. Yet, it is certain that the Internet will continue to change the way we think.

Conclusion and Predictions

The Internet is radically altering the structure and potential of the human brain. It is changing how societies and users perceive, evaluate, and utilize knowledge. The Internet has become a constant companion, a portal for mental extension, and a deliverer of cognitive fatigue. In this final section, we propose four potential outcomes for current developing global trends related to the Internet and knowledge.

In the future, *the size of our digital world will continue to increase*, perhaps even faster and to a greater extent than has been predicted. This growth of information will make information even more accessible to Internet users. The amount of data gathered through digital means will be increasingly utilized by governments, corporations, and individuals. In the case of individuals, many everyday decisions such as when to wake up in the morning and what to eat will be made or provided for them based on computer-generated analysis of information.

The suppression of the linear thought process will continue as Internet users' minds become increasingly habituated to the nonlinear consumption patterns favored online. Expedient and diverse multitasking will become the preferred method of media consumption, as "single-tasking" and linear consumption becomes perceived as rudimentary, inefficient, sluggish, and restrictive.

As digital technologies continue to integrate themselves into society while outprocessing the human mind without bias or exhaustion, *people will seek means to enhance their organic minds (self-perceived and media-promoted as slow or inferior) through a physical union with digital systems.*

The synthetic boundaries between our intellectual technologies and organic intelligences are converging, and primed to continue converging. Toward this end, Google has developed commercial eyeglasses providing constant connection to the Internet while digitally augmenting physical environments and interactions (www.google.com/glass/start). Likewise, Apple is considering wearable Internet devices that will provide users with constant connection (Ovide & Rusli, 2013). The appeal and ease of merging the organic brain completely with digital technologies will further intensify with the continued evolution of artificial intelligence and the commercialization of connected digital interfaces. Taking this interconnection of the digital and organic in the opposite direction and a bit further into the future, it is possible that a digital copy of a person's brain may be transferred to a nonorganic carrier in the manner of cyborgs (Segel, 2013). And, on a truly molecular level, it is already possible to store digital data in DNA (see Naik, 2013).

Time, and perhaps the Internet, will inform us as to whether these and other predictions will come to fruition.

Key Terms

Big Data	Institutions of Knowledge
Bit	Outsourced Brain
Cognitive Load	Techno-Brain-Burnout
Credibility	

Chapter Exercises

Current Issues: When it comes to searching the Internet, companies are increasingly striving to provide the answer before it is asked. Google Glass promises to provide a certain amount of this information (www.google.com/glass/start). Also, Apple has applied for a patent for an augmented reality system that will provide information about surroundings seen through the screens of mobile devices (Murphy, 2013). Have students search for and report on the most current media reports regarding advances in search engines and technologies. Do they consider such news to be more focused on the discussion of technological advances or more focused on portraying the companies in a positive manner? Does this news have an impact on how they search for items on the Internet?

Debate It: Split the class into two groups. Select one of the resolutions below. Have one group support the resolution and have the other group oppose the resolution in a debate.

Resolved: Wikipedia should be considered a legitimate academic source.

Resolved: Wikipedia should not be considered a legitimate academic source.

Ethical Consideration: Have students consider whether it is ever ethical for someone to purposely leak classified or private information from companies, schools, or governments. Are there any occasions when this could be considered legitimate? If so, what makes a person qualified to determine when such actions should be done for a greater good than privacy?

Everyday Impact: Ask student to consider how they evaluate information located on the Internet. What specific sites and general types of sites do they find credible? What specific sites and general types of sites do they not find credible? If they are unfamiliar with a site, what factors do they use to determine whether or not they consider that site to be credible? Have students ever been fooled into thinking a site was credible only to find out that it was not very credible? Why do they think they were fooled and what could be done to prevent such mistakes?

Future Focus: We believe that humans will incorporate digital devices into their brains in the future. Ask students whether they think we are correct. Would they be willing to have a digital device placed in their brain, if doing so was a relative minor and simple procedure and meant being able to do everything they now do with cell phones and their computers without the need for a physical device? What would they consider to be the pros and cons for such devices?

8

Social Networking Sites

The Internet has profoundly altered how people initiate personal relationships, how those relationships are maintained, and how those relationships are terminated. Although previous technologies have influenced how people go about relating, none have had such a major influence on not only how relationships are conducted but also how they are perceived, understood, valued, and categorized. Moreover, no previous technology has been so completely integrated into relational processes.

Social networking sites are exclusively examined within this chapter in part because of their prominence in people's lives and their impact on the Internet as a whole. They are also examined because of their influence on personal relationships and because of their role in the integration of online and offline worlds, knowledge, and realities.

In their most general form, **social networking sites** enable people to construct a profile, assemble a list of connections, and explore the profiles of users. Currently, more than 200 social networking sites are in existence, with billions of registered users throughout the world. The most popular social networking sites, based on the number of registered users, are listed in Figure 8.1.

Although not social networking sites as they are now recognized, early forms of online social networks can be traced to various online communities connected through bulletin board systems. Perhaps the earliest online community site was

Well.com, which was launched in 1985. In their current form, Classmates.com and SixDegrees.com, launched in 1995 and 1997, respectively, were the first social networking sites.

Facebook	Facebook.com
Twitter	Twitter.com
Google+	Plus.google.com
Qzone	Qzone.qq.com
SinaWeibo	Weibo.com
Formspring	Formsrping.me
Habbo	Habbo.com
Linkedin	Linkedin.com
Renren	Renren.com
Vkontakte	Vk.com

Figure 8.1. Top social networking sites.

Founded in February 2004 and based in the United States, Facebook is presently the world's largest social networking site. According to statistics provided by Facebook (facebook.com/press/info.php?statistics), there are more than one billion active users, more than half of whom log in each day. Prior to becoming a publicly traded company in May 2012, Facebook reported $1 billion in yearly profits from $3.71 billion in revenue, 85% of which coming from advertising revenue, with 15% coming from social gaming and other fees (Raice, 2012). In spite of a problematic initial public offering, the following year, Facebook's revenues had increased, and it had begun experimenting with numerous revenue-generating endeavors beyond advertisements and gaming (Rusli, 2013).

Of course, the use of social networking sites is a global phenomenon and is not limited to the United States. In fact, the United States came in second place in the use of social networking sites in a survey conducted by Pew Research Center's Global Attitudes Project. Britain had the highest number of adult social network site users at 52%, while 50% of adults in the United States used social networking sites, which is tied with the amount of users in Russia. Rounding out the top five countries when it comes to social networking site usage are the Czech Republic and Spain, both with 49% (Kohut et al., 2012).

However, much of the research conducted about the Internet has focused on its use in the United States. The Pew Internet & American Life Project has provided a detailed view of Internet use in the United States. In 2011, the number of adults using social networking sites in the United States reached 50% for the very first time, up from only 5% of all adults when social networking site data was first gathered in 2005 (Madden & Zickuhr, 2011). When the data is limited to Internet users rather than the entire population, the number of United States adults using social networking sites increases to 67%, up from just 8% in 2005 (Duggan & Brenner, 2013).

Teenagers continue to use social networking sites more than the adult population. Seventy-six percent of all teenagers use social networking sites. When data is limited to teenage Internet users, who compose 95% of the entire teenage population, that number increases to 81%. By way of comparison, 55% of online teenagers were using social networking sites when data was first gathered for this population in 2006 (Madden et al., 2013).

Social networking sites are globally transforming business and finance, politics, health, education, civic action, and entertainment. They have also been heralded as tools in the toppling of regimes (Ghonim, 2012). However, in this chapter we focus on the intended purpose of social networking sites, which remains the primary reason that people are drawn to them: *relationships*. Specifically, we examine the use of social networking sites in the maintenance of relationships. We also examine the use of social networking sites in the development of identities, which is a secondary but no less important reason people are using these sites. Ultimately, the following questions are answered by this chapter.

Questions Previewing This Chapter

1. *Who uses social networking sites?*
2. *How are relationships being maintained through social networking sites?*
3. *How does the nature of online communication explain the relational benefits of online relational maintenance?*
4. *How does the nature of social networking sites explain their success in maintaining relationships?*
5. *How are social networking sites being used in the construction of identities?*

Maintaining Relationships

Social networking sites are used to accomplish a variety of relational needs and functions. However, the primary relational function of these sites is the maintenance of existing relationships, which in turn increases social capital. In this section, we will specifically examine the use of social networking sites in the maintenance of relationships. We will consider how the nature of online communication and relating may enhance relational maintenance. We will also consider how the nature of social networking sites may explain their use and effectiveness in the maintenance of relationships.

Relational Maintenance

Relational maintenance involves keeping the relationship in existence and keeping the relationship in a satisfactory and desired condition. Relationships are maintained through communication. The very act of communicating with someone reinforces the existence of a relationship, while strategic communicative acts involving intimacy, support, avoidance, and other sorts of behavior are used to keep relationships healthy and at desired levels of closeness (Duck & McMahan, 2015).

The majority of social networking site users utilize these sites to maintain contact with people they already know rather than to meet new people, and to develop closer connections with acquaintances (Baym & Ledbetter, 2009; Craig & Wright, 2012; Ellison, Steinfeld, & Lampe, 2007; Houser, Fleuriet, & Estrada, 2012; Kujath, 2011).

These findings are consistent with a study of what types of relationships comprise the average Facebook user's group of friends. The largest category of friends listed by Facebook users is high school friends, followed by extended family, co-workers, college friends, immediate family, people from voluntary groups, and neighbors. While there do exist friends-of-friends, dormant (nonactive) relationships, and strangers included in Facebook users' lists of friends, the majority of Facebook friends have met at least once. Indeed, the average Facebook user's list of friends includes only 7% of people whom the user has never met in person at least once (Hampton, Goulet, Rainie, & Purcell, 2011).

The value of social networking sites in the maintenance of relationships is further strengthened by findings suggesting that social networking site use is positively associated with other forms of interaction such as face-to-face communication and cell phone usage (Jacobsen & Forste, 2011). It appears that social networking site use supplements and enhances offline interaction rather than replacing or limiting such interaction.

Social Capital

Maintaining the existence of relationships enables people to compile greater levels of social capital. **Social capital** entails resources available from one's social network, consisting of friends, family, romantic partners, coworkers, neighbors, acquaintances and others with whom a relationship is shared. Resources might include support, advice, feelings of self-worth, employment, or any other need people may have in their lives. The amount of social capital possessed can be determined in large part by the number of people in a person's social network and the unique value that each member can potentially provide.

Boase, Horrigan, Wellman, and Rainie (2006, p. 5) classified two types of connections in social networks. **Core ties** include people with whom a close relationship is shared, those with whom important matters are discussed, with whom frequent contact is made, and from whom help is most likely to be sought. **Significant ties** include people with whom there is a somewhat close connection, those with whom someone is less likely to discuss important matters, with whom infrequent contact is made, and from whom help will probably not be sought. Significant ties are greater than mere acquaintances and can provide valuable support if needed but are weaker than core ties.

Internet users in general tend to have greater numbers of significant ties in their social networks than nonusers, while tending to have the same number of core ties as nonusers (Boase, Horrigan, Wellman, & Rainie, 2006). Internet users overall have much greater diversity within their social networks than nonusers (Hampton, Sessions, Her, & Rainie, 2009).

Social networking site users are not more likely to have either a larger or a smaller social network when compared to Internet users as a whole or to non-Internet users. However, social networking site users have been found to possess a larger number of close social ties, feel less isolated, and are more likely to receive social support when needed (Hampton, Goulet, Rainie, & Purcell, 2011).

Ultimately, the use of social networking sites has been found to be positively associated with the development of social capital (Ellison, Steinfeld, & Lampe, 2007, 2011; Aubrey & Rill, 2013; Valenzuela, Park, & Kee, 2009).

The Nature of Online Communication

Overall, then, Internet users tend to have larger and more diverse social networks than nonusers. Social network site users, specifically, feel more connected and more supported. The reasons for these findings encompass and can be explained in part by the nature of online and digital communication, which overcome constraints of distance and time imposed by other forms of human communication.

Face-to-face communication naturally requires people to be within the same proximity in order for interaction to occur. Accordingly, there must be limited distance between people, which limits the geographic diversity and tends to limit the overall diversity of people with whom relationships can be maintained. Telephone, telegraph, letters, smoke signals, and other forms of communication enable people to overcome the limitations of distance but not always the limitations imposed by time.

Synchronous and Asynchronous Communication

Online communication can be either synchronous or asynchronous, which benefits the maintenance of relationships. **Synchronous communication** is that in which people interact in real time, such as through video chat or when playing online games. **Asynchronous communication** is that in which there exists a delay in time. This delay can be slight, such as that experienced with instant messaging, or prolonged, such as that experienced when sending and receiving email. Sometimes instantaneous communication is required or desired, and this synchronous interaction is available online.

Asynchronous types of online communication have the added benefit of enabling the user to better evaluate and consider what they wish to convey. This benefit is especially valuable when engaging in strategic relational maintenance. More than strategic communication, though, the asynchronous nature of online communication does not demand that people coordinate their interactions. Interaction can take place when most convenient for them, making it much easier and more likely to take place.

Richness

Those increasingly few individuals believing face-to-face communication to be the superior form of interaction frequently point to a lack of richness in online communication. **Richness** refers to the number of verbal and nonverbal cues and symbols available when communicating. Face-to-face communication, with numerous nonverbal cues and symbols available in addition to verbal symbols, is considered a rich medium. An email or text message would be considered less rich because of a lack of nonverbal cues and symbols. People use both verbal and nonverbal communication to assign meaning to symbolic activity. Accordingly, it is possible that a lack of richness may lead to misunderstanding.

While difficulties assigning meaning due to a lack of richness may be true at times, emoticons and other activities attempt to assist in overcoming the lack of nonverbal cues when interacting online. Further, the increasingly visual and auditory nature of online activities and interactions is enhancing the richness of online communication.

The Relational Nature of Social Networking Sites

While overcoming distance and time constraints may explain why Internet users tend to have larger and more diverse social networks, it does not completely explain why social network users tend to feel more connected, receive more support, and have more close ties. Like the nature of online communication, the nature of social networking sites can help explain these findings.

In some regard, interacting and relating through social networking sites simply accomplishes the traditional activities in a new way. For instance, a personal message is sent rather than a letter, a wall post stating "Happy Birthday" is given rather than sending a card.

In other ways, however, social networking sites are creating new relational activities and radically altering previous ways of interacting. In what follows, we will discuss some of the ways social networking sites are transforming relationships and how the nature of these sites may explain the benefits users tend to receive. Factors associated with the relational nature of social networking sites include the following: (a) list of connections, (b) ease of interacting, (c) relating is point, (d) sharing everyday life, (e) positive experience, (f) social and civic engagement, (g) relational learning, and (h) social presence.

Lists of Connections

Social networking sites commonly include a list of connections, users who, through mutual agreement, have agreed to be connected. A person's list of connections makes it easier to keep track of existing connections and may provide existential comfort in that this list makes relationships seem more real. The ability to see the connection lists of others with whom a connection is established also reinforces the potential reach and resources of a person's social network.

These lists of connections not only reinforce the existence of relationships but also assist in the maintenance of relationships. Sigman (1991) introduced the term **relational continuity constructional units** to refer to communication that keeps the relationships in existence when people are separated. Lists of connections on social networking sites serve as a form of these units. They reinforce the existence of relationships in spite of absence or lack of interaction. Relationships are reinforced not only for both parties but also for others in respective social networks.

Alternative View

Connection lists are not always positive for users. Tokunaga (2011) found them to be responsible for two of the top three most common negative experiences when using social networking sites, with ignoring or denying requests most common and disparity in the ranking of top connections the third most common.

Ease of Interacting

Social networking sites make interactions and accompanying relational maintenance easier. Indeed, these sites actually encourage interactions with others such as by suggesting potential contacts, encouraging the posting of information and comments, and even reminding users of people's birthdays. Prompted or not, it takes very little effort to post a comment or respond to the contributions of others.

Relating is the Point

Beyond a maintenance role, the primary purpose of social networking sites is to connect and relate with others. People desire to communicate and relate with others, fulfilling such needs as intimacy, security, entertainment, knowledge, and self-worth. Social networking sites are constructed in ways that enable these connections to take place and that enable these needs to be fulfilled. While the popularity of social networking sites is remarkable, what would actually be astonishing is if they did not attract a vast number of people.

Sharing Everyday Life

An ongoing criticism of social networking sites is that users frequently share unimportant and/or shallow information through updates and posts. Social networks occasionally receive criticism for appealing to and encouraging narcissistic tendencies (Carpenter, 2012; Kelly & Duran, 2012). However, the sharing of everyday, seemingly mundane information is actually the primary means through which relationships develop and through which they are maintained (Duck & McMahan, 2015; Wood & Duck, 2006). The supposedly ordinary experiences of life are actually quite extraordinary, not just in their frequency but in their employment in the composition of a person's life and relationships.

Noting the sharing of mundane information through communication technologies such as social networking sites, Tong and Walther (2011) observed that these sites "[normalize] discussing these otherwise unremarkable events (as the system asks, 'what are you doing right now?') as well as inviting reflection and responses." Accordingly, social networking sites "allow individuals to broadcast mundane narratives and reflections to both highly intimate and less intimate partners [providing] more relational maintenance 'bang' for the message-sending 'buck'" (p. 113).

Positive Experience

In spite of the potentially negative experiences and encounters, the use of social networking sites tends to generally be positive and favorable. As a result, the likelihood of their continued use in the maintenance of relationships and to provide relational needs is amplified. Sixty-nine percent of teen users and 85% of adults believe that people are mostly kind to one another on these sites. Naturally, negative experiences will occur on social networking sites, but positive experiences are more common. Examining teen users specifically, 41% of users reported at least one negative outcome from using a social networking site, while 78% reported at least one positive outcome. Further, 65% of users reported at least one experience on a social networking site that made them feel good about themselves (Lenhart, Madden, Smith, Purcell, Zickuhr, & Rainie, 2011).

Alternative View

Although we have highlighted the predominantly positive experiences with social networking sites, keep in mind that there are also negative experiences associated with social networking sites. For instance, unprecedented access to information increases the opportunity to monitor behavior. Some users utilize these sites to keep track of the behaviors of both romantic partners and potential romantic partners (Darvell, Walsh, & White, 2011; Moreno, Swanson, Royer, & Roberts, 2011; Stern & Taylor, 2007). It has been posited that monitoring partners through social networking sites may lead to increased jealousy and distrust, which may in turn result in increased monitoring (Muise, Christofides, & Desmarais, 2009). Chaulk and Jones (2011) found social networking sites a bit more threatening, arguing about the potential for "online obsessive relational intrusion," which can include monitoring and surveillance (see also Spitzberg & Hoobler, 2002).

Social and Civic Engagement

Social networking sites also encourage group activities and civic engagement. According to Raine, Purcell, and Smith (2011), Internet users in general are more likely than non-Internet users to attend meetings or events for a group, volunteer time to a group, contribute money to a group, and take a leadership role in a group.

> Perhaps reflecting their higher levels of participation, Internet users are also more likely than non-users to say that, in the past 12 months they have felt really proud of a group they are active in because of something it accomplished or a positive difference it made (62% v. 47%) and that they have accomplished something as part of a group that they could not have accomplished by themselves (48% v. 35%). (p. 5)

Overall, social networking site users are more likely to engage in group activities online and view the Internet as having a major impact on group engagement.

Relational Learning

Social networking sites represent a new form of relational learning. People have historically learned about relationships through personally engaging in them and though the observation of others. Media products have long been used to learn about relationships, informing people about such matters as how relationships should look and how to behave in relationships (Cohen & Metzger, 1998; Duck & McMahan, 2015).

Social networking sites present another avenue for learning about relationships. However, these sites offer opportunities unavailable and unimaginable through prior means. For instance, the public nature of social networking sites provides users with more opportunities to observe the interactions and relationships of others. Social networking sites also enable users to track the development of their own relationships and to review and evaluate their interactions. It is possible that these opportunities can result in a better understanding of relationships and increase the likelihood of engaging in relationships. It is also possible that increased relational satisfaction and feelings of support are a result of these abilities.

Social Presence

The presence created by social networking sites can be understood as generating social presence. The notion of social presence draws in part from Goffman's (1963) copresence, the mutual awareness that people are "accessible, available, and subject to one another" (p. 22). Lee (2004) moved social presence from the implied shared physical location of copresence, defining social presence as "a psychological state in which virtual (para-authentic or artificial) social actors are experiences as actual social actors in either sensory or nonsensory ways" (p. 45). Tong and Walther (2011) simplified this concept, describing social presence as occurring when "social partners are at least mildly cognizant of one another and feel as though they are in present or potential interpersonal contact" (p. 112).

Social networking sites offer a sense of social presence through a variety of features. The lists of connections and ease of contacting these connections provide a sense of accessibility and availability heretofore unrealized through any previous technology, with the possible exception of mobile phones. These sites also enhance feelings of social presence through real-time updates, photos, videos, and other materials. Photographs and videos make this feeling even more genuine, in essence creating the feeling of being virtually present.

Social Networking Sites and Identity Construction

Beyond their relational utility, social networking sites have become an important means of identity construction. In what follows, we will examine what is meant by the term *identities*. We will then examine the characteristics of social networking sites identity construction.

Identities

Identities are symbolic representations of the self. These symbolic representations are based in part on how we see ourselves, how we want other people to see us, and cultural standards or norms. People do not possess a core self or identity that remains stable and guides their behaviors and actions. Rather, multiple identities with varying degrees of continuity or stability that are continuously being created and re-created as people go about their lives (Duck & McMahan, 2015).

Characteristics of Social Networking Site Identity Construction

Social networking sites have become a common and significant means of constructing identities, profoundly transforming how identities can be understood and studied. In what follows, we will explore the characteristics of identity construction through social networking sites. In doing so, elements of social networking sites specifically used in the construction of identity will be introduced and examined along with the general nature of identity construction through social networking sites. Characteristics associated with social networking site identity construction include the following: (a) photographs, (b) friends, (c) media preferences and texts, (d) static and dynamic, (e) strategic, (f) constrained, and (g) massive public disclosure.

Photographs

The ability of users to display photographs is a common characteristic of social networking sites and is utilized in the construction of identities. Examining the use of Facebook by college undergraduates, Pempek, Yermolayava, and Calvert (2009) found that the majority of those surveyed believed that posting photographs "helped to express who they are to other Facebook users" (p. 233). It is not surprising, therefore, that studies have shown that people are very strategic in their selection of photographs posted online, desiring to convey themselves in a favorable and often very specific manner (Pempek, Yermolayava, & Calvert, 2009; Siibak, 2009; Whitty, 2007a).

Friends

The people with whom a person is connected also assist in the construction of identities. In some ways, connections with people on social networking sites are not unlike connections people highlight offline. For instance, people may introduce or mention people they know to increase perceptions of social capital or popularity. At the same time, connections on social networking sites influence identity construction in very unique and possibly unexpected ways.

Number of Friends. Friends on social networking sites differ from friends offline in that the number of these connections is listed for everyone to see. Someone may be perceived to have a large (or small) social network offline, but social networking sites provide an actual number that can be used by others to evaluate that person. The numbers of friends listed on social networking sites has been found to be a significant cue through which social judgments are made (Tong, Van Der Heide, Langwell, & Walther, 2008).

The number of friends a person lists on social networking sites does create perceptions of social connectedness, which is generally viewed in a positive manner. However, an excessive number of friends may diminish perceptions of genuine social connectedness (Zweir, Araujo, Boukas, & Willemsen, 2011). When assessing social attractiveness and extraversion, "individuals who have too few friends or too many friends are perceived more negatively than those who have an optimally large number of friends" (Tong, Van Der Heide, Langwell, & Walther, 2008, p. 545). When the number of friends listed on a social networking site gets too high, it may cause people to believe that those connections are not genuine or that these connections were gathered for some reason other than sociability.

Physical Attractiveness of Friends. The physical attractiveness of someone's friends will influence others' perceptions of that person's attractiveness. Walther, Van Der Heide, Kim, Westerman, and Tong (2008) discovered that the physical attractiveness of friends listed on Facebook does influence perceptions of the profile owner's attractiveness, and this attractiveness varied in the same direction as friends' attractiveness. Accordingly, people with good-looking friends on social networking sites tend to be evaluated accordingly. Additional research has shown that photographs of friends are also used to determine and evaluate extraversion of the profile owner (Utz, 2010).

Wall Postings by Friends. Comments left by friends will also impact identity construction on social networking sites. In the study mentioned above, Walther, Van Der Heide, Kim, Westerman, and Tong (2008) also discovered that "complimentary, prosocial statements by friends about profile owners improved the profile owner's social and task attractiveness, as well as the target's credibility" (p. 44). When it comes to prosocial statements and physical attractiveness, there appears to be a double-standard in regard to the sex of the profile owner. Female profile owners are perceived to be more physically attractive when comments by friends were positive and less attractive when they were negative. The opposite occurs when it comes to male profile owners. Negative comments involving such socially undesirable behavior as promiscuity and drinking actually increase perceptions of male physical attractiveness.

Warranting. Friends also have an indirect influence on identity construction by increasing confidence in the accuracy of profile information. The **warranting principle** maintains people "place greater credence in information about the personal characteristics and offline behaviors of others when the information cannot be easily be manipulated by the person who it describes" (Walther, Van Der Heide, Hamel, & Shulman, 2009, p. 229; see also Walther & Parks, 2002). The warranting influence of friends on social networking sites is twofold. First, friends will be able to confirm or disconfirm information included on one's profile. Therefore, others are more likely to judge that material as being accurate than they would if that information could not be seen or validated by third parties. Second, the actual comments and profiles of friends may support or negate the identities constructed through one's profile. Indeed, comments and information from friends may actually have a stronger impact than self-generated comments and materials (Utz, 2010).

Media Preferences and Texts

Profile information conveying taste in media products is another element in constructing identity through social networking sites. Media preferences and interests are often discussed in everyday talk with others in the construction of personal identities (Duck & McMahan, 2015). Social networking sites allow these interests to be explicitly displayed to others. Media preferences such as favorite music, movies, and books were found to be a more important marker of identity for profile owners than such "classic identity markers" as political views, religion, work, hometown, relationship, status, sexual orientation, and gender (Pempek, Yermolayava, & Calvert, 2009, pp. 232–233).

Static and Dynamic

Identity construction through social networking sites is both static and dynamic. In other words, some avenues of identity construction on these sites are seemingly fixed while other avenues of identity construction are more likely to change. It is possible for a person to change every feature of his or her profile each day so that the profile and identities conveyed are continuously altered, but this activity is not likely. Accordingly, a person's profile and self-descriptions generally remain unchanged, capturing the identities a person attempts to convey at a particular moment in time. In this sense, identity construction on social networking sites is very different from identity construction offline, which demands that identity construction be continuous. Of course, profiles and self-descriptions are changed from time to time but generally do not undergo frequent dramatic changes.

Some avenues of identity construction on social networking sites are more likely to undergo frequent updating and change. For instance, status updates and comments written on the profiles of other users enable people to continuously construct, develop, and renew their identities. Updating profile photographs and uploading additional photographs assist in the construction of identities, as does uploading such media products as songs and videos, playing online games, and engaging in other activities through social networking sites.

Strategic

Social networking sites also provide people with more time to consider how to develop their identities, enabling them to engage in thoughtful and strategic identity construction. When interacting face-to-face, people do not always have enough time to carefully construct a particular identity. As a result, people interacting face-to-face often do not feel as if aspects of their selves they wish to present have been successfully conveyed to others. Interacting online, however, provides more time to determine how best to construct identities (Walther, 1996). Research indicates that people do carefully and strategically construct online profiles (Ellison, Heino, & Gibbs, 2006). Furthermore, people feel that they are more successful at conveying identities when communicating online (Bargh, McKenna, & Fitzsimons, 2002).

Implicit and Explicit

Identity construction through social networking sites is both implicit and explicit. In some ways, identity construction is conducted explicitly through obvious attempts at presenting oneself in a particular manner. However, Zhao, Grusmuck, and Martin (2008) noted that Facebook users may be more likely to implicitly illustrate their selves as social actors through wall postings and photographs rather than explicitly narrating self-descriptions. This finding is based on discovering limited elaboration generally takes place when completing the "About me" entry on the self-description section of the profile, while the majority of users include numerous wall posts and pictures. Accordingly, users can be understood as presenting themselves as busy social actors, continuously developing identities that can be uncovered by others by visiting their profile.

Constrained

Identity construction is also constrained to various degrees by social networking sites. Some sites offer more freedom than others in the design of one's profile and the content that can be included. However, there are limitations imposed, and users must work within the confines offered by these sites. For some people, utilizing and potentially pushing the boundaries imposed by social networking sites in the creation of their profiles, such as through the incorporation of media texts, conveys particular identities in and of itself.

Overall, however, current social networking sites provide limited versatility for emoting and expressing individual uniqueness. The Facebook interface, for instance, is a predetermined box with finite windows and options. Computer engineers and designers such as Lanier (2010) see this as an explicit design flaw and criticize social networks for their closed systems and restrictive interfaces. Social networking sites also constrain identity construction and presentation through hegemonic categories pertaining sexuality, ethnicity, and gender. When completing a profile using categories provided by sites, users may not find applicable designations. For instance, gay men and lesbians with domestic partners may only find "marriage" as a possible category for the relationship status in some social networking sites. Likewise, a limited number of ethnic and gender categories or designations are generally offered by social networking sites. Accordingly, users are being defined by social networking sites rather than being able to define themselves.

Massive Public Disclosure

Social networking sites are characterized by the large amounts of personal information users provide through their profiles. This characteristic influences privacy, public evaluation, and self-comparisons.

Although some social networking sites enable restricted access to profiles, a massive amount of information about a person can be learned when visiting his or her profile. Further, regardless of restrictions and privacy barriers imposed by users, a person can never be fully certain that the information will not be viewed by unintended others. As a result, social networking sites are impacting how people deal with privacy.

The **Communication Privacy Management** theory explains how people deal with the need for privacy in their relationships (see Petronio, 2002). According to this theory, people perceive ownership of information about themselves and maintain boundaries to control this information. Some information may be revealed while other information remains concealed, and the decision to reveal or conceal information is often based on the relationship shared with another person. The public nature of social networking sites makes it very difficult to maintain boundaries and control private information (Child & Petronio, 2011).

Social networking sites also provide a forum for public evaluation of disclosure and other forms of identity construction though one's profile. Manago, Graham, Greenfield, and Salimkahn (2008) maintained that such public performances of identity may be provided

> virtual applause through the public comment wall … . Because self-displays are available to an audience, the feedback provides social verification. This verification may authenticate the performance by endowing it with social legitimacy. Thus, the portrayal may be more likely to be incorporated as a convincing and attractive aspect of the self. (p. 451)

Of course, positive evaluations of one's behaviors and disclosure will have the same result in offline interactions with others. However, offline performance does not occur with as many people observing. Further, the approval, disapproval, or lack of acknowledgment garnered is not as direct or as public as that delivered on social networking sites, which may serve to reinforce the significance of the feedback (or lack of feedback) provided. Finally, the massive public disclosure on social networking sites may intensify social comparison with others. Manago, Graham, Greenfield, and Salimkahn (2008) further noted that just like social interactions offline, people on social networking sites develop a sense of self in relation to what others are doing and evaluate themselves in comparison with other people. Social networking sites not only provide a massive amount of information to compare with, this information tends to be quite positive in nature. Accordingly, people evaluate themselves in comparison with idealized images of others, which may cause them to evaluate themselves negatively and to feel the need to further enhance their own identities conveyed through their profiles.

Conclusion and Predictions

Used by billions and billions of people throughout the world, social networking sites are a pervasive part of Internet experience. The characteristics of social networking sites along with the nature of online communication help explain their importance in relational maintenance and in the development and strengthening of social capital. Further, these sites are increasingly used in the development of identities. There is little doubt that social networking sites will continue to develop and expand, in terms of their actual numbers, their number of users, and their global influence.

This brings us to predictions about the future of social networking sites. Undoubtedly, *the number and scope of social networking sites will continue to expand.* Although some sites such as Facebook and Twitter will likely continue to have a dominant presence, newer sites will emerge to challenge their dominance with various degrees of success. Also, numerous smaller sites will emerge, catering to specific audiences and providing outlets for people wishing to avoid mainstream sites. Further, *the number of social networking site users will continue to increase,* eventually encompassing not only the majority of Internet users but also the majority of the entire population.

The characteristics and nature of social networking sites will also continue to expand and evolve. We believe that *social networking sites will become increasingly virtual,* enabling people to interact in a virtual space as themselves or other avatars.

Social networking sites will become increasingly vital in the maintenance of re-lationships and the construction of identities as the number of users continues to increase.

However, we also believe that while maintaining a relational focus, *many social networking sites will expand beyond their relational focus, becoming focal points for Internet activity, commerce, self-marketing, and entertainment.* Doing so will enable social networking sites to increase their influence, attract new users, and develop revenue beyond advertisements. This expansion is already evident with gaming on social networking sites, and will eventually include searching for online content with friends, shopping with friends, listening to music with friends, watching video with friends, and engaging in numerous other activities with friends. Facebook ventured into commerce in September 2012, enabling users to send both physical and digital "Gifts" to friends through the site (Isaac, 2012). Further, sites such as Amazon and Spotify increasingly encourage users to share their purchases and media use through social networking sites.

The passing of time will reveal whether these predictions are correct and what other transformations take place involving social networking sites. In the meantime, they remain a powerful online presence and demand attention.

Key Terms

Asynchronous Communication

Communication Privacy Management

Core Ties

Identities

Relational Continuity Constructional Units

Relational Maintenance

Richness

Significant Ties

Social Capital

Social Networking Sites

Social Presence

Synchronous Communication

Warranting Principle

Chapter Exercises

Current Issues: Have students search for and report on the most current media stories regarding social networking sites. Discuss how such reports impact social networking site users. Could such reports influence who becomes or remains users of these sites? Could such reports change how these sites are developed and presented to users? Are there positive or negative slants to these reports?

Debate It: Split the class into two groups. Select one of the resolutions below. Have one group support the resolution and have the other group oppose the resolution in a debate.

Resolved: Face-to-face communication is superior to online communication.

Resolved: Online communication is superior to face-to-face communication.

Ethical Consideration: Materials posted on social networking sites have resulted in people being fired from jobs and being passed over for employment or promotion. Have students consider whether they think it is ethical for social networking sites to be used in making employment decisions. When might an employer be justified in using such posts to judge an existing or potential employee? When might an employer not be justified in doing so? Among other factors, when answering these questions, have students consider the type of job or business, the content of posts, and whether the content of those posts deals with the business.

Everyday Impact: Ask students to consider what identities are being constructed through their social networking site profiles and their use of these sites? What identities are being intentionally constructed? What identities are being unintentionally constructed? If they use more than one social networking site, do the identities being constructed differ with each site? If possible, given the technological capabilities of the classroom, and if students are willing, have a few of the students provide the class with a "tour" of their social networking site pages.

Future Focus: We anticipate that social networking sites will develop into virtual spaces in the future, enabling users to interact virtually and increasing the potential for greater social presence. Ask students whether they think we are correct. What do they think social networking sites will look like in the future? What will these sites provide users in the future?

9

Connecting on the Internet: Pornography and Dating

Since its beginnings, the Internet has been continuously transforming into a means through which people connect with others. And, in doing so, the Internet has been continuously transforming how people connect with others. In the previous chapter, we discussed how people maintain relationships and create identities through social networking sites. Within this chapter, we explore two other areas through which people establish and develop relational connections on the Internet: pornography and dating.

A book on Internet communication would be incomplete without an exploration of pornography. Pornographic materials have been central to the development and expansion of the Internet, in terms of technological advancement, commercial activities, use, and attempts at regulation. As Paasonen (2011a) rightly noted,

> there is little doubt as to the centrality of pornography in terms of Internet history, its technical development, uses, business models, [and] legislation ... it is said to comprise a major part of websites and downloads, to take up the most bandwidth, and to generate the most profit of all web content. (p. 424)

Just as with pornography, the development and use of the Internet can be traced in large part through the development and use of Internet dating sites. Pornography may have been more influential in necessitating increased bandwidth and download speeds along with other transformations. However, dating sites quickly took advantage of the changing nature of the Internet, fully incorporating new messaging, photo, and other capabilities as they became available. Financially, dating sites are also indicative of the Internet's developments. As revenue from these sites rose, some sites succeeded while others failed, and successful sites were often purchased by even more successful sites or large corporations.

In what follows, we begin by discussing the size and scope of Internet pornography as well as what types of pornographic materials users are most likely to seek out. Finally, we present the primary characteristics of Internet pornography.

Then, turning our attention to Internet dating sites, we will examine the scope and influence of these sites. Next, we will discuss the unique characteristics and capabilities of Internet dating sites. We will then explore the construction of Internet dating site profiles. Along the way, the following questions will be addressed:

Questions Previewing This Chapter

1. *How pervasive is pornography on the Internet, and why is the answer difficult to determine?*
2. *What do people search for when seeking pornography on the Internet?*
3. *What are the primary characteristics of Internet pornography?*
4. *What options do people have when using Internet dating sites?*
5. *How are Internet dating sites influencing the way people meet romantic partners?*
6. *What are the primary characteristics of Internet dating sites?*
7. *How do people construct and assess Internet dating site profiles?*

Internet Pornography

Pornography is generally understood and defined as a *depiction*, which means that its very existence is dependent on technology and symbolic activity. The Internet continues the long tradition of the inextricable connection of media and pornography. Yet, beyond previous technologies, the Internet has transformed and continues to alter the understanding and use of pornography.

As mentioned above, we will begin our exploration of Internet pornography by examining its size and scope. As will be discussed, however, determining the actual magnitude and influence of Internet pornography is a challenge and may be impossible to do. We will then discuss what types of pornography users search for when online. Finally, we will examine the characteristics of Internet pornography.

Size and Scope of Internet Pornography

Pornography has a pervasive presence on the Internet. However, determining the size and scope of pornography on the Internet is difficult. One reason for this difficulty is the continuous creation of new sites and material. A second reason is that people are often less than forthcoming about their genuine use of pornography in surveys and other studies. A third reason is that companies producing pornography are often less than forthcoming about their profits. A fourth and final reason is that figures on the amount of pornography frequently come from sites and software devoted to protecting people from the *massive* amounts of pornography on the Internet.

Internet filtering software such as CYBERSitter, Cyber Patrol, McAfee Safe Eyes, and Net Nanny are dedicated to keeping children and those with sensitive eyes safe from Internet pornography. According to statistics offered by FamilySafe-Media (2013), every second "$3,075.64 is being spent on pornography," "28,258 Internet users are viewing pornography," and "372 Internet users are typing adult search terms into search engines." Also according to this site, there exists "4.2 million (12% of total)" websites containing pornography, "68 million (25% of total)" daily pornography search engine requests, and Internet pornography sales are reported as "$4.9 billion." Fortunately, there are books and videos available to assist those addicted to Internet pornography and software to block such content for sale on the site. Somewhat ironically, these services actually increase the amount of money earned as a result of Internet pornography.

Determining the Numbers and Popularity

Truthfully, it is impossible to legitimately calculate the exact amount of pornographic material on the Internet or the amount of money spent and earned from such content. Most numbers are mere speculations and best guesses based on available data. Plus, new content appears on the Internet every second.

Making the process particularly challenging are disagreements concerning what counts as pornography or even what counts as a pornographic site, something that continues to challenge Internet pornography researchers (see Short et al., 2012). For instance, Tumblr.com may be viewed and used by many people as a blogging platform, but there exists a great deal of sexually explicit content on the site. Indeed, nearly 12% of the top 200,000 domains on the sites feature adult content, and 22% of referral traffic to the site originates from pornography sites (Perez, 2013). Further, such sites as Adultfriendfinder.com and Gayromeo.com provide nude pictures of members and opportunities to find sexual partners, but some people may not consider this to be pornography, while other people may consider it to actually be pornography.

Alexa, a site that tracks website traffic, can be used to determine the popularity of pornography sites based on the number of visitors to those sites. Of course, there is the issue of what counts as pornographic materials and pornographic sites. Nevertheless, selecting the Adult category to filter results, the sites listed in Figure 9.1 represent the top 10 most-visited adult websites according to Alexa.

1. Livejasmin.com
2. Youporn.com
3. Xnxx.com
4. Adultfriendfinder.com
5. Streamate.com
6. Freeones.com
7. Literotica.com
8. Adam4adam.com
9. Gayromeo.com
10. Payserve.com

Figure 9.1. Top 10 adult websites. Based on Alexa's Adult category global ratings.

While the Adult category is useful, a complete review of all global ratings according to Alexa ranking provides a different list of pornographic websites. Figure 9.2 provides a list of the top 10 sites known to focus on pornographic material, if defined as that depicting sexually explicit nudity and sexual activity. While rankings continuously change, all sites on the list generally rank globally within the top 250 websites, with Xhamster.com generally ranking within the top 50 websites.

1. Xhamster.com
2. Xvideos.com
3. Pornhub.com
4. Livejasmin.com
5. Redtube.com
6. Youporn.com
7. Tube8.com
8. Xnxx.com
9. Youjizz.com
10. Cam4.com

Figure 9.2. Top 10 pornography websites. Based on Alexa's global rankings.

In addition to ranking of individual sites, it is also possible to examine the percentage of Internet traffic dedicated to pornography. According to Similar-Web, another web traffic site, the worldwide average of pornography-related Internet traffic is approaching 8% (Buchuk, 2013). Figure 9.3 provides national averages of pornography traffic.

Country	Percentage of Traffic
Germany	12.47%
Spain	9.58%
United Kingdom	8.50%
United States	8.31%
Worldwide Average	**7.65%**
Ireland	7.45%
France	7.34%
Australia	7.01%

Figure 9.3. Percentage of Internet traffic dedicated to pornography (Buchuk, 2013).

Alexa rankings and SimilarWeb traffic percentages provide a more complete view of the popularity of Internet pornography in regard to the number of visitors to popular sites. However, we will never know how many sites are actually available, how many people actually use these sites, or how much money is generated by these sites. It is safe to speculate that the answer to such questions is *a lot*, but specific numbers are ultimately impossible to accurately generate.

Searching for Content

While the precise amount of Internet pornography is unobtainable, we do have a good idea of what they are searching for, as a result of research conducted by Ogas and Gaddam (2012). These researchers collected 400 million searches conducted on the search engine Dogpile.com over a one-year period of time by around 2 million people. Of these searches, 55 million (roughly 13%) were searches for erotic content (p. 14). Figure 9.4 presents the top 10 most popular sexual search categories as determined by this research.

Among the findings of this research, it was discovered that men are more likely to search for pornographic images and videos on the Internet, while women tend to seek out pornographic stories. Furthermore, men are very specific about what they want to see. Age and body size are especially prevalent in searches by males. When it comes to age, youth reigns supreme, but a number of searches encompass older categories. Figure 9.5 presents a list of popular age-related searches by heterosexual male users.

Category	Percentage
1. Youth	13.5
2. Gay	4.7
3. MILF	4.2
4. Breasts	3.9
5. Cheating Wives	3.3
6. Vaginas	2.8
7. Penises	2.4
8. Amateurs	2.3
9. Mature	2.1
10. Animation	2.1

Figure 9.4. Top 10 most popular sexual search categories (all users) (Ogas & Gaddam, 2012).

Category	Percentage
1. Teen	5.8
2. Youth	2.0
3. Mom	1.4
4. MILF	1.3
5. Grannies	0.5

Figure 9.5. Top five most popular age-related search categories (heterosexual male users) (Ogas & Gaddam, 2012).

Ogas and Gaddam (2012) also discovered that body size is an important factor in heterosexual male searches, with *larger* women being much more popular than *smaller* women. Adjectives describing larger women (i.e., BBW, chubby, plump) were three times more common than adjectives describing smaller women (i.e., thin, skinny). This finding was supported by an Alexa search of the million most popular websites, which found 504 pornographic sites specifically focused on larger women compared to 182 pornographic sites specifically focused on smaller women.

Characteristics of Internet Pornography

Although determining the actual size and scope of pornography on the Internet is problematic and speculative, we are able to establish its distinctive qualities. In what follows, we will explore the primary characteristics of Internet pornography.

Unlimited, Concrete, and Interactive

Internet pornography is unlike any previous mediated pornography due to its unlimited, concrete, and interactive nature. First, Internet pornography is essentially unlimited in its availability and scope. Because of the large and relatively inexpensive storage capacity of the computer systems used to create pornographic websites, a virtually unlimited amount of photographs, movies, stories, games, and other content is available.

Second, compared to pornography through other media, the Internet provides pornography that is concrete, specific, and vivid. Rather than a small number of pictures of a model in a magazine, for instance, Internet sites can provide users with a large number of pictures of that model from a variety of perspectives.

Finally, pornography on the Internet is interactive. The Internet enables users to create a more personal experience through the vast selection of pictures and video options just mentioned. Beyond this capability, however, the Internet enables users to interact with models and performers using live webcams (see Richtel, 2013). Further, Internet users can engage in sexual activity though online avatars, with avatars controlled by other users and with avatars programmed through websites. As such, the appearance of participants and activities can be totally determined and controlled by users. Making the experience truly interactive are **teledildonics**, devices used to simulate sexual activity and manipulated remotely and digitally. These can be controlled by an online sex partner or by an individual user interacting with online content.

Affordable, Accessible, and Anonymous

In the late 1990s, Cooper (1998) described pornography as Triple A: affordable, accessible, and anonymous. This description has remained accurate through the ensuing years. Pornography on the Internet is definitely affordable. In fact, a great deal of pornographic material is available on the Internet at no cost. Some of this free material is provided by pay sites as samples in attempts to entice users to purchase a membership in order to access additional materials. Paid sites also band together, featuring links to others sites and enabling users to access multiple sites for a single fee.

Pornography on the Internet is also highly accessible. This accessibility is due in part to the ease of moving from one site to another made possible through cooperation among sites just addressed. It is also due to the unlimited nature of Internet pornography addressed above. Quite simply, pornography on the Internet is constantly available. At any moment in time, pornography can be accessed with essentially no delay.

Finally, pornography on the Internet is anonymous. As long as the privacy settings have been accurately established on one's browser, it is likely no one else will know that pornographic materials have been accessed. The anonymous nature of Internet pornography has resulted in personal changes among users. As Cooper (1998) presciently and fully explained,

> the belief that one is unknown, (both real and perceived), is found to have a powerful effect on sexuality in a number of ways including: A sense of freedom, and increased willingness to experiment, a faster pace of self-disclosure, and an ability to talk openly about one's sex life, questions, concerns, and/or fantasies. (p. 188)

Alternative View

As will be discussed in Chapter 13, anonymity and privacy on the Internet are illusory. Pornography can now be accessed in the "privacy" of homes and on carefully obscured screens when amongst other people. However, through tracking, surveillance, and the creation of clearly distinguishable digital trails, the use of pornography might now be more public than ever.

Peripheral Pornography

Another characteristic of Internet pornography has been its influence in the increase and proliferation of **peripheral pornography**, which is pornography dedicated to areas outside the mainstream or traditional offerings. Limited only to the imaginations of producers and consumers, such material is dedicated to a host of sexual cultures, fetishes, sexual preferences, and tastes.

The rise of peripheral pornography can be explained through increased awareness and use of these materials and through decreased financial risk for producers. Increased awareness and use of peripheral pornography has been made possible in part by search engines. Prior to advanced search engines, it was difficult to locate certain categories of peripheral pornography online, even if a user was aware of its existence and the possibility of accessing such materials. Because of search engines, still other users discovered this material's existence through key word searches. As more people sought this material, the number of sites offering the material increased, which in turn increased the likelihood that the material would be discovered by someone new, which in turn increased demand for more sites.

Coupled with increased awareness and use of peripheral pornography is the relatively minimal financial risk for producers of this material. When profits were based on sales or rental of physical items, which cost a great deal of money to physically produce, store, and ship, producers of pornographic material were hesitant to invest money in products that might interest only a limited number of people. As a result, they tended to focus on traditional mainstream pornographic material rather than on material of a peripheral nature, with a comparatively smaller, and at the time, unknown market. Now, once it has been created, the Internet enables producers to supply this material with minimal costs. As a result, more peripheral pornography has been created, which has increased consumer awareness and use and resulted in the demand for more material, which has increased the likelihood that it will be discovered by new consumers, in turn increasing demand for material.

Participatory

Pornography on the Internet is also participatory in nature. A host of issues and competing perspectives arise from this characteristic of Internet pornography. Among these matters, and what enables participation, is the production and distribution of one's own materials (see Richtel, 2013). Pornography in general has a history of incorporating personal submissions into content. Like many aspects of traditional pornographic materials, however, this activity has been transformed through the Internet.

Making this endeavor particularly unique is that the Internet enables users to increase production values of these materials and become their own distributors of materials. These capabilities have led Coopersmith (2006) to note that the Internet has resulted in the "democratization of pornography." This democratization of pornography through the Internet developed from

> technologies that encapsulated the expertise and skills necessary to record, edit, and distribute, thus enabling anyone to be a producer….This technology can be seen as liberating and empowering, allowing individuals to actively create their own pornography, not just passively consume the work of someone else. (pp. 10–11)

Arising from the capability of producing and distributing personal pornographic materials is the issue of whether it leads to empowerment or subjugation. Pornography has often been viewed by some as oppressing and exploiting women (see Long, 2012). In the case of Internet pornography, though, a number of pornographic sites have been developed by sex performers themselves, and these sites have become among the most visited and profitable sites on the Internet. On the other hand, it might be argued that while these performers have taken more control, women in general are still being subjugated.

While this issue will never be fully resolved or agreed upon, many scholars have focused on empowerment through these sites. Podlas (2000) suggested "that far from being complicitous in the oppression and exploitation of women, [these pornography site] webmistresses may reflect a degree of emancipation from male-dominated female imagery and economic control" (p. 847). It was further argued by DeVoss (2002) that

> these sites work against condescending attitudes that view women sex workers always already as victims, thus denying any alternative possibilities.... Instead, [these sites] are spaces where they own and control their bodies and sexualities, spaces where they appropriate stereotypical notions of pornography. . . [transgressing] expected norms of sexuality and [upsetting] established conventions of representation. (pp. 76–78)

Perhaps strengthening the subjugation argument, participation in Internet pornography also takes place through rating sites. On these sites, people submit nude photographs of themselves, frequently concentrating on certain parts of their bodies, depending on the focus of the site. These photographs are posted and then evaluated by users of the site, using a numbered rating system with the possibility for comments. Such sites as *RateaBBW*, *MeNude*, and *RateMeNaked* focus on overall looks, while such appropriately named sites as *BoobCritic*, *RateLegs*, *FloppyDicks*, *BootyVote*, and *RateMyMelons* focus on obvious body parts.

While these sites can be viewed as further subjugation, it can also be argued that contributing to such sites can be viewed as empowering as well as sexually and personally fulfilling. For Waskul and Radeloff (2010), contributors can be viewed as

> "artisans". . . maintain[ing] creative control over the products of their erotic labor.... [and] playfully deploy[ing] the technology to uniquely fashion creative images and see[ing] themselves through the eyes of others—to manipulate erotic looking-glasses to create a sense of their own bodies as erotic generators. (p. 215)

Accordingly, as Waskul (2002) previously observed in regard to cybersex, for contributors to these sites, "the excitement that others receive from seeing them nude is repaid unto the self by the indirect yet comforting knowledge that one's body is appealing and desirable" (p. 213).

The participatory nature of the Internet has led to an increase of amateur pornography. In the same vein as rating sites, sexual and personal pleasures may be gained from sharing pictures and videoing sexual encounters. Beyond personal motivations, however, this increase in amateur pornography has been accompanied by an increase in authenticity.

Paasonen (2011b) noted that the rise in amateur pornography on the Internet assists in creating an aura of authenticity that "translates as realness, immediacy, and presence" (p. 78). Far from airbrushed photographs of models and properly lit studios, pimples, bruises, stretch marks, and other imperfections are often visible though amateur pornography. Further, the sexual prowess demonstrated by professional porn actors and actresses is not necessarily evident in amateur videos. Moreover, in spite of technological advances, amateur pictures and video often do not reflect the skill and quality of the professional. As Paasonen further maintained, "these notions of realness and directness are mapped into the notions of directness that are associated with both pornography as a genre and the Internet as a medium" (p. 80).

Although yet to be determined, it is possible that this authenticity may combat long-held arguments that pornography promotes unrealistic expectations of performance and body image. Aside from unrealistic expectations of instant gratification through constant accessibility and unrealistic beliefs that everyone else in the world is having sex, this authentic amateur pornography might be less likely to promote other unrealistic expectations.

Social

Internet pornography is also social. Internet pornography users are often portrayed as loners sitting in a darkened room, much like customers of pornographic movie theaters were portrayed as wearing sunglasses with their heads down and trench-coat collars turned up to obscure their faces. However, pornography in general has an equally long history of social behavior, with individuals frequently exchanging and sharing pornographic materials with other users. An early study on the use of pornography discovered that men were using it not only for sexual stimulation but also for male bonding (Hite, 1981).

Additionally, through digital tablets and smartphones, Internet pornography can now be accessed and used publically, enabling both intentional and unintentional sharing of materials. This capability has prompted disputes between users and passersby as well as the passing of laws in Tennessee, Louisiana, and Virginia, in addition to proposed laws in other states, banning public displays of pornographic material (Richtel, 2012).

Pornography was especially social in the early history of the Internet. Prior to advanced search engines and the World Wide Web, pornography on the Internet was essentially a completely social endeavor. Users shared pornographic materials on bulletin boards, Usenet groups, and Internet relay chat networks (see Mehta & Plaza, 1997; Slater, 1998). Peer-to-peer file sharing services were also a common way for users to access and distribute pornography. Such services have come under tremendous pressure and regulation due to sharing on child pornography, however (Coopersmith, 2006).

The social aspects of pornography are particularly evident through forums and fan cultures converging on the Internet. Some pornography sites enable users to discuss performers, share information, categorize, review, rank, and engage in other interactions with one another. Freeones.com, a massive and multi-purpose pornography site, includes, among other features, an extensive forum and message board. As Lindgren (2010) noted "almost as much energy and space is devoted to cultivating the 'We' of the view collective, as to discussing the female porn stars who are the main object of the viewers' attention" (p. 178). Accordingly, the development of community and the establishment of personal connections are vital elements in the discourse of the users of this and other Internet pornography sites.

Internet Dating Sites

While pornography is completely dependent on technology, romantic relationships can be initiated and exist without technology. However, technological change has historically resulted in changes in courtship rituals and forms. As with these other technologies, the Internet has profoundly influenced the ways in which romantic relationships are initiated, maintained, and dissolved.

In what follows, we will explore the scope and influence of Internet dating sites. In doing so, we will examine what types of sites are available and the prevalence of the Internet in relationship formation. We will then discuss the characteristics and capabilities of these sites. Finally, we will turn our attention to the construction of Internet dating site profiles.

Scope and Influence of Internet Dating Sites

Internet dating sites provide people searching for romantic partners with a wide range of options and are increasingly used as a method of matchmaking. Figure 9.6 lists the top 10 dating sites based on Alexa rankings.

1. Match.com
2. Datehookup.com
3. Blackpeoplemeet.com
4. Rsvp.com.au
5. Cupid.com
6. Afrointroductions.com
7. Singlesnet.com
8. Friendfinder.com
9. Fropper.com
10. Plentyoffish.com

Figure 9.6. Top 10 dating websites. Based on Alexa rankings.

Internet dating sites generally enable users to focus searches to find people with desired characteristics and backgrounds. However, an increasing number of sites are being developed that focus on a very specific characteristic. Not surprisingly, given the importance of faith for many people when dating and considering serious romantic relationships, many sites offer opportunities to seek out potential romantic partners with shared religious beliefs. Figure 9.7 offers a list of some of these sites.

Sites	Religious Community
Catholicmatch.com	Catholic
Catholicmates.com	
Catholicmingle.com	
Bigchurch.com	Christian
Christianmingle.com	
Singlec.com	
Hinduconnections.com	Hindu
Hindufaces.com	
Imilap.com	
Jdate.com	Jewish
Jewishcafe.com	
Sawyouatsinai.com	
Muslima.com	Muslim
Qiran.com	
Salamlove.com	

Figure 9.7. Religious dating sites.

While religion is a prevalent Internet dating site category, other categories are also popular among these sites. These categories are often based on body type, race/ethnicity, age, profession, and other characteristics. Figure 9.8 provides a list of some of the most common Internet dating site categories.

Sites	Category
Bbwcupid.com Findmybbw.com Largefriends.com	Big Beautiful Women (BBW)
Active-singles.org Fitkiss.com Fitness-singles.com	Fitness Enthusiasts
Encountersdating.co.uk Itsjustlunch.com Ivorytowers.net	Professionals
Amigos.com Interracialmatch.com Shaadi.com	Race/Ethnicity
Agematch.com Maturefreeandsingle.com Seniormatch.com	Seniors
Hmates.com Mpwh.net Stdmatch.net	Sexually Transmitted Diseases
Datingforparents.com Justsingleparents.com Singleparentlove.com	Single Parents

Figure 9.8. Common dating site categories.

Some users seek romantic partners based on even more specific characteristics, and a number of **niche Internet dating sites** are available for these users. One site, Passionsnetworks.com, offers pages based on such categories as music preferences, food and drink preferences, and sports preferences, among many others. There are even sites for lovers of live action role playing (Larppassions.com), clowns (clownpassions.com), mustaches (stachepassions.com), and the mighty mullet (mulletpassions.com). Figure 9.9 provides a list of other niche Internet dating sites.

Sites	Users
420dating.com	Marijuana Users
Alikewise.com	Book Lovers
Ashleymadison.com	Married People Seeking Affairs
Beautifulpeople.com	Attractive People
Cupidtino.com	Apple Product Fans
Cougarlife.com	Cougars/Cubs
Datemypet.com	Pet Lovers
Democraticpeoplemeet.com	Democrats
Farmersonly.com	Farmers
Gk2gk.com	Geeks
Inmate.com; Womenbehindbars.com	Convicts
Datealittle.com	Little People
Nolongerlonely.com	Mentally Ill
Republicanpeoplemeet.com	Republicans
Seekingarrangement.com	Sugar Daddies/Sugar Babies
Tallfriends.com	Tall People
Theuglybugball.com	Unattractive People
Veggiedate.org	Vegetarians
Wewaited.com	Virgins

Figure 9.9. Niche Internet dating sites.

The Internet and Relationship Initiation

As we discussed in Chapter 8, social networking sites are primarily used for maintaining relationships rather than initiating relationships. Internet dating sites, of course, are dedicated to the initiation of relationships and the initiation of romantic relationships specifically. And, just as social networking sites seem to be a pervasive and growing means through which relationships are maintained, Internet dating sites are a pervasive and growing means through which relationships are initiated.

However, claims of success in finding long-term romantic partners through Internet dating sites are difficult to confirm. Most data and positive claims about these sites come from the sites themselves (Bialik, 2009). Comparatively, and as discussed earlier in this chapter, these data should be considered in the same vein as data conveying the ubiquity of Internet pornography offered by companies offering protection from such content.

At the same time, the popularity of these sites and the dramatic increase in the role of the Internet in the initiation of relationships should not be overlooked or dismissed as less than impartial. Beyond whether or not Internet dating sites are successful tools in the initiation of long-term romantic relationships, they are used by millions of people and have resulted in relationships involving people who likely would not have met otherwise.

The Internet has become a predominant means through which romantic partners meet. Though not limited to dating sites, longitudinal research conducted by sociologists Michael Rosenfeld and Reuben Thomas (2012) examined the rise of the Internet in the initiation of romantic relationships. As might be expected, family and school, the previously dominant sources for being introduced to romantic partners, have been steadily declining for decades. Within the past 15 years, the Internet has "partially displaced…neighborhoods, friends, and the workplace as venues for meeting partners" (p. 523). Furthermore, within this period of time, the percentage of people meeting and forming relationships among people with no previous social ties has increased. Additionally, "individuals who face a thin market for potential partners, such as gays, lesbians, and middle-aged heterosexuals, are especially likely to meet partners online (p. 523). Indeed,

> meeting online has not only become the predominant way that same-sex couples meet in the United States, but meeting online is now dramatically more common among same-sex couples than any way of meeting has ever been for heterosexual or same-sex couples in the past. (p. 531)

Characteristics and Capabilities of Internet Dating Sites

Internet dating sites have unique characteristics and capabilities that make them different from face-to-face and other technology-based matchmaking and partner selection. Within this section, we will examine these characteristics and capabilities, along with their advantages and disadvantages when searching for a romantic partner.

Time and Space

Perhaps the most obvious advantage of Internet dating sites is their ability to enable users to overcome constraints of time and space when searching for romantic partners. Beyond providing users with additional time to strategically develop their own symbol presentations and to strategically assess the symbolic presentations of others, issues that will soon be examined, these sites enable users to search at their own convenience. Time searching for a potential romantic partner is also spent more efficiently when using Internet dating sites. Internet dating sites provide a much greater number of potential romantic partners, the personal information of whom can be gained quickly through a review of their profiles.

Greater Selection

Internet dating sites also provide users with a greater selection of potential romantic partners than do other means of matchmaking. Beyond providing greater numbers in general, Internet dating sites provide a greater selection for users searching for romantic partners with specific characteristics and for users in a challenging dating market. As discussed in the previous section, there exists a host of Internet dating sites focused on very specific populations. These sites assist users searching for partners with specific interests, physical features, religious beliefs, and any number of other characteristics. Further, as confirmed by Rosenfeld and Thomas (2012), these sites also assist users in thin dating markets such as those encompassing sexual orientation or age.

Alternative View

Although access to greater numbers of potential romantic partners might appear to be advantageous, Finkel, Eastwick, Karney, Reis, and Sprecher (2012) argued that online daters are likely to find the process of browsing so many profiles cognitively laborious, which may decrease their level of interest in any specific potential partner during the browsing process—and it might ultimately undermine their levels of happiness with and commitment to a potential partner once a relationships moves offline. (p. 34)

Greater Description

When searching for a romantic partner, such attributes of attraction are generally discovered through talk, which results in some potential romantic partners being filtered out and removed from consideration and other potential romantic partners moving forward and remaining under consideration (Duck, 1998; Duck & McMahan, 2015). Internet dating sites provide users with the added advantage of receiving a wealth of information about potential romantic partners. Such information, both in terms of amount and type, may not be readily available through other means of matchmaking and requires a great deal of effort to receive. In a single location there exists not only information concerning those attributes of attraction listed above but also information concerning what attributes that person desires in a romantic partner.

Strategic Creation and Strategic Assessment

An upcoming section will be devoted to the ways in which profiles on Internet dating sites are created. These sites are unique from other sorts of matchmaking in that users are provided additional time and additional tools to create the most favorable impression for potential romantic partners as possible. When creating a profile, however, people are better able to strategically create the person they wish to convey. For those viewing these profiles, Internet dating sites also provide additional time for scrutiny and assessment. Many users take advantage of the extra time and asynchronous interactions by utilizing online tools to search for public records, compare profiles on other sites, and simply run random searches about a potential partner (McKenna, 2008; Gibbs, Ellison, & Lai, 2011).

Greater Comfort

Searching for a romantic partner and the intricacies of dating are frequently accompanied with anxieties and discomfort. These negative feelings can involve fears of rejection, loss of control, identity issues, and related phobias, among many other factors (Chorney & Morris, 2008; Hope & Heimberg, 1990).

Alternative View

It might be assumed that people high in dating anxiety would be more likely to use Internet dating sites than people low in dating anxiety. However, the opposite has actually been found to be true (Valkenburg & Peter, 2007). Further, people who are more sociable in general are more likely to use these sites than people who are less sociable (Kim, Kwon, & Lee, 2009). While Internet dating sites may assist people who are particularly anxious about dating, they do not level the playing field. Rather, more sociable and less anxious people view these sites as yet another means of finding romantic partners.

Internet dating sites can alleviate some of the concerns associated with matchmaking and dating. For those individuals who fear or feel uncomfortable in public settings, these sites eliminate the need to venture into crowded areas in search of a romantic partner. Concerns associated with making a good initial impression are somewhat reduced through the ability to strategically create an acceptable profile. Likewise, concerns associated with initiating interaction to engage in information seeking are reduced through the ability to review the profiles of others.

Internet dating sites also alleviate fears associated with rejection. Concerns associated with rejection and with unrequited love are among the leading causes of dating apprehension, and involve not only personal rejection but also the possibility of having to reject others. In fact, it has been argued that the rejecter is often in a more difficult position that the person being rejected (Baumeither & Wotman, 1992; Baumeister, Wotman, & Stillwell, 1993).

Fears associated with rejection are alleviated through Internet dating sites in a variety of ways. First, users of these sites are more likely to have zero shared history or shared social network connections. Second, rejection through digital means is less difficult than when rejection occurs face-to-face, which might involve numerous negative nonverbal cues, conversational leave-taking decisions, and other interactional factors. Finally, Internet dating sites provide options unavailable through other matchmaking means, such as automated rejection responses or simply remaining unresponsive to requests for correspondence (Tong & Walther, 2010).

Constructing Profiles

As alluded to previously, the ability to strategically create a profile is one of the characteristics that make Internet dating sites distinct from other attempts to find romantic partners. In this section, we will examine the common components of these profiles along with discussing the meanings and strategies used when creating a profile. We will specifically examine the use of deception when crafting a profile and how other users evaluate and determine the accuracy of these profiles.

Profile Characteristics

Internet dating sites naturally vary in regard to components included in user profiles. However, there are a number of common elements which tend to be included in these profiles. Some of the most common of these features are shown in Figure 9.10.

Photographs	User Name	Location
Sex	Age	Height
Body Type	Hair Color	Eye Color
Race/Ethnicity	Description of Self	Ideal Partner
Faith/Religion	Political Views	Education
Professions	Salary	Interests
Children (Status/Desire)	Smoking/Drinking Habits	Languages

Figure 9.10. Components of Internet dating site profiles.

When creating their profiles, users of these sites generally have the option to complete some areas while disregarding others and to provide varying degrees of details in their responses. This ability provides them with opportunities to strategically create how they want to be perceived by potential romantic partners. Of course, as will be discussed, users are not always completely honest when developing their profiles.

Photographs and Physical Appearance

People tend to be very strategic about the photographs of themselves posted on the Internet (Pempek, Yermolayava, & Calvert, 2009; Siibak, 2009; Whitty, 2007a). Given the importance of appearance and physical attraction in the selection of romantic partners (Feingold, 1990), especially initially and when briefly encountering the other person as would be the case when scanning photographs (Luo & Zhang, 2009), photographic representations on a person's profile become that much more important. Indeed, photographs have been found to be the strongest predictor of attraction for those searching for romantic partners through Internet dating sites (Fiore, Taylor, Mendelsohn, & Hearst, 2008).

In spite of the importance of physical appearance, however, it has previously been posited that interactions on the Internet enable the playing field to be leveled, providing people who do not fit the stereotypical notion of beauty to attract partners through other means or decreasing the importance of physical attraction (e.g., Levine, 2000). Unfortunately for those who are less than good looking, though, this does not appear to the case when it comes to Internet dating sites.

Brand, Bonatsos, D'Orazio, and Deschong (2011) discovered that "photos rated as physically attractive had profile texts that were rated as more attractive, even though photos and texts were rated by different judges" (p. 166). It is suspected that people who are generally perceived to be attractive have greater confidence when constructing other portions of their profiles and write more appealing texts as a result. Good-looking people seem to be at an advantage both offline and online, then, when it comes to attracting potential romantic partners.

Screen Names

The use of a screen name is a component unique to matchmaking through Internet dating sites. A person's actual name has been found to impact how that person is perceived and treated by others (Mehrabian, 2001), so it is reasonable to expect screen names to accomplish the same thing. In fact, screen names can be used in the construction of identities (Duck & McMahan, 2012), and people tend to be fairly adept at judging personalities based on names used in e-mail addresses (Back, Schmukle, & Egloff, 2008).

Whitty and Buchanan (2010) found that Internet dating site screen names generally fall into one of the following categories: (1) looks, (2) sexual, (3) personality, (4) wealthy, (5) classy/intellectual, (6) humorous, and (7) neutral. Among the findings of these authors, sex differences were discovered related to attraction and motivation to contact in regard to screen names. Men were more likely than women to find screen names conveying favorable physical appearance to be attractive, while women were more likely to find neutral screen names and those conveying intelligence to be attractive. Both men and women were also more likely to contact users with these respective screen names.

Overcoming Potential Limitations

The strategic construction of Internet dating site profiles includes deliberately addressing personal characteristics that may be perceived negatively. Some characteristics involving body type, age, marital status, and employment among others may be less than attractive depending on the target audience of potential romantic partners. As will be discussed below, some users of these sites deal with such issues through deception. Other users avoid discussing these issues in their profiles. Still other users directly address these issues through various discursive means (Alterovitz & Mendelsohn, 2013; Jonson & Siverskog, 2012; Young & Caplan, 2010).

Deception

Users of Internet dating sites are often deceptive when completing their profiles and when communicating online with potential romantic partners. Toma et al. (2008) found that 81% of participants provided at least one inaccurate bit of information on their profile. In general, women tend to lie about their weight

(DeAndrea et al., 2012; DeHall, Park, Song, & Cody, 2010; Toma, Hancock, & Ellison, 2008). Men tend to lie more about their height (DeAndrea et al., 2012; Toma et al., 2008), along with personal assets, relationship goals, personal interests, and personal attributes (Hall, Park, Song, & Cody, 2010).

Interestingly, people using these sites do not necessarily believe they are guilty of deception but believe other users of these sites are engaged in deception. Gibbs, Ellison, and Heino (2006) discovered in their research that "94% of respondents strongly disagreed that they had intentionally misrepresented themselves in their profile or online communication, and 87% strongly disagreed that misrepresenting certain things in one's profile or online communication was acceptable" (p. 169). Correspondingly, users of these sites tend to rate their profile pictures to be relatively accurate, while independent judges rate one-third of these photographs as inaccurate (Hankock & Toma, 2009). At the same time, even though they do not always recognize their own deceptive practices, users indicate deception to be common among other users (Gibbs et al., 2006; Lo, Hsieh, & Chiu, 2013).

Presenting an Ideal Self

Accordingly, it appears that people tend to be deceptive when completing Internet dating sites profiles and communicating with potential romantic partners online. However, users of these sites do not believe they are being blatantly deceptive. This inconsistency likely encompasses how people view the creation of these profiles and their desire to present an ideal self.

When completing Internet dating site profiles, users must balance between accuracy and desirability (Ellison, Heino, & Gibbs, 2006; Whitty, 2007b; Whitty & Joinson, 2009). This balance is even more tenuous given the strategic capabilities these sites provide users, which were discussed above. Users of these sites may not "expect the profile to be an exact digital representation of a corporeal being, but rather a consciously constructed amalgamation reflecting a fluid sense of identity drawing from past, present, and future selves, subject to daters' limited self-knowledge" (Ellison, Hancock, & Toma, 2012, p. 51). Thus, when creating their profiles, users of these sites are presenting an idealized version based on their idealized past, distorted present, and idealized and anticipated future self. When inaccurately describing such a physical characteristic as body type or when inaccurately representing their income, users may be describing the person they wish to become and believe these characteristics to be attainable in the near future and, importantly, by the time a face-to-face encounter will occur.

Conclusion and Predictions

More than essentially any other content or activities, that associated with pornography and relationship initiation can be used to trace the development of the Internet. Sites dedicated to these endeavors are a pervasive presence. As discussed throughout this chapter, both Internet pornography and Internet dating sites are massive in size and scope. Further, each has unique characteristics that make its development and use distinctive among other sites and online activities.

As we look toward the future, certain predictions can be made concerning Internet pornography and Internet dating sites. First, *the number and scope of both types of sites will continue to expand.* This expansion will likely include the further development and presence of peripheral pornography and niche dating sites respectively. Likewise, *the number of users of these sites will continue to increase.* A reciprocal relationship will continue to exist between the increased numbers of sites, range of content, and users. In the case of Internet dating sites specifically, the number of users will continue to increase as the stigma attached to using these sites continues to lessen in significance.

As with other relational activities on the Internet we believe that *pornography and online dating sites will become increasingly virtual,* enabling people to interact in a virtual space as themselves or other avatars. Similar to its past influence on the development of the Internet, pornography, specifically, will be influential in advancing the creation and presence of virtual reality through the Internet. Sexual activities will increasingly take place using teledildonics and other means. Although Internet dating sites have never lead in the development of Internet capabilities, they are quick to take advantages of this progress. Accordingly, there will be opportunities for virtual interaction when using these sites. Like other predictions, the passing of time will determine their accuracy. And, like other relational activities on the Internet, connecting through pornography and Internet dating sites will continue to be influential and worthy of attention.

Key Terms

Niche Internet Dating Sites
Peripheral Pornography
Teledildonics

Chapter Exercises

Current Issues: Have students search for and report on advertisements for Internet dating sites. Ask student to consider how these advertisements might impact how these sites are perceived and whether people searching for romantic partners will use them. Are some of these advertisements more persuasive than other ones? If so, why might that be the case?

Debate It: Split the class into two groups. Select one of the resolutions below. Have one group support the resolution and have the other group oppose the resolution in a debate.

Resolved: Makers of Internet filtering software are most dedicated to protecting children from objectionable material.

Resolved: Makers of Internet filtering software are most dedicated to making money.

Ethical Consideration: When people attempt to meet potential romantic partners in a public space, they often endeavor to look better and dress nicer than usual. However, this behavior is usually not considered deceptive. Ask students whether they would consider selecting a particularly favorable photograph when constructing an Internet dating site profile to be different.

Everyday Impact: Ask students who are searching for a romantic partner whether Internet dating sites have impacted their search, regardless of whether or not they have personally utilized them. Ask students not searching for a romantic partner how Internet dating sites have impacted people in their social networks.

Future Focus: We anticipate that Internet pornography will become more virtual, immersive, and sensory in the future, once again being a driving force for the development of these features and characteristics online. This development will be assisted by such immersive devices as Google Glass. Ask students to consider whether we are correct. What do they think Internet pornography will offer users in the future?

10

Transcending Space, Time, and Class: Video Sharing, Video Gaming, and Praying Online

Scholars from a variety of disciplines, from sociology to communications, from Turkle (2011) to Bauerlein (2011), regularly assault, deconstruct, and celebrate the social implications and transformations produced by the modern evolution of computer mediated communication. Without question, computer mediated communication online continues to nurture "dramatic, unprecedented changes in our personal lives and social relationships" (Kempers, 2002, p. 118). Indeed, access to entertainment, education, community, family, and religion is increasingly liberated from pure face-to-face interactions and other physical constraints.

Online, we see diminishments in a host of variables that previously limited one's access to numerous sociocultural systems and experiences. For instance, a lack of suitable transportation or financial support, the remoteness of a location or one's physical disabilities have all become less obstructive or discouraging when communicating digitally in virtual contexts (Loane & D'Alessandro, 2013). Indeed, digital technologies appear to be redefining the parameters and contexts available for socializing while adding new layers, pockets, and levels to society and popular culture.

Opportunities for education, base entertainment, and intellectual enlightenment are no longer solely tethered to prodigious textbooks, stadiums, and "invitation only" fluorescent classrooms. Lectures, software tutorials, product tests, live shows, and more thrive online as free, tradable, amateur, and professional videos. Guided meditations, a shaman's blessings, or a pastor's pontifications are also no longer solely confined to specific physical moments, persons, and contexts. Online, an array of seamlessly engaged online forums, smartphone applications, and web-based virtual communities of all breadths and types are committed to religious-based exchanges. Similarly, manic and heated videogame challenges between friends and neighbors have moved beyond shared physical spaces like the darkly lit, sticky arcade and the sweaty, suburban neighborhood playpen. Today, through online portals and game consoles, players can engage players from nearly anywhere at anytime, from the rural to the urban, from the elite to the subordinate; kids in Russia can cross swords with teenagers in China, boys in Canada can race girls in Korea, North Americans in Indiana can plan a virtual raid with Mexicans in Guadalajara, and on it goes with near infinite possibility.

Preview

In part, this chapter (and unit) extends our commentary on an ongoing popular cultural shift away from a social dependency on physical space and physical interactions for social experiences. Toward this end, in this chapter, we explore three sociocultural systems and their relationships with the Internet and America: First, we describe the video sharing explosion prevalent online across demographics. Here, we describe the penetration of online video in America (and abroad) while placing emphasis on the notable applications and functions these videos have assumed in society and popular culture. Second, we introduce the video game industry as a near universal and lucrative enterprise offering users a social and emotional outlet. We highlight the diffusion of an eccentric variety of video games across populations before transitioning to their appeal for users and researchers. Third, we end with a portrait of how various religious agencies are repurposing the web and digital communication technologies for spiritual purposes. Each sociocultural system outlined and analyzed in this chapter adds to our portrait of the Internet as an inclusive, unbiased, and convergent system uniting bodies, minds, preferences, and ideas in multiple forms and experiences.

Questions Previewing This Chapter

1. *How have video-sharing websites been adopted and applied in popular culture?*
2. *How are video games becoming a universally adopted experience?*
3. *What is contemporary research saying about the video-game experience and video-game users?*
4. *How have religious organizations and interests exploited computer mediated communication technologies?*

The Video-Sharing Explosion: YouTube.com and Beyond

On April 23, 2005, the first video was uploaded to YouTube.com. The video, titled "Me at the Zoo," was a 19-second offering featuring the spontaneous musings of the site's cofounder, Jawed Karim. In October 2006, Google absorbed the website for $1.65 billion.

Since its introduction, consumption of online video continues to rampantly increase. For example, in 2011, YouTube averaged well over three billion **unique views** a day (Moore, 2011). By 2012, it averaged nearly four billion unique views a day with some one billion unique visitors each month (Grossman, 2012; youtube.com).

Today, professionals and amateurs representing a plentitude of demographics and regions appear to consume and produce online video daily. Every month in the United States, some 150 million Internet users watch over 14.5 billion videos, averaging 97 videos per viewer (Stelter, 2009). YouTube.com has evolved into a seemingly bottomless and anarchic vessel overflowing with millions and millions of videos. Indeed, in 2013, YouTube announced users were uploading 100 hours of video to the site every minute (or 6,000 hours of video every hour; 144,000 hours of video every day) (youtube.com).

Alternative View: Fleeting Engagements Online

YouTube is often accused of operating at a loss (Gladstone, 2013). Despite its popularity and wide adoption, the average user still spends a mere 15 minutes on YouTube per day (Grossman, 2012). In contrast, the average American spends nearly three hours a day watching television. However, time spent watching online video as a whole appears on the rise. For instance, users spent 49% more time watching videos online in May 2009 versus May 2008 (Mindlin, 2009).

Beyond YouTube.com, a vibrant array of video-sharing websites have emerged online, nationally and internationally, including Vimeo.com, Vine.co, Tudou.com, Hulu.com, Dailymotion.com, and Youku.com. As of 2011, 71% of all online adults use video-sharing sites with 28% using these sites daily (Moore, 2011). Without question, video sharing has become a significant and popular component of the Internet experience. Indeed, *The Pew Research Center* declared videos a "key social currency online" (Raine, Brenner, & Purcell, 2012). In this section, we describe and deconstruct four significant uses and gratifications of these prevalent video-sharing websites.

1. A Platform for Entertainment and Artists

Over half of adult Internet users watch humorous videos online (Purcell, 2010). In 2011, four of the five most watched videos online were short bursts of professional and amateur comedy. For example, the fifth most watched online video in 2011, "Nyan Cat," featured a crudely animated cathead with the body of a Pop-Tart racing through space and time, creating a rainbow trail. The cat passively soars to the rhythmic chants of a Japanese pop song for some three minutes. By early 2013, "Nyan Cat" had amassed over 90 million unique views (youtube.com). Videos of this whimsical and eccentric nature are common online and commonly consumed.

When video sharing began online, short-length home videos and low-budget, amateur offerings often found massive audiences, earning favor and exposure as the most shared videos of that year (Dawsey, 2012). Indeed, many lucky and talented amateurs have often found themselves propelled forward in popular culture due to their exploitation of video-sharing services. Across 2012, however, a shift in these trends became evident. The most shared and popular videos of 2012 appeared more polished and better funded. Oftentimes, the top viewed videos we see now cost millions of dollars to produce (Dawsey, 2012). Savvy professionals, government agencies, production companies, and media entrepreneurs have penetrated the blossoming medium, capitalizing on a now proven market. Indeed, in 2013, YouTube.com partnered with select producers to beta launch a paid subscription service featuring longer, more professional content (youtube.com). To a degree, the cheaply generated clip-based video experience is being replaced and dampened, but not lost (Stelter, 2009).

Alternative View: Amusing Ourselves to Death

Nearly 30 years before the Internet delivered Miley Cyrus's bizarre and osten-tatious party anthem music video, "We Can't Stop," 10.7 million views in 24 hours, Postman (1985) published Amusing Ourselves to Death: Public Discourse in the Age of Show Business, *an inspired critical assault on rising cultural consumption and production patterns in the age of television. In his text, Postman argued the public was slowly adjusting to "incoherence" and being amused into "indifference" (p. 111). Postman fretted we were collective-ly entering an Aldous Huxley–styled prophecy where culture is nothing more than a burlesque. Postman warned, "when a population becomes distracted by trivia, when cultural life is redefined as a perpetual round of entertainments, when serious public conversation becomes a form of baby-talk, when, in short, a people become an audience and their public business a vaudeville act, then a nation finds itself at risk; culture-death is a clear possibility" (pp. 155–156). For some, such as Keen (2007), author of* The Cult of the Amateur: How Today's Internet is Killing Our Culture, *Postman's arguments have become exponentially more poignant when examining current cultural consumption patterns in the age of the Internet.*

2. A Tool for Education

Online video content is not limited to cats, comedy, and digital art. Entre-preneurs are also producing vivid and inspiring educational videos online. In 2010, 38% of adult Internet users consumed educational videos, exposing another viable market online (Purcell, 2010). In 2007, the TED organiza-tion responded to trends online, launching their own video-streaming service, TED Talks. TED Talks offers hundreds of free videos featuring academic pre-sentations highlighting a variety of innovative ideas from around the globe. For instance, these videos educate viewers on topics such as robotics, the hu-man brain, the human orgasm, the upcycling of "obsolescent" technologies, the nature of the universe, the science of happiness, and the influence of big data, to name a few.

In 2011, TED's YouTube.com channel became the "number one" subscribed nonprofit channel with over one million subscribers (youtube.com). As of 2013, more than 1,400 TED Talks are now available online with select collections addi-tionally available through Netflix.com (ted.com). *Business Insider* called the TED Talks platform a place for the brightest minds to spread ideas and engage a global audience (Blackstone & Groth, 2012). Indeed, in 2012, TED Talks had over 800 million page views internationally (Blackstone & Groth, 2012).

Outside of large organizations like TED, smaller organizations and thought producers also effectively exploit online video sharing. For instance, Annie Leonard of Free Range Studios produces fast-paced animated lectures educating viewers on environmental and social issues. Leonard's first online video, "The Story of Stuff" (2007), a 20-minute animated lecture deconstructing human production and consumption patterns and distributed for free, has amassed over 15 million views (storyofstuff.org).

3. A Political Instrument

Political campaigning and political musings inevitably pervade every medium. As covered in more detail in Chapter 12 the Internet has become a compelling and effective tool for modern campaigning and political transparency. Indeed, various politicians regularly produce online video for YouTube.com in an effort to capitalize on its ongoing popularity and accessibility. For instance, during the 2008 U.S. presidential campaign, BarackObama.com uploaded more than 1,800 videos to YouTube.com. Videos are generally shared and disseminated socially via email, Facebook, and Twitter. Indeed, across the 2012 U.S. presidential campaign, 55% of all registered voters reported consuming political videos online (Smith & Duggan, 2012). Videos centered on U.S. presidential hopefuls Barack Obama and Mitt Romney amassed more than two billion unique views in 2012 (Dawsey, 2012).

After securing the 2012 U.S. election, the Obama administration continued their involvement in video-sharing platforms. For example, the 2013 State of the Union address was live-streamed on YouTube.com.[1] Further, in 2013, YouTube.com announced all members of the U.S. Congress would have the opportunity to access enhanced features on their platform, including the ability to live-stream content from their offices and beyond (youtube.com). Efforts such as these show potential to enhance political transparency while increasing the accessibility of our politicians, who often reside out of reach.

Alternative View: Digital Dissidence

YouTube.com, Vimeo.com, and other live-streaming services have also become successful means for political dissidence. With limited resources, frustrated citizens operating on the fringes of popular culture and mainstream politics often turn video sharing platforms into their personal global podiums. For instance, on March 5, 2012, "Kony 2012," a 31-minute video aimed at raising awareness on war criminal and African militia leader Joseph Kony, captivated the online multiverse. In one single day, the audacious and politically charged "Kony 2012" collected some 30 million unique views, earning the record for most views in a 24-hour period (Dawsey, 2012). By 2013, "Kony 2012" had over 97 million views. (The radical applications of YouTube are further explored in Chapter 14: International, Corporate, and Radical.)

4. A Platform for Socialization

Video sharing online offers families, friends, schoolmates, and colleagues embodying different times and spaces the opportunity to connect, reconnect, share, and unite with one another. In 2010, research indicated home videos were one of the most common videos posted online, shared by 62% of video uploaders (Purcell, 2010). Home videos span large spectacles, small moments, and spontaneous blunders; capturing wedding ceremonies, a child's first words or slide down the stairs, late-night karaoke debacles, and more. Home videos have transcended living rooms and family reunions to thrive in wider views, untethered to constraining and fleeting contexts.

In early 2013, Vine.co partnered with Twitter and launched online to offer consumers a video recording "app" for free for their mobile devices. Made for motion instead of stills, Vine.co appeared as the indirect offspring of the heavily adored Instagram "app." Vine.co enables users to seamlessly capture and send video of their lives from anywhere, to anyone, using only their smartphone. Vine.co promotes the app as an opportunity to create "windows into the people, settings, ideas and objects that make up [the user's] life" (vine.co). Within four months, the "app" ascended the charts to become the number one free app in the United States (Souppouris, 2013). In spring 2013, Vine.co and YouTube.com helped make HBO's *Game of Thrones* episode, "The Rains of Castamere," the "most social episode of any HBO show ever," with more than 700,000 mentions online (Hernandez, 2013). Users of vine.co and YouTube.com uploaded and shared hundreds of videos from living rooms and bedrooms showing excited viewers reacting to a momentous and surprising climax in the HBO show.

The Video Game Industry: An Expanding and Immersive Enterprise

The video game enterprise is vast and multiplying (see Table 10.1). Gaming revenues recently surpassed the global revenues of both the music and film industries (Wallop, 2011). The average U.S. household owns at least one game-playing device (Entertainment Software Association, 2012). Persons without traditional game-playing consoles can (and do) download and stream interactive entertainment through various portable devices, such as a touch tablet or a mobile phone. Indeed, 33% of surveyed households play video games on their smartphones (Entertainment Software Association, 2012).

Today, game playing appears nearly universal. 97% of surveyed 12–17-year-olds report playing video games on the web, a console, or a portable device (Lenhart, Kahne, Middaugh, & Evans, 2008). Further, dampening notions of gender bias, males and females increasingly participate in near-equal numbers. In 2012, 47% of game players were female (Entertainment Software Association, 2012).

Table 10.1. Video Game Sales Rising.[2]

Year	Total Sales
2004	$9.9 Billion
2007	$13 Billion
2011	$24.75 Billion

In this section, we offer an overview of the video-game industry, with an emphasis on video-game content, video-gaming online, and video-game research. First, we describe the dynamic scope of video-game content. Second, we introduce the lucrative diffusion of **social games** online. Virtual social gaming has evolved into a significant market and phenomena, nationally and internationally. Finally, we present video games as a complex and compelling venue for emergent research. Two contemporary research studies on video games are summarized.

The Dynamic Breadth of Video-Game Content

The video-game experience is endlessly diverse, embodying manic colorful explosions, surreal alternate realities, and muted, action-charged reflections of history. Consumers appear to embrace this breadth with a hint of insatiability. A typical teen plays at least five different categories of games and nearly half play eight or more different game types (Lenhart et al., 2008). These findings reflect both the immense popularity of gaming and the vast range of compelling video game content available.

Across video game libraries and platforms, an eclectic and ever-growing assortment of environments, characters, structures, puzzles, and interactive narratives vie for downloads, dollars, minds and fingers. There are role-playing games such as World of Warcraft and Final Fantasy, first-person shooter games such as Halo and Call of Duty, action and adventure games such as Silent Hill and Uncharted, fighting games such as Mortal Kombat and SuperSmash Brothers, sports games such as NFL Madden and NBA Jam, racing games such as Mario Kart and Gran Turismo, and puzzle games such as Portal 2 and Super Meat Boy. Indeed, every day, new games and new types of games are born while popular titles are updated and sequels are demanded.

Alternative View: The Rise of Lifestyle Games

Scholars, families, and governments continually voice concern and frustration over trivial, addictive, and violent video games distributed to children and teenagers (Carvajal, 2007). Entrepreneurs have responded, directly or indirectly, by developing and releasing lifestyle games designed to enhance the user's brain health, physical health, world health, and social relationships. For instance, My Life Coach, designed for the portable Nintendo DS, analyzes walking and rewards exercise and a hearty diet with game play (Carvajal, 2007). Another Nintendo DS offering, My Word Coach, challenges users with vocabulary-building exercises. According to its website, My Word Coach is intended to teach users to be more expressive and confident with their vocabulary. Through these and other offerings, video games appear to be transcending their base entertainment roots to appeal to an even more diverse global audience.

Baxter (2011) proposed,

> *The always-connected nature of the online gaming medium is ideal for health-based content that tracks progress, such as brain fitness or calorie tracking. These games can be connected to sophisticated database systems that track health progress, offer targeted recommendations, and even provide personalized, healthy gaming experiences that adapt and scale to the appropriate level of challenge for each user. (p. 109)*

Social Games: Competitive, Cathartic, and Lucrative

Mirroring trends in video sharing (discussed above), video gaming is also heavily social. Seventy-six percent of gaming teens play games with others at least some of the time (Lenhart et al., 2008). Fueled by the popularity of social networks and our increasingly mobile connectedness, social games have evolved into a significant market in the video game industry. Across 2010, 2011, and 2012, social gaming became a national phenomenon with the ravenous adoption of games such as Farmville, Words With Friends, Draw Something, and Letterpress.

Social games encourage collaborating, engaging, meeting, and competing with friends virtually through various social networks and/or mobile phones. Oftentimes, engaging friends inside these virtual environments is the best (or only) method for personal ascension within the game. Friends trade resources, compete for shared resources, and compete using the same resources. Games can be synchronous, played live and finished in moments, or games can be asynchronous, taking hours, days, or weeks to complete, depending on each participant's particular habits. The top three reasons supplied for playing these social games are "fun and excitement," "a competitive spirit," and "stress relief" (Loechner, 2011).

In 2011, research found 118.5 million people were playing social games at least once a week, up 71% from the previous survey (Loechner, 2011). In April 2013, one single social game title, *Farmville 2*, boasted 40 million monthly users (Murphy, 2013). Further, many of these users frequently spend real-world money within these virtual, social environments. Thirty-one million players purchased virtual currency in 2011, up 86% from the previous survey (Loechner, 2011). Four years after its launch, Zynga, producer of the endlessly adored Farmville series and other such social games, was worth some $4 billion.

The Study of Video Games

Rabid success and popularity often introduces public introspections and evaluations from journalists, scholars, and critics. In this regard, video games are no exception. Game studies have become an exciting object for emergent research and researchers. Doctorate programs focusing primarily on gaming have emerged at Southern California University, MIT, and the Georgia State Institute of Technology. Further, as video gaming entrenches itself in popular culture, we see an outpouring of books on gaming related topics, including Ryan's (2003) *Narrative as Virtual Reality: Immersion and Interactivity in Literature and Electronic Media*, McGonigal's (2011) *Reality Is Broken: Why Games Make Us Better and How They Can Change the World*, Bogost's (2007) *Persuasive Games: The Expressive Power of Videogames*, and Kutner and Olson's (2008) *Grand Theft Childhood: The Surprising Truth About Violent Video Games and What Parents Can Do*.

Academic interest in gaming appears to multiply in tandem with the expansion of the gaming enterprise. *The Wall Street Journal* (2008) observed, "the number of gaming thinkers— and outlets for their work—is mushrooming" (p. W3). We often see five common approaches to the study of video games: the historical approach, the reception approach, the aesthetics approach, the cultural approach, and the social approach.

In this section, we present two research contributions framed primarily from the reception approach and the aesthetics approach. Roughly, both studies described in this section analyze the cognitive and psychological effects and appeal of video games and video game aesthetics. First, we discuss Bowen's (2005) research on video game **immersion** and its impact on emotion. Further, we identify five significant variables affecting video game immersion. Second, we summarize Grodal's (2000) analysis of the video game narrative experience when compared to traditional forms of entertainment.

Video Games and Emotion

The Bowen Research Firm surveyed 535 gamers to explore how important an emotional experience and response was to the gameplay of any particular game. Specifically, Bowen (2005) asked gamers to describe how deeply their favorite games trigger various emotions on a scale of 1 to 5 (with 5 being the most intense). Bowen concluded that a user's favorite video games trigger various notable emotions, some very deeply. Here, we summarize Bowen's core findings.

Over one-third of those surveyed reported video games create an engaging emotional experience and response. Half of all game players surveyed described an emotional response as important to their reception of any particular game. More than two-thirds felt games delivered as much emotional impact as movies, music, or books. The top emotions inspired by video games were competitiveness, fear, and a sense of accomplishment. Other highly ranked emotions included honor, loyalty, integrity, awe and wonder, sadness, delight, and beauty.

Role-playing games were ranked as the most emotionally affective, with first-person shooters ranked as the second most affective. Oftentimes, surveyed users reported physically weeping during dramatic moments (featuring character deaths or romantic unions). Here, players reported intense attachments to the narratives and characters of role-playing games due to the depth of the stories, the poetic musical scores, and the dozens and dozens of hours invested into gameplay (Bowen, 2005). These findings support the notion that the level of immersion, or **telepresence**, induced by a video-game environment is, in part, directly influenced by the frequency of gameplay and the perceived quality of the narrative. Some gamers were so deeply affected by specific video game experiences, they reported being depressed in their physical lives, outside of the virtual confines of the game reality.

Variables affecting video game immersion. Several variables affect the level of video game immersion, or the level of involvement, attachment, and commitment of the video game player to the video game. For instance, multiple levels are designed to make a video game more involving by increasing the commitment of the video game player. A player will often feel increasingly immersed in a game reality as they engage more levels and spend more time inside that game reality (as evident in Bowen's findings outlined above). Five additional variables significantly affecting video game immersion include:

1. *Environmental vividness* of the game environment in terms of breadth and depth.
2. *Interactivity* in terms of speed, range, and mapping.
3. *Number of players*, including both the range and number of discrete roles as well as potential collaborations.
4. *Number and kinds of choices* for each player in terms of acts, scenes, tools, goals, and roles.
5. *Creativity possible on each level*, including the range and types of innovations in rules and context, the ability to implement new strategies, use tools in a wide variety of ways, find and employ new resources, and recast established artifacts in new artistic ways.

Video Games and the Pleasures of Control

In *Media Entertainment: The Psychology of Its Appeal*, Grodal (2000) deconstructed the various gratifications and positive effects derived from the gaming experience, contrasting the video game experience against traditional forms of entertainment, highlighting the video game experience as both an extension and separate experience to film and television consumption. Compared to film and television, video games offer users more control over their particular narrative. Grodal observed, "the hallmark of most video games is that they transform traditional forms of entertainment into an interactive form that enables the player actively to participate in shaping the [experience]" (p. 197). Video game users generally control the action, the perspective (or point of view), and the ways in which the game world is represented.

Increased control helps further bond the player to the game world, enhancing the cognitive connections between the narrative, the characters, and the user.

Echoing Bowen (2005) and Steuer (1992), Grodal reported heightened interactivity and the player's strong sense of control over the narrative will elicit various potent psychological responses, including catharsis effects, equilibrium effects, and boosts to the user's self-esteem and sense of accomplishment.

Religion and the Internet

For centuries, religions of varying complexities, shapes, and colors have projected and carved meanings and structures into our physical, mental, and social environments. Sociologists observe, religions oftentimes emerge and solidify to respond to an inherent human need to belong, luring individuals into structured communities that become essential to identity and identity formation (Kempers, 2002). From the printing press to the television, religious entrepreneurs often tap popular mediums to communicate and recruit. For example, in the 1980s, religion was frequently broadcast on television and televangelism soared with characters such as Jimmy Swaggart, Pat Robertson, Jim Bakker, Billy Graham, Jerry Falwell, and Robert Schuller, each pontificating on our bubble screens. In the 1990s, as the Internet coalesced and Web 1.0 spread, religious agents began using the new medium to promote and disseminate sermons, church newsletters, religious texts, and other information.

Savvy religious pundits quickly saw the Internet as an exciting new method for capturing more eyes, minds, and souls. Indeed, across the last several decades, online religious communities have steadily multiplied in number while becoming increasingly more interactive (Campbell, 2004, p. 87). Today, the Pew Research Center (2004) reports, 64% of "wired Americans" have used the Internet for a spiritual or religious reason, at least once (Hoover, Clark, & Rainie, 2004). From streamingchurch.tv to lifechurch.tv to stpixels.com, organizations have formed online committed to hosting and building virtual religious collectives where members can both passively consume sermons and participate in constructing (or remixing) content.

In the sections that follow, first, we describe the dynamic array of digital options and tools employed by these online religious communities. Further, we highlight one specific and relatively new tool, the virtual confessional. Arguably, the virtual confessional vividly reflects an ongoing shift in religious preferences, from formal to more informal. Finally, we transition to a discussion of the social uses and functions of online religious communities and tools.

Cyberchurch: The Virtual Congregation

Hewitt (1998) defined cyberchurch as a "church without walls in which people literally gather together via the Internet." Campbell (2005) extended this definition, highlighting that a cyberchurch generally has no offline equivalent, differentiating them from churches with a presence online and offline. Members of a cyberchurch communicate using various media channels. Figure 10.1 summarizes the various digital options commonly and currently employed by online religious communities.

Figure 10.1. Digital options employed by online churches.

The Confession Drop Box Phenomena

The virtual confessional is a unique and emergent digital tool deserving a more detailed explication. Virtual confessionals are designed to offer participants a sense of personal and social catharsis (Banerjee, 2006; Gross, 2011). Unspoken secrets can often induce alienation from friends, daily social anxiety, and relational avoidance. Anonymous virtual confessions enable participants to unload potentially toxic and poisonous baggage within a compassionate and inviting community.

The Internet currently offers a plethora of portals for personal confessions, including www.absolution-online.com, www.confessionjunkie.com, www.grouphug.us, postsecret.blogspot.com, mysecret.tv, and "Confession," a Roman Catholic application available on the iPhone, iPod, and iPad. Users can often find and experience a certain absolution in confessing secrets through these services (Banerjee, 2006; Gross, 2011). Indeed, months after its initial launch, mysecret. tv had amassed more than 1,500 unique confessions (Banerjee, 2006). In the following section, we further explore the social uses and functions of online religious communities and tools.

Social Uses and Functions: Expansion, Inclusion, and Personalization

Online religious communities and tools can service a variety of social needs beyond the cathartic experiences available through virtual confessionals. For instance, Rev. Craig Groeschel, founder of LifeChurch.tv, one of the largest church organizations in America, promotes religious websites and applications as a fruitful means for extending messages and outreach beyond physical presentations and churches (Banerjee, 2006). In this section, we summarize several additional and notable functions of online religious communities and tools.

Typically, offline church experiences are heavily structured, linearly organized, and intrinsically tethered to particular times and geographies. In contrast, religious outlets online give participants more limitless opportunities to share their particular voices and angst on a variety of topics, at any moment during the night or the day, from any place, from the bedroom to the barstool. Further, navigating specialized forums, websites, and engaging users inside chat rooms with particular ideas give participants more unique control over their individual experience. Arguably, online religious engagements can be more directed, or personalized, to an individual's immediate needs and wants.

A 2011 survey conducted by the Pew Research Center offers extended insight into the impact of the Internet on religious communities. The report concluded, the Internet has had a significant impact on religious groups to "communicate with members," "draw attention to an issue," "connect with other groups," "organize activities," "raise money," "impact local communities," and "recruit new members" (Jansen, 2011).

Conclusion and Future Possibilities

For many, sharing videos, playing video games, and interacting online is still a relatively young and intensifying practice. Indeed, YouTube.com is not yet 10 years old. For reference, the television set was still broadcasting primarily in black and white in its first 10 years on the scene. As our online communication technologies grow into adulthood, significant technological and social transformations brew on the incoming horizon. Here, we offer two thoughtful predictions for the future of video sharing, video gaming, and virtual gatherings.

The Massive Anarchic Proliferation of Online Video

As of 2013, more than 500 million photos are shared daily while 100 hours of video are uploaded to YouTube each minute (Fiegerman, 2013). Beyond 2013, on the wings of Vine.co and similar accessible video-recording applications, we anticipate billions and billions of video fragments will be uploaded online and shared in loads similar to photos, with an ongoing and increased frequency. The cataloging and tagging systems applied to this gargantuan influx of content will become increasingly important and vital. Without explicit, imposed organization, chaos and anarchy will reign across our plump and ostentatious virtual libraries.

The advancement and probable mainstream adoption of Google's "Google Glasses" (and similar wearable computers) will further help Internet users feed nearly infinite and anarchic video archives. Heavy users will record and upload every tiny, trivial, and momentous occurrence happening across their waking and sleeping lives, from childhood into death. (For a slanted look at the potential relational implications of these devices and compulsive recording practices, see Armstrong and Brooker's (2011) *Black Mirror* episode "The Entire History of You.")

The Lucrative Progression of Immersion Technologies

The physical layers separating the user from the virtual gaming environment are dissolving and will only continue to dissolve into the future. For example, in 2011, the Xbox 360 released Microsoft's Kinect, a hands-free sensor device that captures and integrates a player's physical motions into virtual environments. Here, a user's natural physical movements translate into virtual action inside a game. As a result, the perceivable boundaries between the physical player and the virtual character blur, necessarily increasing immersion levels.

Today, the consumer appears to desire, and demand, the enhancement of variables determining and constraining immersion. Indeed, within 60 days on the market, more than eight million Kinect devices were sold, making the device "the fastest selling consumer-electronics device in history" (Schmidt & Cohen, 2013, p. 18). Further, in 2012, the innovators behind the "Oculus Rift," a wearable virtual-reality headset, crowd-funded their research and development efforts through Kickstarter, raising $2 million over their $250,000 goal. The "Oculus Rift" is intended to enable users to step into the game. By 2013, video snippets were shared online showcasing the "Oculus Rift" as it delivered mesmerized test subjects horrific guillotine simulations and transcendent trips to the ocean.

The social impact and transformations motivated by these emerging technologies will be vast inside and beyond the gaming medium. For instance, in the near future, users will be able to seamlessly and virtually attend meetings with colleagues and spiritual leaders live from anywhere on the planet. Geographic constraints placed on collaborations and congregations will only continue to lose relevancy and power as these devices become more complex, immersive, and commercially viable.

Classroom Exercises

Current Issues

For this exercise, instruct students to browse popular video sharing services online, from YouTube to Vine to Vimeo. Considering the four uses and gratifications described in this chapter, what types of videos have been most viewed and shared this month and week? Are these videos being shared and traded on students' social network feeds? What potential variables made these videos so popular and why?

Everyday Impact

As discussed above, social games and their creators have built a notable market online. For this exercise, encourage students to explore how their own gaming experiences and preferences might be impacting this trend and their social relationships. Do they play social games? What types of games do they play and why? Do these games impact their social relationships? How?

Debate It

Resolved: Face-to-face communication is necessary to produce shared religious meaning.

Resolved: Face-to-face communication is unnecessary to produce shared religious meaning.

Religion permeates the online multiverse. In response, various religious leaders voice concern that synchronous and asynchronous forms of virtual congregation may be impoverished when compared to a more traditional face-to-face model (Dawson, 2004, p. 80). For instance, McGillion (2000) argued flesh-and-blood relationships, mostly absent when interacting primarily through digital media, are an indispensable condition of shared religious meanings. For this exercise, divide students into two opposing groups to debate the significance or lack of significance of flesh-and-blood relationships in manufacturing shared religious meanings.

Key Terms

Unique Views	Cyberchurch
Lifestyle Games	Personalization
Social Games	
Telepresence	
Immersion	

Endnotes

1. In contrast to "Nyan Cat," however, the "The 2013 State of the Union Address" has received significantly less views, with well under a half-million views (youtube.com).
2. Figures provided by Ritchtel (2005), Alexander (2008), and Entertainment Software Association (2012).

11

The Pragmatics of Communication—The Internet as a Coherent and Complete Societal System

The Internet has developed into a complete and coherent societal system. This statement is made in recognition of the Internet's major position in individual lives and societies as a whole. Nearly every aspect of a person's life is touched directly or indirectly by the Internet. Nearly everything that can be accomplished physically can be accomplished digitally. Those things in life that cannot be accomplished digitally have been transformed through the Internet.

Within this chapter, we will examine four key areas that have been transformed through the Internet: (1) health care, (2) banking, (3) employment, and (4) education. These areas were isolated because of their prevalence online, their unique qualities, and their fundamental position within society. Many additional areas, some of which will be addressed in the conclusion, are emerging and expanding online. As such, there is little doubt that the Internet will continue developing as a complete and coherent societal system.

Questions Previewing This Chapter

1. *How is the Internet transforming health care?*
2. *How is the Internet transforming banking?*
3. *How is the Internet transforming employment?*
4. *How is the Internet transforming education?*

Internet and Health Care

The Internet maintains a prominent position in health care. Seventy-two percent of Internet users have looked online for health information (Fox & Duggan, 2013). Patients are playing a more active role in their own treatment based on information discovered through the Internet, and medical researchers and providers are using the Internet to uncover medical trends and to assist in their treatment of patients. Further, growing numbers of people are using the Internet for support when dealing with a variety of medical issues (Loane & D'Alessandro, 2013). The top online health care sites according to Alexa.com rankings are listed in Figure 11.1.

Nih.gov

Webmd.com

Myfitnesspal.com

Mayoclinic.com

Weightwatchers.com

Mercola.com

Drugs.com

Menshealth.com

Focusoncroahnsdisease.com

Cdc.com

Figure 11.1. Top online health care sites; based on Alexa rankings.

Within this section, we will examine the use of the Internet as a source of health care information, the use of the Internet in health care research and treatment, and the use of the Internet for health care support.

Internet and Health Care Information

As mentioned above, obtaining health care information on the Internet is a common activity among Internet users. In many cases, the Internet is ranked as the most frequent source of information about a medical condition, a more frequent source than even health care providers (Google/Compete, 2011a, 2011b). Searches for information encompass a variety of categories, including general information about a medical issue, treatment options, others' experiences, and drug information, among other categories (Fox & Duggan, 2013).

Three factors make obtaining health care information through the Internet especially unique. First, accessing this information has enabled people to play a more active role in their health care needs. Patients increasingly arrive at medical appointments with specific questions, suggestions, and treatment options based on materials found on the Internet. In some cases, information obtained through the Internet enables people to bypass health care providers altogether (Google/Compete, 2011c).

A second factor making the obtaining of health care information on the Internet unique and so widespread is the desire for secrecy and privacy. Medical issues are often topics people want to avoid talking about or disclosing to others, especially if there is a social stigma attached to a medical condition or issue. The Internet is the most frequent source of information about depression, for instance (Google/Compete, 2011a).

Finally, half of all health-care-related searches on the Internet are done on behalf of someone else (Fox & Duggan, 2013). Unlike most other forms of information seeking on the Internet, searching for health care information is likely to involve the needs of others.

Treatment and Research

Health care treatment and research are also being transformed through the Internet. These transformations include those involving patients and providers along with the medical profession in general.

As mentioned above, patients are playing a more active role in their treatment. The information discovered through the Internet enables them to ask questions of health care providers and participate in developing treatment plans. In preparation for treatment, patients also use the Internet to determine which health care providers and treatment facilities to utilize. The Internet is used much more frequently as a source of information than members of one's social network or such media systems as television, magazines, or newspapers (Google/Compete, 2012a, 2012b).

The Internet also assists medical providers with the treatment of patients. Internet material is used twice as much as print material by medical personnel when conducting research. Further, the majority of Internet research conducted by providers is based on interactions with patients (Google/Manhattan Research, 2012).

Beyond searching for information, the Internet is also being used by providers in the actual treatment of patients. **Virtual doctor visits** are those in which patients and providers meet through online video. Such visits tend to be cheaper than visits to a provider's physical office—which is beneficial for patients, business, and insurance companies (Mathews, 2012)—and are certainly more convenient, especially for patients geographically distant from medical facilities (Abrams, 2012). Virtual doctor visits have yet to become widespread, primarily due to state regulations and governing medical bodies. However, signs point to dramatic future growth of such treatment options. A study conducted by Cisco (2013a) found that 74% of consumers were open to virtual office visits.

Finally, the Internet is making it possible for researchers to track diseases and discover previously unknown medical correlations. For instance, **Google Flu Trends** (http://www.google.org/flutrends) uses billions of search queries to monitor flu activity in over 25 countries. In a similar vein, researchers are attempting to track flu epidemics and other public health issues through such social networking sites as Twitter. Real-time comments using key words connected to the flu and other health issues could be used to track health outbreaks as they occur. And, as with Google Flu Trends, this information could possibly be available many days prior to reports from the Centers for Disease Control and Prevention (Jarvis, 2013).

The use of search engines and social networking sites to track health issues requires massive amounts of data. These large quantities of data have also become available through the digitization of patient records. The aggregation, or collection and combining, of data from numerous patients provides medical researchers with new insights into treatments. These researchers are able to uncover correlations involving symptoms and treatments that would otherwise be undiscovered (Jaret, 2013).

Researchers can also share data more easily through the Internet. For example, the **1000 Genomes Project** (http://www.1000genomes.org) posts the DNA of over 1,000 people from all over the world to be used by researchers for study. Although researchers are provided with a wealth of data that can potentially be used for medical discoveries, there are privacy concerns about this and similar uses of medical information.

Support

The Internet also connects people dealing with health concerns with others with similar experiences. Eighteen percent of Internet users have searched online for people with similar health concerns. Percentages increase for those who are caregivers, those with chronic conditions, and those who have experienced recent medical issues (Fox, 2011).

Online support groups connect people who are experiencing similar health concerns. A host of online support groups exist for a variety of health concerns. These groups are available for those who are personally dealing with a health issue and for those with a relational connection to someone dealing with a health issue. These groups also facilitate the sharing of information and mutual support among users (Chung, 2013a; Loane & D'Alessandro, 2013). And they provide users with a sense of empowerment (Campbell, Coulson, & Buchanan, 2013).

Online support groups are similar to offline support groups in their purposes. However, there are many differences between them. An obvious difference is the ability of users to overcome space and time; another difference is the constant availability of most groups. There is no need to travel to a physical location, and support is available whenever needed. Beyond convenience and availability, some users prefer the support provided online to that provided offline (Chung, 2013b). Another major difference between online and offline support groups is the degree of structure and professional support. Online groups are developed by participants themselves and are less likely to follow a pre-established structure. Any structure that develops is created over time by participants themselves. Also, moderators tend to be fellow members rather than a professional moderator or medical provider (Coulson & Shaw, 2013).

Internet and Banking

Banking through the Internet is becoming increasingly common, utilizing the services of both traditional banks and Internet-only banks. In fact, as growing numbers of customers move toward online and mobile banking, many physical bank branches have closed and more are expected to close in the near future (Passy, 2013; Sidel, 2013a). Presently, 61% of transactions occur online or through mobile devices, with only 14% taking place at physical branches, and the remaining taking place using automated teller machines or other methods (Sidel, 2013b).

Alternative View

Banks do not offer services through the Internet to be kind. One bank reports saving $300,000 each year for every physical branch they are able to close (Sidel, 2013a). It also enables banks to keep customers. It has been estimated that using the Internet to pay bills, deposit money, and deduct money reduces customer turnover by 95%. A customer may become angry with his or her existing bank and want to move to a new one. However, moving multiple bill payment arrangements and direct deposit arrangements is seen as overwhelming (Schwartz, 2011).

In terms of customers, a study conducted by the Federal Reserve (2012) found that 68% of U.S. consumers with a bank account and Internet access had used online banking within the past year. When bill paying is isolated as a single category, the percentage of consumers using the Internet for such transactions reaches nearly 74% (IDC Financial Insights, 2012). Internet banking overall is even more common in other countries, with 88% of customers in Sweden, 76% in France, 70% in Canada, 74% in Australia, Poland, and South Africa, respectively, and 73% in Belgium banking through the Internet (Ipsos, 2012a).

Within this section, we will discuss factors influencing Internet banking adoption. We will then explore the rise of mobile banking. Finally, we will examine Internet-only banks.

Internet Banking Adoption

The majority of customers use the Internet for their banking needs. Some people simply have no interest in Internet banking (Patsiotis, Hughes, & Webber, 2012), but the remaining population is still deciding whether or not to embrace this type of financial service. Those consumers already using the Internet for their banking needs and those who are still contemplating this use of the Internet usually base their decision on the following five factors: (a) ease, (b) marketing, (c) risks, (d) price, and (e) Internet access/familiarity (Clemes, Gan, & Du, 2012).

The *ease* at which customers can utilize banking services will determine whether or not they will engage in Internet banking. With the possible exception of cashing in a jar of coins, the Internet has the possibility of making all banking easier and more convenient. Banking websites and mobile apps vary in their usability, and this will influence the decision to utilize these services (Akhlaq & Ahmed, 2013; Akhlaq & Shaw, 2011; Google, 2012).

Marketing by financial institutions also influences whether Internet services are utilized by customers. When searching for a bank, the vast majority of people use the Internet. Some people talk to bank representatives or utilize brochures or other materials provided by banks, and a few rely on advertisements on television or in print media (Google/Shopping Sciences, 2011). In all avenues, and especially through Internet advertising, financial institutions must make consumers aware of what Internet services are available (Alsajjan & Dennis, 2010).

Consumers also evaluate *risks* when determining whether to engage in Internet banking. Comfort levels of banking consumers concerning Internet banking have been shown to vary a great deal (Yousafzai & Yani-de-Soriano, 2012). Reports of bank hacking (Perlroth & Hardy, 2013), for instance, may cause some people to wonder whether using the Internet for financial transactions is safe. Lee (2009, pp. 131–132) found such fears involving security risks to be among four additional concerns related to Internet banking. Consumers also consider financial risks (monetary loss due to transaction error), social risks (adopting or not adopting Internet banking may be perceived negatively by others), time risks (slow Internet speeds; delays in transactions), and performance risks (bank sites malfunction or go down). Strengthening confidence in personal abilities and in Internet banking may be decreasing some of these concerns. And, in spite of worries about hackers and identity theft, nearly 70% of consumers indicate a willingness to provide more private information to banks if it means more personalized service and convenience (Cisco, 2013b).

Price is also a factor when customers consider Internet banking. Negative price issues include the cost of Internet access and limited data-packages available to some users. Positive price issues include lower fees charged by banks for Internet transactions and conversely higher fees charged by banks for transactions involving phone or branch representatives.

Finally, *Internet access/familiarity* is a factor in Internet banking adoption. This factor is somewhat related to some of the risks listed above. Some people do not have confidence in their abilities to use the Internet. Other people simply do not have Internet access. For many in this latter group, especially in developing countries (O'Brien, 2010), but also in industrialized nations, the answer is mobile banking.

Mobile Banking

In many ways, banking through mobile devices is following the path of Internet banking in general. Mobile banking adoption encompasses many of the same concerns just discussed in regard to Internet banking (Luo et al., 2012). Further, although there are significantly fewer mobile banking users than Internet banking users, their numbers are also rising. For instance, there have been recent one-year increases of anywhere between 100%–200% in Brazil, China, Kenya, and the United States (Cellular-news, 2011).

Alternative View

The exact number of users of anything depends on who is asked and what is asked, and the same holds true for mobile banking. In the United States, the Federal Reserve (2012) found that 21% of consumers with a bank account and mobile phone had used mobile banking within the past year. This percentage is slightly lower than other studies. However, confining use to the past year might miss out on consumers who have used it in the past. The percentage might also be higher if only consumers with mobile phones with Internet access were included rather than including everyone with a mobile phone, some of whom do not have a phone with such capabilities. A study conducted by Compete/Kantar Media (2012) isolated consumers who owned mobile phones with Internet access, and the percentage was around 58%. A third study (Inmobi, Viggle, & IAB, 2013) surveyed Viggle.com users, who tend to be tech-savvy and between 25 and 34 years of age. This study asked whether respondents were using mobile apps or mobile-optimized websites for banking, and it was discovered that 58% use the former and 50% use the latter.

As with Internet banking in general, consumers using mobile banking generally do so for practical convenience (Google, 2012). The vast majority of customers utilize mobile banking services to check account balances and recent transactions. Other uses include transferring money, bill payments, locating in-network ATMs, managing investments, and depositing checks using mobile phone cameras (Federal Reserve, 2012).

Online-Only Banks

Bank customers are also moving toward **online-only banks**, those banks with no, or at least very few, physical branches or locations. Online-only banks are a small part of the banking industry overall, holding less than 5% of deposits. At the same time, they have experienced significant growth in the past few years, reaching $364 billion in 2012, an increase of 400% since 2004 (Passy, 2013).

These banks have significant advantages over banks with physical locations. They tend to offer customers greater analysis of transactions and spending. Their sites are more likely to be user-friendly than those of traditional banks. And a lack of expense for staff and maintenance of physical branches means customers usually earn about 5% more interest on deposits (Passy, 2013; Wortham, 2013).

Internet and Employment

The Internet has become a vital employment tool. Job seekers use such job-search sites as Indeed (www.Indeed.com), Monster (www.Monster.com), and Career-Builder (www.Careerbuilder.com) to gain employment, and job seekers are increasingly using mobile devices as part of their job search (Weber, 2013). People in need of assistance for the completion of small jobs or short-term tasks and those people wishing to offer their services can use such sites as Fiverr (www.Fiverr.com), Taskrabbit (www.Taskrabbit.com), and Thumbtack (www.Thumbtack.com), with $5, of course, being the going rate for services listed on Fiverr. Many employees utilize the Internet in the workplace, both for work and for personal reasons.

Within this section, however, we will focus on the use of the Internet to **telecommute**, either by working from home or from a location other than a physical office. Telecommuting is a growing and sometimes prevalent method of work globally. Ipsos (2012b) reported that 35% of employees in 24 nations surveyed report telecommuting at least some of the time. Seven percent of employees in these countries report telecommuting exclusively. Table 11.1 lists the nations surveyed, the percentage of employees telecommuting at least some of the time, and the percentage of employees telecommuting exclusively.

Table 11.1. Telecommuting in 24 Countries.

Nation	Percentage of Employees Telecommuting	Percentage of Employees Telecommuting Exclusively
India	82%	32%
Indonesia	71%	16%
Mexico	58%	6%
South Africa	56%	12%

Table 11.1. *Continued*

Nation	Percentage of Employees Tele-commuting	Percentage of Employees Telecommuting Exclusively
Turkey	56%	6%
Saudi Arabia	48%	14%
Argentina	45%	9%
Russia	42%	9%
China	41%	6%
Australia	33%	7%
Poland	31%	6%
Brazil	28%	9%
Great Britain	28%	6%
Sweden	26%	6%
United States	26%	4%
Belgium	24%	2%
Spain	24%	3%
Japan	18%	5%
Canada	17%	3%
South Korea	17%	5%
Italy	14%	4%
France	12%	5%
Germany	12%	4%
Hungary	8%	2%

Note. By Ipsos, 2012b.

United States Census Bureau (2012) data show that nearly 7% of people in the United States work from home. This number is slightly higher than the number of telecommuters discovered by the Ipsos (2012b) survey. However, the Census Bureau data includes people who are self-employed and not telecommuting to an employer's physical offices. Still, in the majority of cases, self-employed individuals rely on the Internet in order to work from home.

Within all nations, the percentages of people working from home through the Internet depends on such factors as type of employment (United States Census Bureau, 2012), Internet availability and speeds, as well as employer willingness

(Swisher, 2013). From an employee perspective, the majority of employees would work from home, if they were able to do so (Ipsos, 2012b).

In what follows, we will examine the reported advantages and disadvantages of using the Internet to work from home. As we explore these pros and cons, keep in mind that this discussion entails the present state of working from home using the Internet. As virtual reality (Blascovich & Bailenson, 2011) and virtual gamification or gameplay (Reeves & Read, 2009) become more prevalent online and within workplace environments, these advantages and disadvantages will likely be altered.

Advantages of Telecommuting

Telecommuting comes with both advantages and disadvantages, some of which are contradictory. We will begin by examining some of the advantages.

One advantage of telecommuting is an increase in *productivity*. Perhaps the result of greater life and work satisfaction discussed below along with fewer distractions, the productivity of telecommuters has consistently been shown to be greater than employees at a physical office. For instance, a study examining "call-center" employees at CTrip, a Chinese travel agency, saw tremendous performance increases from those working from home. Such performance increases as working-minutes-per-shift and more calls-per-minute were attributed to fewer breaks or sick days and a quieter work environment, respectively (Bloom et al., 2013). Correspondingly, a study conducted by Cisco (2009) found that 69% of employees cited higher productivity when working remotely, and 75% believed their work had improved.

Perhaps as a result of this increased productivity, another advantage of telecommuting is the potential for *increased revenue*. In studies conducted in multiple nations, businesses that allow telecommuting have reported greater revenue growth than those businesses that do not allow telecommuting (Hendy, 2013; Silverman, 2013). In addition to increased employee productivity, businesses can possibly save substantial money when not requiring physical space for employees (Silverman, 2012).

A third advantage of telecommuting is the ability to *attract and retain top employees*. Not being bound geographically, businesses have a larger talent pool from which to draw (Silverman, 2012). Offering the ability to telecommute may also influence employee retention. The student survey conducted by Cisco (2009) cited above noted that 91% of teleworkers named being able to work remotely to be important to their overall satisfaction.

A final advantage of telecommuting is an *increased quality of life*. Multiple factors explain findings that telecommuting employees view working remotely as improving their quality of life (Cisco, 2009). One reason for this view is feeling better about the quality of their work and productivity mentioned previously. Workers can save money on transportation, clothing, and other work-related expenses. The vast majority of employees working from home (83%) also report having less stress due to not having to commute to a physical workplace. The vast majority of these employees (78%) also report a greater balance between work and family by working from home (Ipsos, 2012b).

Disadvantages of Telecommuting

There also exist disadvantages to telecommuting. And the issue of work and family is indicative of the somewhat contradictory nature of the advantages and disadvantages.

One disadvantage of telecommuting is a *reduction in boundaries between work and home*. Fifty-three percent of telecommuting employees note that there is greater conflict within their families because of reduced boundaries between work and home (Ipsos, 2012b). Again contradictory, however, some research has shown this spillover between work and family to have both favorable and unfavorable results (Troup & Rose, 2012; Wolfram & Gratton, 2012). Beyond family issues, employees may also feel as if they have to always be working or find themselves working more, given remote access to the workplace.

A second disadvantage of telecommuting is some people's *perceptions and fears of employee inactivity*. Studies indicating increased productivity should disprove perceptions and fears of employee's taking advantage of telecommuting by not working as hard or not working at all. Nevertheless, these perceptions exist in the minds of some employees who do not telecommute, and these fears exist in the minds of some supervisors with employees who do telecommute. However, just as some employee computers are monitored for activities in physical workplaces, the computer activities of telecommuters are often closely monitored by employers (Shellenbarger, 2012).

Another disadvantage of telecommunication is a *decreased chance for promotion*. In most nations, employees still primarily work in a physical location, and being physically present and available makes these employees more visible to supervisors than telecommuting employees. It also enables employees in the physical workplace to be more aware of activities taking place within the business and enhances their ability to cultivate relationships with other employees and supervisors (Silverman & Fottrell, 2013). A decreased chance for promotion is a concern for the majority of employees who do telecommute (Ipsos, 2012b).

A fourth and final disadvantage of telecommuting is a *lack of interaction* with other employees. Relationships in the workplace are important (Duck & McMahan, 2015), and telecommuters tend to feel socially isolated from colleagues (Ipsos, 2012b). These feelings of social isolation make interactions that do take place among employees to be extremely valuable in terms of organization identification and commitment (Fay & Kline, 2011, 2012). A general lack of interaction may also lead to conflict when employees must work together in virtual teams (Ayoko, Konrad, & Boyle, 2012). At the same time, some telecommuters see their communication with other employees to be as good as or better than that taking place face-to-face (Cisco, 2009).

Internet Education

As with other areas of life, the Internet is transforming education. Within this section, we will examine the present state of online education by examining the numbers of students enrolled, the numbers of courses and degrees available online, and the views of administrators and faculty. Next, we will examine types of online courses. Finally, we will examine advantages and disadvantages as well as concerns encompassing online education.

The State of Online Education

Online instruction is increasingly taking place at the K–12 level (Banchero & Simon, 2011; Hollander, 2012). However, it is much more prevalent in higher education, and it is there where online instruction is being developed, shaped, and studied before moving to lower levels of education. Accordingly, we will primarily focus on online education at this level.

The Babson Research Group has tracked online education in the United States since 2002 and issued a report examining its first 10 years of tracking (Allen & Seaman, 2013). According to this study, 86% of higher education institutions offer online courses, which increased from 72% of institutions offering online courses in 2002. The percentage of institutions offering complete online degrees has risen more dramatically in this time period, from 35% in 2002 to 63% in 2012.

Thirty-two percent of higher education students took an online course within the previous year. Less than 10% of higher education students did so in 2002.

Seventy-seven percent of academic leaders believe online course learning outcomes to be the same or superior to face-to-face courses. This percentage has risen from 57% of academic leaders who held this view in 2002.

Professors tend to take a more negative view of online education. Only 30% of academic chairs believe that faculty in their departments accept the value and legitimacy of online education. This percentage has only slightly increased from 28% of faculty holding this view in 2002.

Alternative View

According to ICEF Monitor (2012), the United States leads the world in online education both in terms of numbers and in development. However, other countries offer online instruction, and they put forward seven who are leading the way.

India is experiencing rapid growth in online education, and online programs are expected to reach US $1 billion in revenue by the end of the decade.

China boasts nearly 70 online colleges, but has been plagued by limited Internet access in rural areas and diploma mills.

South Korea has 17 online colleges, but the high value placed on face-to-face education in that country has been a hindrance.

Malaysia's Asia e University offers online education in 31 Asian nations.

The United Kingdom's Online Learning Taskforce has recommended a £1 million investment in online education.

Australia has experienced a 20% growth in online education in the past five years.

South Africa has developed nationwide online resources and offers online courses in high schools and institutions of higher education across the country.

Types of Internet Courses

There are six types of Internet courses in existence now or in development: (1) Internet-only course, (2) Hybrid course (Blended course), (3) Internet-enhanced course, (4) Virtual course, (5) Virtual-reality course, and (6) Massive Open Online Course (MOOC).

Internet-Only Courses

When thinking about online education, **Internet-only courses** likely come to mind first. These are courses taking place completely online with no, or minimal, face-to-face interaction among instructors and students. In many cases, there is no face-to-face interaction at all, and participants may be scattered geographically. In other cases, especially those in which students are enrolled at and taking face-to-face courses at the same university, an opening meeting may be required or students may be able to meet with the instructor during face-to-face office hours. In both types of cases, nearly all learning and interaction takes place through the Internet.

We will examine the benefits and drawbacks of online instruction below. Here we will discuss how students choose which courses to take online. Three factors ultimately influence this decision. First, students consider the suitability of the subject area. Some students do not view public speaking, laboratory-based, or language courses as more suitable for face-to-face instruction, for example. Second, students consider the difficulty of the course. Students who believe the material will be difficult tend to prefer face-to-face instruction. Third, and finally, students consider the importance of the course and their interest in the material. Courses students find personally valuable and interesting are more likely to be taken face-to-face (Jaggars, 2013).

Blended Course (Hybrid Course)

Some courses provide a mix of Internet and face-to-face instruction. **Blended courses (Hybrid courses)** take place both online and with face-to-face interaction among instructors and students. These courses, naturally, require the relatively close physical proximity of the instructor and students. Courses using the blended or hybrid approach may require students to physically attend class to take examinations or to complete certain assignments.

These courses are likely to include those indicated by students to not be suitable for a completely online format. For instance, performance-based courses such as public speaking may require students to deliver presentations in the classroom. Laboratory-based courses such as chemistry may require students to attend lab sessions in person and complete other course activities online. Language courses may require students to speak to one another physically rather than interacting through text.

Internet-Enhanced Course

Some face-to-face courses are supplemented by the Internet. **Internet-enhanced courses** are those taking place primarily face-to-face but that use the Internet to supplement course instruction and for course management. In addition to Internet-based materials referenced and used by instructors, these courses may use such course-management programs as Blackboard, Canvas, and Moodle. Features of such programs include access to the course syllabus, schedule, assignments, grade information, testing, and online contact with the instructor and other students. Internet-enhanced courses may also use social networking sites to post course information and encourage interaction among students (Tess, 2013). Even if instructors are not enhancing their courses through the Internet, many students are doing so. Eighty-six percent of students have supplemented course material with digital content they have discovered on their own (Google/TNS & Tru, 2012).

Virtual Course

In many ways, **virtual courses**, those taking place with instructor and students interacting primarily through webcams, are like traditional face-to-face courses or telecourses using video. The differences include the geographic and physical separation of students and instructors. And, in the case of the latter, a number of students tend to be in the same physical location, with others taking part through video feeds. For some virtual courses, though, the Internet is used as more than just a means of transferring audio and video. Course materials, videos, and materials located online can be integrated into instructions and lessons. Further, examinations are often taken and assignments are submitted online.

Virtual-Reality Course

This next type of course is not yet available in any significant way. However, it will likely become more widespread in the future and has some intriguing possibilities. **Virtual-reality courses** take place with instructors and students interacting through avatars. Discussing the possibilities of infinite realities, Blascovich and Bailenson (2011) conveyed some of the possibilities of virtual interactions and instructions. For instance, students can receive individualized instruction based on their preferences for instructor closeness, eye contact, and other communicative activities. They will also be able to decide the instructor's appearance and voice. The avatar instructor's communication can essentially be molded to each individual student's learning preferences. And, the avatar instructor's communication can be modeled based on best classroom communication practices, such as the most effective distance from students and eye contact with students.

Also, consider how virtual reality can be used to enhance instruction. History courses, for example, could transport students to key moments in time and enable them to interact in this virtual setting. Medical students can practice on virtual patients before dealing with real patients, and teachers can teach a classroom of virtual students during their training. Such instruction is actually being developed and incorporated to a certain extent (e.g., Banker & Corbeil, 2013; Radechi, et al., 2013). However, costs are still too high for this method of instruction to be a widespread (virtual) reality. Still, this may very well be what instruction looks like in the future.

Massive Open Online Course (MOOC)

Finally, **massive open online courses (MOOCs)** are large-scale courses offering free enrollment and open to anyone online. The three major MOOC providers are Coursera (www.coursera.org), edX (www.edx.org), and Udacity (www.udacity.com). MOOCs include a range of content taught by instructors from a variety of institutions. Lectures are available for viewing online and students can also take quizzes and examinations to test their learning of the material. There is generally no contact between students and instructors, and some courses begin with instructors discouraging students from attempting to make contact (Jacobs, 2013). Students cannot usually receive credit toward degrees for these courses. However, some universities are allowing students to earn course credit for these courses or are blending MOOC content with existing courses (Korn, 2013a; Lewin, 2013a).

Advantages and Disadvantages of Online Education

Online education has been met with both praise and apprehension. In what follows, we will examine both the advantages of online instruction along with disadvantages of and concerns about online instruction. This discussion will primarily focus on Internet-only courses and MOOCs rather than blended and other types of courses.

Advantages

Perhaps the most common and obvious advantage of online education is its convenience for students. Courses can be largely completed around a student's schedule and can be completed regardless of a student's proximity to a physical campus. Correspondingly, it has been argued that online learning provides nontraditional student populations and groups less likely to attend college greater access to instruction and academic resources (Morse, 2012). Online instruction may also solve problems involving classroom overcrowding, as student populations continue to grow (Lewin, 2013b). Especially in the case of MOOCs, students have the opportunity to learn from great instructors from around the world. As more students enroll in online classes, tracking their activity online will enable instructors to compile this data and analyze what works best pedagogically (Mayer-Schonberger & Cukier, 2013, p. 115). Finally, with more than 50% of students requiring remedial courses, it has been suggested that online courses, especially MOOCs, could ease their transition into college work with minimal or no cost (Lewin, 2013b).

Alternative View

It has been suggested that online instruction may ease students requiring remedial work into college classes. The argument rests primarily in the findings that these students are likely to drop out of school after their first semester, wasting time, money, and other resources. However, remedial students require more personalized instruction than college-ready students. They are also more likely than college-ready students to drop out of online courses. Further, achievement gaps are more pronounced between remedial and college-ready students when taking online courses (What We Know About Online Course Outcomes, 2013).

Disadvantages and Concerns

Among the disadvantages and concerns surrounding online instruction is a potential lack of access to and interaction with instructors. As mentioned above, many faculty members view online courses as lacking value and legitimacy, which is in opposition to the views of many administrators (Allen & Seaman, 2013). Based on findings that students avoid online courses when viewing the material as important and valuable (Jaggars, 2013), students may also question the learning opportunities available from online courses.

A second concern about online courses, while some instructors utilize the Internet in order to engage with online students, the majority of online instructors do not. When instructors do engage with online students interpersonally, though, it has been found to positively enhance academic performance (Jaggers & Xu, 2013).

Correspondingly, there are significant differences in student and instructor expectations for online courses (Creating an Effective Online Environment, 2013, p. 3). In terms of *responsibility*, students expect online instructors to guide and motivate them personally through various learning activities. Online instructors view students as independent learners. In terms of *instructor presence*, students expect instructors to have an active presence, while instructors see themselves in the role of content manager on the side. In terms of *course materials*, students expect a variety of content delivery, while instructors deliver text-based content. Finally, in terms of *communication*, online students expect instructors to provide quick feedback and responses at any point. Online instructors do not see themselves as being constantly on call whenever students desire information.

Another concern connected with online instruction is the high failure and withdrawal rates among students. Failure and withdrawal from online courses are significantly higher than for face-to-face courses. Students in need of remedial work upon entering college are especially likely to struggle with successful completion of an online course. Further, online courses have also been found to be negatively correlated with degree attainment (What We Know About Online Course Outcomes, 2013).

Finally, there are concerns among faculty that their jobs will be outsourced and that instruction will take place by a few instructors around the world or by instructors who are not qualified (Belkin & Korn, 2013; Korn, 2013b). Accompanying these concerns, are worries that budgets may be slashed at existing physical colleges and universities in favor of online universities and instruction (Jacobs, 2013).

Conclusion and Predictions

The Internet has become a complete and coherent societal system, directly or indirectly touching nearly every aspect of society and everyday life activities. In some ways, following the path of previous technologies, the Internet has had a greater and more complete societal impact than any previous technology due to its characteristics and capabilities available through its use. Health care, banking, employment, and education are just some of the areas transformed through the Internet. And the Internet's influence will likely increase as more people gain access to high-speed connections, become more confident in conducting activities through the Internet, and become more confident in their own ability to conduct activities through the Internet.

As we look to the future, the *Internet growth rate of some areas will slow*. This prediction is not because of an overall lack of use. Rather, since the Internet is already being used by a large percentage of people, dramatic growth in a short period of time will not be possible. There will no longer be one-year increases of 100%, 200%, or more users in some areas, for example.

On the other hand, the *Internet growth rate of some areas will increase.* Such areas as telecommuting through the Internet, which are not as prevalent as other areas, will see dramatic increases in the future.

Further, the *Internet presence and demand of other areas will emerge.* As with virtual doctor visits, the emergence of such areas as law advice and practice and gambling on the Internet face regulatory and legal obstacles. At the same time, these and similar areas show a growing presence and demand online, which is influencing regulators and lawmakers (e.g., Pratt, 2013).

The Internet's presence will reach beyond mobile devices and computers. Specifically, *there will be a rise in smart homes and telematics.* **Smart homes** are those in which climate control, lighting, and other features in the home are connected to and can be controlled by the Internet. Present smart-home technology includes climate control and lighting mentioned above as well as security systems, all of which are becoming more advanced and sophisticated. In the near future, most appliances will be connected to the Internet, so that a hot-water heater can inform homeowners of leaks, refrigerators will provide grocery lists and coupons, washing machines and driers will run at the most efficient times, and all appliances will be able to monitor their performance (Wingfield, 2013). **Telematics** involves the connection of automobiles to the Internet. Many vehicles already come equipped with diagnostic tools, security features, and navigation that are connected to the Internet. Traffic data, parking-space locators, driverless cars will be arriving in the near and perhaps not-so-distant future (Bennett & White, 2013; Markoff, 2013). Both smart homes and telematics will assist in making the Internet a more accepted, comfortable, and, eventually, domesticated and unremarkable part of life.

Finally, *virtual reality will become a common characteristic of many areas in life and society.* Still in early development and availability, avatars and virtual reality have been shown to be useful in health care, including exercise, physical therapy, and psychological wellness. Travel could, of course, take place virtually. Potential home buyers could take a true virtual tour of homes, providing a more accurate evaluation than photos and videos. Customers missing physical bank branches can conduct transaction in virtual banks. And, as mentioned within the chapter, virtual reality will completely transform education and work (Blascovich & Bailenson, 2011; Reeves & Read, 2009).

Key Terms

1000 Genomes Project

Bitcoin

Blended Courses (Hybrid Courses)

Google Flu Trends

Internet-Enhanced Courses

Internet-Only Courses

Massive Open Online Courses (Moocs)

Online-Only Banks

Online Support Groups

Smart Homes

Telecommute

Telematics

Virtual Courses

Virtual-Reality Courses

Virtual Doctor Visits

Chapter Exercises

Current Issue: In recent years, the most visible telecommuting policy made by a company was perhaps that of Yahoo. Somewhat paradoxically given the company's focus and contrary to the policies of the majority of technology companies, Yahoo announced that employees working from home would have to return to their offices and employees would no longer have the option to work from home (Swisher, 2013). Have students search for and report on the most current media reports regarding telecommuting. Have other companies followed Yahoo's lead? Or, have companies continued to expand telecommuting options for employees? Is telecommuting portrayed as more positive or more negative in these pieces?

Debate It: Split the class into two groups. Select one of the resolutions below. Have one group support the resolution and have the other group oppose the resolution in a debate.

Resolved: Online instruction is superior to face-to-face instruction.

Resolved: Online instruction is inferior to face-to-face instruction.

Ethical Consideration: Even though the DNA samples in the 1000 Genomes Project are anonymous, one study illustrated that a participant's identity could be revealed using the sequence, the participant's age and region (both available in the study), a genealogy website, and general Internet searches. This study discovered not only the identity of participants but also relatives of the participants who did not take part in the project (Gymrek et al., 2013). There are similar privacy concerns surrounding the use of data from social networking sites and Internet searches. Ask students to consider whether it is ethical or unethical for companies and researchers to use such data. Does the fact that this information is being used for medical tracking and research make a difference, in their view?

Everyday Impact: Have students consider the types of Internet courses examined within the chapter. Which ones would they prefer to take and why might that be the case? If they have taken Internet courses, have them describe their experiences. Aside from content, what did they most enjoy about these courses and what did they most dislike about these courses?

Future Focus: There are growing attempts to promote digital payment through mobile devices at physical stores, restaurants, and other businesses. Google Wallet (www.Google.com/wallet) and PayPal's Cash for Registers (www.Paypal.com/webapps/mpp/merchant-cash-for-registers) are among the few notable examples of such services. Another type of payment system bypasses financial institutions all together. Bitcoin (www.Bitcoin.org) is a virtual currency system that can be exchanged without any transaction fees and can be used at a growing number of online merchants as well as among Internet users themselves (Needleman & Ante, 2013). Additional digital currencies have been created, including Litecoin (www.Litecoin.org), Freicoin (www.Freco.in), and Ripple (www.Ripple.com). A study conducted by PayPal found that an average of 83% of consumers in Australia, Canada, Germany, the United Kingdom, and the United States would want to use such a service (Business Wire, 2013). Likewise, a survey of tech experts conducted by the Pew Research Center found the majority agreed that by 2020 "most people will have embraced and fully adopted the use of smart-device swiping for purchases they make, nearly eliminating the need for cash or credit cards" (Smith & Rainie, 2012, p. 3). Ask students if they have ever used such payment options or would be willing to use such payment options in the future. Depending on their answers, what do they see as the benefits of such systems? What are the drawbacks to such systems, and what might cause them to bypass their use? Do they believe similar concerns arose when credit and debit cards were first introduced? Do they believe there will come a time when physical money will no longer be in use?

12

The Internet and United States Politics

This chapter provides a review and analysis of the Internet's use in United States politics. Specifically, we examine the rise and development of Internet politics from its first use in a presidential campaign in 1992 to its use in the 2008 presidential campaign. We then examine five key functions of the Internet in political campaigns.

Before proceeding, however, we need to discuss our approach to this chapter. When writing the history of anything, certain elements and events will be included while others will be excluded. In this case, our review and analysis of Internet politics focuses on presidential campaigns.

Of course, campaigns for congressional, state, and now even local positions utilize the Internet. Notable early events include Jesse Ventura winning the Minnesota gubernatorial race as a Reform Party candidate in 1998, beating major party candidates in part by relying heavily on the Internet. And a number of congressional candidates were developing campaign websites by that 1998 midterm election (see D'Alessio, 2000).

However, the use of the Internet in campaigns has developed primarily through presidential campaigns. Beyond the greater efforts extended for such campaigns in general, this development is due in large part to funding issues. In the early days of the Internet, the return on one's investment by campaigning online was not significant. The money of campaigns less well funded than those of presidential candidates could be better spent elsewhere. More recently, heavily funded presidential campaigns have been able to hire well-equipped teams to develop their Internet campaigns. Ultimately, though, campaigns for other political positions have benefited from, learned from, and emulated the efforts and successes of presidential Internet campaigning by presidential candidates.

Also, our overview of the Internet's rise and development ends at the 2008 presidential campaign, even though this book is being published after the 2012 re-election of Barack Obama. Because that election has occurred too recently, numbers, events, issues, and activities related to the use of the Internet by voters and campaigns are still being uncovered. Rather than provide an incomplete analysis of that campaign, we have purposefully not included it within this chapter. We do, however, refer to the 2012 campaign at times when discussing the functions of the Internet in political campaigns.

Moreover, at this point in Internet politics, ending our overview at 2008 is somewhat appropriate. That campaign was a banner year in the development of Internet campaigning, and much of what took place in 2012 was initiated in 2008. This statement in no way implies that what happened in 2012 is unimportant, for initial endeavors introduced in 2008 were further developed in 2012. However, with the exception of additional tracking of voters by candidates in 2012, few new developments took place.

With that, we begin our review and analysis of the Internet's use in U.S. politics, starting with 1992, ending in 2008, and leading to a discussion of the five primary functions of the Internet in political campaigns.

Questions Previewing This Chapter

1. *How did Internet politics develop in the United States?*
2. *In political campaigns, what is the informing function of the Internet?*
3. *In political campaigns, what is the engaging function of the Internet?*
4. *In political campaigns, what is the connecting function of the Internet?*
5. *In political campaigns, what is the fund-raising function of the Internet?*
6. *In political campaigns, what is the tracking function of the Internet?*

The Rise and Development of Internet Politics in the United States

When considering politics and the Internet, political discussion and debate occurred on many early bulletin boards, and political discourse subsequently took place through each developing communication platform of the early Internet. However, the Internet was not initially viewed as a viable means of campaigning. If for no other reason than the fact that few people had access to the Internet, which meant that campaign money and resources of any campaign could be best spent elsewhere.

1992 Presidential Campaigns

The first use of the Internet in a presidential campaign took place during the 1992 election by both George H. W. Bush and Bill Clinton. Bush sent some 200 speeches and position papers to a number of bulletin boards (Davis, Baumgartner, Francia, & Morris, 2009), while Clinton distributed speeches, position papers, and other materials to newsgroups as well as creating a listserv (Bimber & Davis, 2003; Davis et al., 2009). Neither candidate truly used the Internet as a genuine campaign tool, however.

1996 Presidential Campaigns

The 1996 presidential campaigns witnessed an increased awareness and use of the Internet, but its actual incorporation and purpose in campaigns was still limited. One candidate, Lamar Alexander, did announce his candidacy on American Online, accomplishing what his aides rightly called "the first cyber-announcement in history" (Berke, 1995). In addition to Alexander, Republican candidates Pat Buchanan, Steve Forbes, Phil Gramm, Richard Lugar, and eventual nominee, Bob Dole, all had websites devoted to their campaign. In another first, Dole announced the address of his website during his closing statement in the first presidential debate, leading to a reported two million visitors during the next 24 hours (Davis et al., 2009). The incumbent, Bill Clinton, continued his online presence as well, but like the websites of other candidates, his offered very little value to users.

The 1996 election also presented the first use of the Internet during a major party's convention. The Republican National Convention was the first to feature chats with delegates, elected officials, and candidates, audio of the proceedings, press releases, full text of speeches, and the party platform document online. The convention website received 500,000 hits during the first day of the convention (Stuebe, 1996).

It has been suggested that the first political Internet campaigns occurred during the 1996 campaign (Graff, 2007; Selnow, 1997). However, candidate websites were little more than billboards or brochures, with information conveyed essentially being universal and unchanging. There was no indication of massive use or any significant effect on outcomes of the elections (Pollard, Chesebro, & Studinski, 2009). Indeed, the number of adults going online during the campaign period of 1995 and 1996 only ranged from 10% to around 20% (Zickhur & Smith, 2012), with just 4% of all adults visiting political sites during 1996 (Media Studies Center, 1996, cited in D'Alessio, 2000).

2000 Presidential Campaigns

While presidential candidates had websites in 1996, the first significant use of websites for campaigning occurred during the 2000 presidential campaign (Pollard, Chesebro, & Studinski, 2009). Presidential contenders Gary Bauer, Bill Bradley, Pat Buchanan, Steve Forbes, Orin Hatch, John Kasich, Alan Keyes, John McCain, and Dan Quayle all had websites, and even potential candidate Elizabeth Dole had a website dedicated to her exploratory committee. Of course, candidate Al Gore and eventual winner, George W. Bush, both had websites as well.

What contrasted the 2000 campaigns with the 1996 campaigns is that these websites were not created and then forgotten. Rather than offering essentially unaltered billboards or brochures, websites for the 2000 campaign featured updated information. Although negligible, especially by existing standards, they also offered various degrees of interactivity. Campaigns "had moved beyond thinking of the Web as an electronic brochure to viewing it as an electronic headquarters" (Foot & Schneider, 2006, p. 10).

The key explanation for the significant use of websites in 2000 is the profound growth in the number of Internet users. By this point in time, nearly half of all adults had access to the Internet (Lenhart, 2000), and an increasing number and percentage of voters were going online for election news. In fact, the percentage of the general public who went online for election news was over four times greater in 2000 than in 1996 (Kohut, 2000). While a greater percentage of these people sought information on news websites rather than candidate websites and, ironically, a smaller percentage visited candidate websites in 2000 than in 1996 (Kohut, 2000), the actual numbers of people visiting these sites had greatly increased. Accordingly, it was in the candidates' best interests to have a significant online presence.

Alternative View

Anstead and Chadwick (2009) maintain that pointing to "technological diffusion" as a reason for an increase in the use of the Internet as a campaign tool only tells part of the story. Further, they note that other nations with high Internet usage such as the United Kingdom show the Internet as having "only a marginal influence on elections." They suggest that unique political institutions of nations influence the impact of the Internet and other technologies. Specifically, "technologies can reshape institutions, but institutions will mediate eventual outcomes" (p. 56). So, while we can point to increases in Internet usage as a reason for the increased use of the Internet by campaigns, we must also recognize that campaigns would only use the Internet if it was successful. A lot of people drink milk, but campaigns have yet to place advertisements on milk cartons. Further, preexisting political structures and cultures are also responsible for the Internet success, not just the fact, however influential, that more people began using the Internet.

Furthermore, it was becoming obvious that certain populations of voters could be targeted with greater results using the Internet. Disparities among certain populations of Internet users were particularly profound in 2000, especially in terms of age and education (Kohut, 2000; Lenhart, 2000).

Coinciding with the developing ability to use the Internet to target very specific populations was the changing nature of campaign advertising. As Binder and Davis (2000) noted, both Bush and Gore

> emphasized targeted advertising crafted to convey specific messages to particular media markets. Rather than running nationwide advertisements on the [television] networks ... Bush and Gore addressed themselves to local television audiences, making appeals and counter appeals that often revolved around quite narrow issues. (p. 30)

It was becoming quite evident that a focused, specifically targeted message would be more effective than a general, widely dispersed message. Using local media markets assisted with this strategy; however, that method would still pale in comparison to what could be accomplished through the Internet.

2004 Presidential Campaigns

When the 2004 presidential campaigns commenced, the position of the Internet in American life had taken on increased importance. By this point, over 60% of adults used the Internet (Zickuhr & Smith, 2012), and even more significant than actual numbers of users, the Internet had become an important source of information for many adults (Rainie, Cornfield, & Horrigan, 2005).

Once again, candidates from both major parties had websites, including the incumbent George W. Bush. On the Democrat side, Carol Mosely Braun, Wesley Clark, Howard Dean, John Edwards, Dick Gephardt, Bob Graham, Dennis Kucinich, Joe Lieberman, Al Sharpton, and eventual nominee John Kerry all had websites with varying degrees of sophistication and utility but with vastly more development than websites in support of previous campaigns. Nevertheless, in spite of increasing recognition of benefits derived from interactivity, especially among younger voters, the interactive capabilities of these sites were still relatively limited (Tedesco, 2006).

Beyond the greater development of some websites and the increased use of the Internet as a source of political information, three items involving these campaigns are especially worthy of note. First, while many users cited convenience as the main reason for taking to the Internet to find political information, over half claimed the Internet provided information that was unavailable through such established media outlets as television and newspapers. Moreover, some of these users specifically noted that they were able to find material sources and materials online reflecting their own interests and values (Rainie, Cornfield, & Horrigan, 2005). These findings help explain why increasingly more people were using the Internet as opposed to other media outlets.

Second, of the eventual nominees, Kerry supporters were more active online than Bush supporters, and they felt as though they got more out of engaging in politics online than Bush supporters (Rainie, Cornfield, & Horrigan, 2005). These findings, of course, could be explained by Kerry supporters in general being more likely to use the Internet, and to use it to greater degrees of satisfaction than Bush supporters in general. However, this discovery is an indication that Democrats had caught up to or surpassed Republicans in terms of using the Internet as a campaign tool.

The third, and arguably the most significant impact of the Internet during these campaigns, was the use of the Internet for fundraising. Four years earlier, John McCain's campaign raised $1 million online during the 24 hours following his surprise victory over George W. Bush in the New Hampshire primary, and amassed contribution totals of over $6 million online. These feats seemed remarkable and perhaps unrepeatable at that time. During the 2004 primaries, however, Howard Dean's campaign raised around $20 million online, a significant portion of its total donations. John Kerry's campaign eventually outdid all others by raising around $82 million online, an amount that far exceeded Al Gore's total number of contributions four years earlier and dominated the George W. Bush campaign's online donations of $14 million (Justice, 2004).

Especially noteworthy concerning these online contributions are their amounts and their sources. The amount of money received from people donating less than $200 increased by four times the amount received in 2000 (Justice, 2004). Further, a report from the Institute for Politics, Democracy, and the Internet (Graf, Reeher, Malbin, & Panagopoulos, 2006) found that the majority of contributions (more than 80%) from people aged 18–34 occurred online. While this age group was less likely overall to contribute money to campaigns, a precedent for online donations had been set as members of this group aged and would find themselves in an age demographic more likely to contribute. This report also found that Democrats were far ahead of Republicans when it came to receiving online donations. Democrats received more than twice the number of online donations, in excess of $500 (64% to 31%), and they had an even greater advantage in contributions of $100 or less (54% to 19%).

As with the above discussion concerning the online activity of Kerry supporters, fund-raising efforts make it clear that Democrats had the online advantage over Republicans by the 2004 campaigns. It has been noted that Bush's website was dedicated more to organizing voters rather than fund-raising (Justice, 2004) and that Kerry's fund-raising overall had to be more aggressive to compete against an incumbent president (Graf et al., 2006). However, Kerry's online contributions at nearly six times more than those of Bush's, in addition to the overall online fund-raising advantage of Democrats, exhibit that this party had embraced the Internet to a much greater extent than Republicans. This advantage was to become even more pronounced in four years.

2008 Presidential Campaigns

The 2008 campaigns have been deemed unprecedented for a number of different reasons. Among these reasons are the number of primary candidates on both the Republican and the Democratic tickets and the phenomenal amount of money raised during the primaries. Aptly describing the early stages of the 2008 presidential campaigns, Trent (2009) noted that there were not only eighteen candidates who, in one way or another formally announced their candidacies and campaigned at least through the Iowa Caucus, there were another eleven who, at one point or another, were serious enough about their presidential aspirations to formally announce the suspension of their campaigns. (p. 1)

Trent (2009) further noted that this group of candidates "raised and spent more money than had ever been raised and spent in the early stages of any other presidential election cycle—more than twice the amount as during the early stages of the 2004 presidential elections" (pp. 1–2).

The campaign committee of one candidate in particular, Barack Obama, a first-term senator perceived as having extremely limited experience, highly partisan positions, and seemingly little chance of beating Hillary Clinton in the Democrat primaries let alone winning the general election, used the Internet as a vital campaign tool. Following Obama's victory in the general election, *Huffington Post* editor-in-chief, Arianna Huffington, was quoted as saying "were it not for the Internet, Barack Obama would not be President. Were it not for the Internet, Barack Obama would not have been the [Democrat] nominee" (Miller, 2008).

McKinney and Banwart (2011) rightly cautioned against maintaining that Obama "became president of the United States simply because he was more skillful than his opponents in adopting the Internet and digital technologies as a campaign communication tool" (p. 3). There are simply too many factors and variables involved in a political campaign. Nevertheless, in support of Huffington's bold statement, it has been suggested that Obama won the presidency through his committee's use of the Internet, specifically, "by converting everyday people into engaged and empowered volunteers, donors and advocates through social networks, e-mail advocacy, text messaging and online video" (Lutz, 2009, p. 2).

As we look closer at the 2008 presidential campaigns, we will examine two particular areas. First, we will examine how the Internet was being used by potential voters. Then, we will examine how the Obama committee specifically used the Internet during the 2008 campaign.

Voter Internet Use

Of course, this effective use of the Internet by the Obama campaign was due in no small measure to an increase in the use of the Internet itself and in the central role it then played in the lives of voters. By this point in time, over 70% of adults used the Internet (Zickuhr & Smith, 2012), and for many people, it had become a pervasive part their lives. In terms of its political role, 74% of Internet users and 55% of the entire voting age population went online when seeking political information or to receive or share political information with others (Smith, 2009).

Engaging Young Voters

Just as it is not legitimately possible to say the Internet or any other strategy won the election for a candidate, it is not legitimately possible to say that one group won the election for a candidate. It can be argued with some degree of legitimacy, however, that certain groups can have a significant influence on the results of an election. In the 2008 presidential election, Obama received an unprecedented percentage of votes among young voters (*The New York Times* Exit Polls, 2008), who voted in larger numbers than ever before (United States Census Bureau, 2012), who were the only age group to show a significant increase in voter turnout (United States Census Bureau, 2012), and who were the most politically active users of the Internet (Smith, 2009). These numbers and trends are displayed visually in Tables 12.1, 12.2, 12.3, and 12.4.

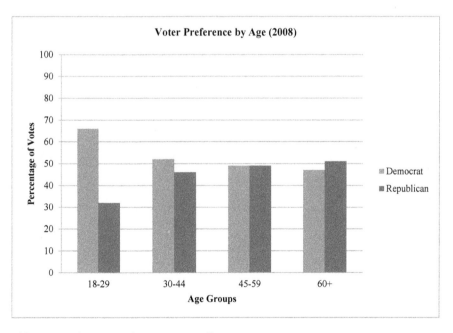

Table 12.1. *The New York Times* Exit Polls, 2008.

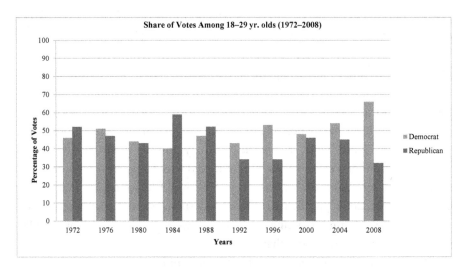

Table 12.2. *The New York Times* Exit Polls, 2008.

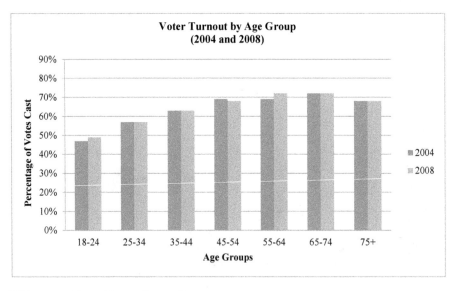

Table 12.3. United States Census Bureau, 2012.

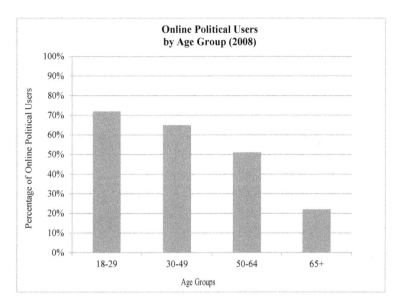

Table 12.4. Smith, 2009.

Connecting, Fund-Raising, and Advertising

The Obama campaign used the Internet to connect voters through online social networking, to raise unprecedented amounts of money, and to advertise. In doing so, the committee did not necessarily do anything unique in terms of their use of the Internet. Rather, they strategically improved existing methods and techniques. Carr (2008) noted that

> like a lot of Web innovators, the Obama campaign did not invent anything completely new. Instead, by bolting together social networking applications under the banner of a movement, they created an unforeseen force … that helped them topple the Clinton machine and then John McCain and the Republicans. (para. 6)

Although we will examine the Obama campaign's use of online social networking, fund-raising, and advertising, efforts individually, they should not be understood as having existed in isolation or without mutual impact. Each of these areas assisted and bolstered one another in effective ways.

Connecting voters. The Obama campaign connected with voters through a dominant online presence in both the primary and the general elections, far outpacing Democratic challenger Hillary Clinton and then Republican nominee McCain. These connections were developed through the candidate's official website

(BarackObama.com) and through activity on a variety of social networking sites. In total, Obama had five million supporters on at least 15 social networks (Vargas, 2008b). Table 12.5 illustrates Obama's online dominance during the general election.

Table 12.5. Obama's Online Dominance in the 2008 General Election.

Website	BarackObama.com
	72% of total traffic
	JohnMcCain.com
	28% of total traffic
Facebook	Obama
	2 million friends
	McCain
	600,000 friends
Twitter	Obama
	112,000 followers
	McCain
	4,600 followers
YouTube	Obama
	18 million channel visits
	McCain
	2 million channel visits
	Obama
	115,000 subscribers
	McCain
	28,000 subscribers
	Obama
	1,800 videos
	McCain
	330 videos

Note. Sources: Fraser & Dutta, 2008; Vargas, 2008a.

These online activities resulted in two very important achievements: connection with the campaign/candidate and connection with other supporters. First, voters gained a sense of personal connection with the campaign/candidate, which possibly increased their likelihood to volunteer, donate, encourage others, and, of course, cast their vote for the candidate. Engendering a sense of personal connection with voters is by no means a new strategy. However, given the number of people using the Internet sites at this point in time, this was the first campaign able to fully achieve these connections online. Furthermore, much campaign information came through the social networking sites. These sources for information are important, because they were the same sites used to gather information about and to interact with friends, which likely increased feelings of personal connection with the campaign. Further, the frequency at which people used social networking sites made it more likely that the information would be encountered.

Alternative View

As is also the case with past campaign efforts, these personal connections were highly artificial. Obama himself was far-removed from the inner-workings of the digital arm of the campaign, updates and interactions were highly scripted, and campaign workers were exceedingly disciplined and unable to speak about the campaign to outsiders without explicit and official approval (see "Barack Obama's Get-out-the-vote Machine, Is Bigger, Faster and Smarter," 2008).

A second achievement through these online efforts was the connection of voters to each other. These connections were made among those already sharing a relationship and among those whose only connection was based on the campaign. As with other candidate's sites during this campaign and during previous campaigns, BarackObama.com provided visitors with campaign information, encouraged visitors to provide personal information, and, of course, donate to the campaign. However, more than any other previous campaign site, BarackObama.com encouraged the connections among voters. Throughout the campaign, two million profiles were created by voters, 400,000 blog posts were written, 200,000 offline events were planned, and 35,000 volunteer groups were created (Vargas, 2008b).

Fund-raising. Candidate Obama originally pledged to accept public funding, which is accompanied by specific spending limits. Candidates were given $84 million in public money but prohibited from spending more than that amount between their party's convention and the actual election (Rutenberg, 2008a). Following his successful fund-raising during the primaries, however, he broke this pledge, becoming the first major party candidate to do so and angering both Republicans and fellow Democrats (Luo & Zeleny, 2008).

While perceived and portrayed as ethically questionable, this reversal was strategically brilliant, given the vast amounts of money eventually raised, the vast amounts of advertising that was then purchased, and the other spending that could take place. In total, the Obama campaign raised $745 million, and spent nearly $730 million (Opensecrets.org, 2008a). In doing so, previous records for fund-raising and advertising were easily broken. Further, both figures are over twice the amount of those of McCain, who was also bound in the final weeks of the campaign by spending limits imposed through his decision not to forgo the acceptance of public funds. In total, the McCain campaign raised just over $368 million and spent just over $333 million (Opensecrets.org, 2008b).

While the total amount of money raised by the Obama campaign was groundbreaking, the amount of money raised through the Internet was even more so. Five hundred thousand dollars, nearly 70% of their total donations, was raised online (Vargas, 2008b).

Advertising. This record amount of funding, which came without the imposition of spending limits, enabled the Obama campaign to engage in a historic advertising campaign. Through Internet-based funding, the Obama campaign was able to increase the use of traditional media. From January 1, 2007 to November 4, 2008, the Obama campaign spent $310,144,381 for television advertisements airing 570,963 times. These numbers can be compared to those of the McCain campaign, which spent $134,792,298 for advertisements airing 274,737 times (CNN, 2008). Overall, the Obama campaign aired over twice the number of advertisements than the McCain campaign. This advantage was even more pronounced in key battleground areas, running seven or eight times the number of advertisements as McCain (Gorman, 2008). Further, as indicated by the amount of money spent, the Obama campaign was also able to air these advertisements at coveted and much more expensive times, ensuring that more people and specific groups of voters would be watching. The Obama campaign also ran a half-hour advertisement/infomercial during primetime on four broadcast networks and the cable channels MSNBC, BET, and TV One shortly before Election Day (Rutenberg, 2008b).

The Obama campaign also spent a great deal of money purchasing Internet advertisements. For instance, in February of 2008, it spent nearly $3 million on Internet advertisements. This amount was 10 times more than Clinton, who was still in the race at that point, and 20 times more than McCain. Throughout the campaign, advertisements were included on political websites, newspapers, and attached to politically based search terms on Google and other search engines (Mosk, 2008). In October 2008, advertisements were included on Facebook homepages, and in a first, a total of 18 Xbox Live video games. These advertisements were included as banners and billboards in a variety of games, primarily targeting voters in battleground areas and those states allowing early voting to take place (Gorman, 2008).

Functions of Internet Campaigning

The Internet is now fully within its second decade as a vital campaign tool. Accordingly, it has become possible to isolate specific functions of Internet campaigning. Foot and Schneider (2006) previously noted the occurrence of common Internet campaigning practices. Specifically, "through the production of Web objects, campaigns' strategies of informing prospective voters, involving supporters, connecting Internet users with other political actors, and mobilizing advocates are manifested on the Web in different ways and to varying degrees" (p. 24). Similarly, Davis, Baumgartner, Francia, and Morris (2009) noted the use of Internet campaigning in communicating with voters, mobilizing voters, and raising funds.

Utilizing some of the terms offered by these previous authors, with some alterations in how they are characterized and understood based on ensuing campaigns and our own perspectives, we will examine primary functions of Internet campaigning. In doing so, we will also explore an additional function that has grown in importance through recent campaigns. In what follows, then, we will examine the following functions of Internet campaigning: (a) informing, (b) engaging, (c) connecting, (d) fund-raising, and (e) tracking.

Informing

While informing voters and controlling information is nothing new, the Internet has changed how these endeavors can be accomplished and has profoundly enhanced the effectiveness of such efforts.

Cost Effective

One benefit of using the Internet to inform voters is its cost effectiveness. Since the 1960s, television advertisements have been a primary means of informing voters. The comparative costs of making campaign information available online or uploading the advertisements themselves is quite reasonable. In some cases, such as uploading videos on campaign sites, social networking sites, and video sites, the expense is nothing more than the cost of campaign staff. Using numbers quoted by Joe Trippi, campaign manager for Howard Dean in 2004, the Obama campaign's YouTube material was watched for over 15 million hours in 2008. Purchasing this many hours on broadcast television would have cost nearly $50 million (Miller, 2008).

Alternative View

Following the success of Jesse Ventura in the 1998 Minnesota gubernatorial race against better-funded major-party candidates, it was tempting to think that the Internet could be a financial equalizer. Candidates with less financial backing could get their message out and reach voters like never before, countering the efforts of candidates who could pay more for advertising. However, the Internet has not proved a great equalizer, and the Jesse Ventura case has not been replicated. The reason this equalization did not come to fruition is that all candidates began using the Internet, including the better-funded candidates who could actually afford to do more with the Internet than those with less financial backing.

Highly Targeted

Another advantage of the Internet is an increased accuracy when targeting voters. The Internet has enabled campaigns to target voters with unmatched accuracy and efficiency. Campaign websites that match content with particular visitors are increasingly being developed. For instance, through determining a user's internet protocol address, websites can present the most relevant information for voters in that particular physical location. Returning and registered visitors to websites will be presented with campaign materials molded to their specific background and activities during previous visits.

Unlimited and Constantly Available

Another advantage of using the Internet to inform voters is the vast amount of information that could be provided and the constant availability of this information. The Internet also enables campaigns to provide a virtually unlimited amount of materials for voters to access. Moreover, these materials are always available when voters want to use them and are generally easy to access. Advertisements using other media are fleeting, brochures and other printed materials are misplaced, but information on the Internet is always accessible to voters.

Alternative View

Given the unlimited and constantly available nature of political information on the Internet and the fact that voters are coming to the information, it might be tempting to envision a scenario in which all voters are flocking to the Internet and becoming well-informed voters. Unfortunately, that does not seem to be the case. Although, as illustrated within the history portion of this chapter, increasing numbers of voters are using the Internet to gain political information, low-informed voters are still rampant. Still, there are some glimmers of hope within the uninformed darkness. Hardy, Jamieson, and Winneg (2009), for example, found that accessing information online was connected with greater ability to distinguish facts from deceptions. Brundidge and Rice (2009) contended that accessing information online can increase political knowledge for some people, but at the same time, these individuals are most likely to already be highly politically informed.

Updated

The Internet enables campaigns to update information quickly. This rapid updating of information, for instance, might enable a campaign to modify materials that have been proven wrong by opposing campaigns or have simply found to be erroneous. Additionally, the campaign can react to information presented by the opposition or other actions by the opposition swiftly and decisively. Finally, the ability to update information quickly enables campaigns to address developing matters of national or even regional interest quickly.

Sharable

Another feature of information provided through the Internet is that it is sharable. Information can be forwarded by users to potential voters in their social networks. This action, of course, benefits campaigns since voters are not only coming to the information but also spreading that information to others. As we shall discuss below, information coming from trusted people in one's social network may be more likely to be consumed and may even be perceived as more important and credible.

Engaging

Engaging voters involves getting them involved with the campaign and getting to vote for the candidate. As with advancing the ability to inform voters, the Internet has enhanced a campaign's ability to engage voters and has provided voters with greater opportunities for engagement.

Notification

At the most basic level, the Internet allows campaigns with a superior means to notify supporters of campaign events and of volunteer opportunities. As with informing voters in general, information pertaining to such events and opportunities is always available and is easily accessible for voters searching for such information. Campaigns can also reach voters through a number of different measures such as e-mail, social network updates, and notifications on mobile devices.

Convenient

The Internet also makes campaign engagement very convenient for supporters. A supporter does not have to remember campaign events, for instance, because helpful reminders will be sent. Supporters do not have to concern themselves with finding campaign locations, because links to turn-by-turn directions can accompany the notifications. If supporters wish to discuss the campaign with others, they need not worry about remembering the positions of the candidate, since such material is readily available for them to access.

Variety

Such time-honored campaign support as filling seats at campaign rallies and stuffing envelopes for campaign mailings is still necessary and evident to varying degrees. However, the Internet provides a greater variety of endeavors for supporters of a campaign. Supporters can assist the campaign in reaching potential voters using tools and guidance provided by the campaign or through their personal efforts. Supporters may also contribute to campaign social networking sites, create blogs, and create an assortment of digital content in support of a candidate. Some supporters endorse candidates on message boards and comment sections while also combating opposition on these sites. Further, campaigns increasingly rely on supporters to provide information on rumors and negative attacks against the candidates.

Get-out-the-vote

Perhaps the most important aspect of a campaign is encouraging potential voters to actually cast their votes for a campaign. Increasingly, campaigns are encouraging voters in early-voting states to cast their votes prior to Election Day. Promoting early voting, naturally, helps ensure that these voters cast their ballot. The Internet also assists campaigns when determining how to manage the final push on the day itself. Although there have been some initial software problems with both Democrat and Republican efforts to track voters, both parties are tracking voters casting ballots efficiently in real time (Jacobs, 2012).

Connecting

Campaigns also endeavor to connect voters both with the candidate and with one another. Although connecting voters through the Internet had been attempted by previous campaigns, its first truly successful implementation came via the 2008 Obama campaign. And, in many ways, this approach served as the foundation for that campaign's Internet strategy and as fundamental to the election victory.

Connecting a Candidate and a Campaign

Connecting supporters with a candidate and a campaign engenders a sense of belonging and a feeling that personal relationships have been established. Once established, supporters will be more likely to work on behalf of a candidate, donate money to the campaign, and cast their vote for the candidate.

Candidates have long attempted to establish feelings of personal connection with voters. Prior to the Internet, however, contact with potential voters was more fleeting and expensive. The virtually unlimited and cost-efficient nature of the Internet enables campaigns to focus more on humanizing a candidate than they may have done in the past. Campaigns post numerous pictures of a candidate in various scenes that may connect that candidate to potential voters, for example. Personal statements about a candidate's family, life experiences, beliefs, and values are also provided to an extent and with an effectiveness that was not fully possible using other methods.

Campaigns have also taken advantage of characteristics of social networking sites used and developed by campaigns. For instance, campaigns pay special attention to the favorites lists included on profiles on some social networking sites. In the 2008 presidential campaign, there was discussion of favorites listed on the Facebook pages of Obama and McCain. Observers noted that Obama's list of favorite songs, books, and other media products seemed to better connect him with potential voters than those listed by McCain (Fraser & Dutta, 2008).

Campaigns can also personalize material presented about a candidate based on specific visitors. Through microtargeting, campaigns can provide information about candidates which match what a specific visitor to a site views as important. For instance, if a visitor views family issues as particularly important, a campaign can highlight information and materials about a candidate's family life and how it impacts policy decisions.

Campaigns also attempt to connect voters with the campaign itself. Connecting voters with the campaign serves to further engage supporters through volunteer efforts and fund-raising as well as ensuring that they actively urge others to support a candidate and cast their vote accordingly. Connecting voters to a campaign can reinforce connections with a candidate. This connection can also be used when voters are less than captivated with a candidate but nevertheless feel strongly about issues associated with a campaign and the efforts of a campaign.

These efforts can be reinforced further through cultivating online relational connections with the campaign. It is now possible for a campaign to maintain personal and frequent contact with supporters, thanking them for their efforts, assuring them how much they mean to the campaign, and urging them to continue working. And, the convenience of supporter engagement mentioned above makes it even more likely that their efforts will continue, further connecting them with the campaign. Campaigns also strive to become part of the everyday digital lives of supporters. Beyond constant updates and personalized correspondence, campaigns offer supporters wallpaper, screen savers, banners, and alerts to further ingrain connections with voters. More than supporting a candidate or a campaign, voters are supporting something in which they are personally invested.

Connecting Voters

Furthering these personal investment and relational connections, campaigns also encourage connections among voters themselves. These connections are cultivated through social networking components of campaign websites, existing social networking sites, and by encouraging personal campaigning and fund-raising on the part of supporters. Indeed,

> sixty-six percent of social media users have employed the platforms to post their thoughts about civic and political issues, react to others' postings, press friends to act on issues and vote, follow candidates, "like" and link to others' content, and belong to groups formed on social networking sites. (Rainie et al., 2012, p. 1)

Twenty-two percent of registered voters let others know how they voted by posting the information on a social networking site (Rainie, 2012).

These connections among voters benefited the campaign in a few different ways. First, connections with other voters likely increased feelings of association with the candidate and the campaign, encouraged activities in support of the campaign, enabled the campaign to maintain a continuous presence through the activities of others, and compelled continuous support and prompting to vote. Second, messages and appeals for donations may be more believable and persuasive coming from someone with whom a close connection is shared rather than from the campaign itself. Third, the creation of profiles and updates through existing social networking sites or social networking established through campaign websites provides unmatched access to the personal information and behaviors of voters.

Fund-raising

Fund-raising has always been a primary focus of campaigns regardless of the media or technology being utilized. So, while the importance of fund-raising is not unique to Internet campaigning, it is nearly impossible to overstate the Internet's significance in fund-raising endeavors. Fund-raising is what attracted politicians to Internet campaigning, and politicians have taken full advantage of raising funds through the Internet.

Fund-raising efforts by candidates, especially those taken advantage of by the Obama campaign in 2008, were outlined when discussing the rise and development of Internet politics. Still, there are three online fund-raising trends to specifically highlight within this section.

Unprecedented Donations

The Internet has enabled candidates to raise unprecedented amounts of money. This ability was evident fairly early in the development of online campaigning, and campaigns continue to increase their revenue though online donations. While not specifically addressed within the rise and development portion of this chapter, with final numbers and analysis still pending, the 2012 election between Obama and Republican nominee Mitt Romney resulted in even greater fund-raising numbers both online and offline than any other previous campaign.

One-Click Donations

The Internet makes it easy for people to donate money to candidates, and campaigns have sought to make it even easier for supporters to do so through one-click donations. The 2012 Obama campaign developed **Quick Donate**, a digital tool that allowed them to store credit card details and other personal information following a voter's initial contributions. After this initial contribution, voters could donate money with an email, text message, or single click on a campaign site or publicity material. Donations were strategically solicited from these previous donors following campaign rallies, in response to speeches from the opposition, and other key events (Fitzpatrick, 2012). The Romney campaign followed the Obama campaign's lead with **Victory Wallet**, their own version of Quick Donate. The

Romney campaign found that once people donated, they were at least three times more likely to do so again (Judd, 2012). In part, this finding was due to the ease with which supporters could now contribute.

Social Donations

In 2008, the Obama campaign linked fund-raising and individual efforts. Through social networking sites, supporters were able to solicit donations from friends, family, and other people in their social networks on behalf of that campaign. Even if these people were not supporters of that candidate, they donated because of their relational connections. That tactic was calculated and controlled, but social donations are also occurring independently and without campaign guidance. The Internet has made it incredibly easy to discuss support for candidates and encourage the support and financial contributions of others. In some cases this is occurring at the suggestion of campaigns, but the actual efforts and methods are those of individual users and result in reaching potential contributors who would not have been solicited otherwise (Judd, 2012).

Tracking

The previous four characteristics of online campaigning have all been enhanced through the tracking of supporters and potential voters. This tracking takes place through the development of databases and other software. It also takes place largely unbeknownst to visitors of campaign websites. Tracking enables campaigns to discover information about specific voters and specific voting blocs, which enables them to tailor personal correspondence, advertisements, funding requests, and even website displays specifically to them.

Tracking Through Software

As mentioned at various points above, campaigns are better equipped to identify and track voters through software systems. The Obama campaigns of 2008 and 2012 in particular were able to compile and share voter information among operatives and volunteer with incredible accuracy and ease. These databases included far more than broad demographic information. Rather, they included details about activities, behaviors, and opinions voters sometimes themselves did not even know they had.

Both the Republicans and Democrats have developed software to track voters on Election Day in order to better get out the vote. In 2008, the Obama campaign used a system know as Houdini, so named because it would magically remove people who had already voted from get-out-the-vote lists. This enabled volunteers to focus their efforts on potential voters who had yet to vote, saving time and resources. Obama's 2008 victory came without Houdini, though, which went down on Election Day. In 2012, the Obama campaign used a system called Narwhal to track voters, which seemed to work when the time came, by all accounts. The Romney campaign, following Obama's lead once again, had a similar system called ORCA, so named because orcas are the only nonhuman predators of narwhals. Unfortunately for the Romney campaign, like Houdini four years before, ORCA went down on Election Day. Especially unfortunate for the Romney campaign, ORCA had been billed as a "secret weapon" to counter the Obama ground game (Jacobs, 2012; Madrigal, 2012).

Tracking Through Websites

Campaigns are also tracking supporters and potential voters through websites. Some examples of such tracking are obvious to users of political websites. The Obama campaign site, which offered users social-networking capabilities, could be used to mine data provided by users through their activities and registration information. When visitors to any campaign site provide such data as their names and contact information, that material will be used by the campaign.

Yet, campaigns are getting more clandestine and active in their tracking of users. This strategy was especially true in the 2012 election and will likely increase in upcoming elections. By dropping cookies and other tracking devices into the computers of visitors, their websites, campaigns are able to track their online activities. In 2012, the Obama campaign site dropped 87 tracking cookies into the computers of users. It was frequently noted in the press when discovered that these were more cookies than used by major retailers. Always seeming to be outdone, the Romney campaign site used 48 tracking devices, just over half of the number used by the Obama campaign (Crovitz, 2012). It has been reported that tracking firms have information on around 80% of all registered voters (Beckett, 2012).

Conclusion and Predictions

The Internet has been used in presidential campaigns since 1992, and its use and significance have grown exponentially during this period of time. The Internet is now a part of congressional, state, and even small, local campaigns. However, campaigns of these levels have benefited from, learned from, and emulated the efforts and successes of presidential campaigns.

As we look toward the future, *the significance and prevalence of campaigning through the Internet will continue to increase.* Fund-raising records were broken once again in 2012, and these records will likely fall with each ensuing presidential campaign. At some point, increases will level off and will not be as dramatic, but that point will be many election cycles in the future. As a result, presidential campaigns will continue to use the Internet to its fullest abilities.

Voter tracking will emerge as a powerful campaign tool. Tracking of voters through the Internet began in earnest in the 2012 presidential campaign, and this campaign strategy will be further developed and utilized in the future. Doing so will assist in the continued breaking of fundraising records and the continued molding of campaign messages to specific voters.

It is also likely that a number of *regulations will be introduced with regard to Internet campaigning.* Some regulations and campaign law decisions have already been put forward. However, regulating Internet campaigning will become a major issue in the future, especially as it pertains to the tracking of voters, fund-raising, and spending.

Finally, *future campaigns will incorporate virtual reality.* Supporters and potential voters will be able to interact with candidates in a virtual environment, eliminating the need for extensive candidate travel and enabling campaigns to reach even more voters. Supporters will be able to attend fully immersive campaign rallies and interact with other supporters in virtual environments. Campaigns will be able to mold candidate avatars to mimic the nonverbal behaviors of individual voters and flawlessly engage in other communicative strategies utilized with various degrees of success and reach by physical politicians.

As with other predictions, time, and in this case, additional election cycles, will confirm the accuracy of these speculations.

Key Terms

Connecting (Function of Internet Campaigning)

Digital Wallet

Engaging (Function of Internet Campaigning)

Tracking (Function of Internet Campaigning)

Fund-Raising (Function of Internet Campaigning)

Informing (Function of Internet Campaigning)

Quick Donate

Chapter Exercises

Current Issue: As students are studying this chapter, it may or may not be a presidential election year. Of course, presidential campaigning never really ends and is usually taking place regardless of the point in the election cycle. Have students examine current or potential candidate websites. If possible, given the technological capabilities of the classroom, display current sites for the class to examine. What features are shared among these sites? Do some sites seem more developed than other sites? If so, what makes these sites stand apart from the others? What changes would they make to these campaign sites if they were in charge of them?

Debate It: Split the class into two groups. Select one of the resolutions below. Have one group support the resolution and have the other group oppose the resolution in a debate.

Resolved: Barack Obama would not been elected President of the United States of America if not for the Internet.

Resolved: Barack Obama would have been elected President of the United States of America even without the Internet.

Ethical Consideration: Have students consider whether it is ethical for campaigns to track voters without their knowledge? Are there limits to how much the tracking of voters is acceptable? Should regulations be established concerning the tracking of voters? If so, what regulations would be most necessary?

Everyday Impact: Ask students whether they have ever donated money to a campaign through the Internet. If so, what motivated them to donate money to the campaign? Beyond perhaps agreeing with a candidate's issue or disliking that candidate's opposition, how do they think the campaign succeeded in receiving a donation from them? If they have never donated money to a campaign through the Internet, they have no doubt been solicited for a donation. Beyond perhaps not having enough money or not liking any of the candidates, what has prevented them from donating money to campaign? If their answer is just not caring about politics, campaigns attempt to get people to care. Why have these campaigns failed with them and succeeded with others? How might campaigns change their approach to them?

Future Focus: Looking back from two decades into the future, it is very easy to recognize the Internet's potential in campaigns. If it were not clear to campaigns in the 1990s, certainly it must have been obvious by 2000. Campaigns did start using the Internet in earnest by 2000. However, even at that point, there were a number of politicians who were not convinced of the Internet's utility in campaigns. And we should not really find fault with them. The Internet of today does not look like the Internet of the 1990s or early 2000s. Ask students to describe campaigning on the Internet 20 years from now. Do they anticipate another technology, yet to be invented or fully realized, being used in political campaigns?

13

Privacy, Transparency, and the Internet in America

Telegraphy, telephony, and radio launched a rippling trend in communications: the sharing and representing of our symbols by, in, and through electricity and machines. Following telephony and the radio, the computing machine crept into existence, aided by hush-hush research on anti-aircraft machinery and other war efforts. From Colossus to Mark I to ENIAC, these machines extended the trend, further mechanizing and digitizing human behaviors and experiences. Initially, these computation devices were bulky and novel, limited in accessibility and scope. In less than a century, however, these devices evolved, commercialized, connected, turned palm-sized and ubiquitous. Computing machines inevitably transformed more and more genres and types of human communication into digitally digestible, electronically transferable, and storable symbols, or binary bytes.

Today, information of all breadths, ranges, and varying intimacies, from all types and classes of persons, is digitally mediated. Evermore, messages are sent and received through smartphones, tablets, and laptops connected to the Internet. As host and railway, the Internet witnesses much. IBM (2013) estimated "the world creates 2.5 quintillion bytes of new data daily" (Risen & Lichtblau). Further, this ongoing flood of data is only primed to increase in quantity and intensity. The International Data Corporation observed (2013), "from now until 2020, the digital universe is expected to double every two years" (Risen & Lichtblau).

Echoing visions of the Wild West, from its inception, the Internet has remained ripe for profiteering and highly unregulated (or "self-regulated"), creating unique opportunities for entrepreneurship and exploitation. Accordingly, digital savvy corporations, governments, and researchers of all sizes have surreptitiously and overtly gained wider and more intimate access to our information. Indeed, the harvesting of our daily data has become common practice, a lucrative and profit-laden business equivalent to trading pork bellies. New industries and new communicative practices dictate new rules, often creating unique shifts in traditional societal norms. Of significant note here, expectations of privacy have been, and continue to be, radically redefined by and through digital communication technologies and practices.

Preview: A Drama of Opposing Interests

Online, we buy, sell, read, browse, gossip, debate, trade, write, share, coordinate, and manage accounts, videos, pictures, friends, articles and business. Immense value surrounds and permeates all these elaborate and impromptu digital exchanges of information. The attributed value of all this data has inspired legions of dedicated systems built to routinely harvest, store, and share our patterns, preferences, and digital whispers. As a result, notions of privacy shift and change online like a chameleon scrambling to change color in a gumball machine. Almost weekly, privacy policies and practices are being tweaked and criticized, enforced and fined. Each change and critique affects the individual differently, both online and offline. Predominantly, however, users continue to seamlessly engage and exploit the online multiverse from anywhere and everywhere. The social and mental gains, popularity, and other appeals of digital communications have maintained an ongoing, albeit capricious, consumer tolerance for a lack of privacy online. To date, the **risk-benefit ratio** appears to favor online communications despite a diminishing sense of privacy and a growing sense of transparency.

In this chapter, we explore, in detail, the complicated and ever-evolving issue of privacy in the modern digital world. First, we offer four contrasting and interlinked perspectives on privacy and information management in the new digital age. Second, we introduce the unregulated origins of the Internet and the value of information, or **big data**, as two key, motivating factors for the current state of privacy online. Third, we describe some of the more popular tools and practices active online nurturing a mutually profitable consumer transparency while limiting privacy. Fourth, in contrast, we introduce several tools promoting more individual privacy or control over digitally mediated communications.

Fifth, we present several common rationales defending the merits of transparency and privacy online. Sixth, we describe various public instances where opposing perceptions and conceptions of privacy and transparency collide. Finally, we end with a prediction of the future landscape of digital communications as it relates to privacy and transparency.

> ### Questions Previewing This Chapter
>
> 1. *How is privacy defined and framed across unique agents and agencies?*
> 2. *Why and how has the Internet changed expectations of privacy?*
> 3. *How are current digital technologies redefining what can be known?*
> 4. *What are some of the negative and positive consequences of data mining, routine surveillance, and political transparency?*
> 5. *How are differing perspectives for privacy and transparency interacting and colliding online and offline?*

Framing Privacy

Privacy is a complex and pluralistic idea in constant flux, tethered to particular contexts and forces (Solove, 2008; Dyson, 2008). Across time, our communications technologies, from the postcard to the smartphone to the social network, introduce novel opportunities for lucrative surveillance and benevolent social transparency. As a result, these new technologies, and their mainstream adoptions, often implicitly redefine and agitate traditional ideologies, practices, and expectations. For instance, today, on the wings of the digitization of everything, traditional conceptions of "a right to be let alone" appear increasingly discarded and publicly endangered, if not ruled irrelevant. Arguably, the digital communicator is never, and never wants to be, alone; and for those wanting to be let alone, they will find only limited and radical options.

More than ever before, the digital communicator lacks a sincere ability (or right) to obscure, control, and understand the distribution and repurposing of their digital communications and personal data. In part, this is due to the rapid diffusion of digital communication technologies and the dynamic perceptions and conceptions of privacy's evolving role in the new digital environment. In the sections that follow, we introduce four frequently conflicting and interlinked perspectives on privacy's role in America: the traditional, federal, commercial, and radical.

Traditional: A Right to Control

Petronio's (2004) popular Communication Privacy Management (CPM) theory proposes the average individual believes she owns her private information and has "a right to control whether the information is disclosed as well as to whom it is disclosed" (Kennedy-Lightsey, Martin, Thompson, Himes, & Clingerman,, 2012, p. 666; Petronio, 2004). Westin (1967) described information privacy further as "the claim of individuals, groups, or institutions to determine for themselves when, how, and to what extent information about them is communicated to others" (p. 7). Rule (2007) later defined privacy similarly as "the exercise of an authentic option to withhold information on one's self" (p. 3). Many scholars reflect these sentiments, presenting privacy, quite simply, as a "form of control" (Rosen, 2010, p. 44).

These voices collectively reflect the traditional conception of privacy, which, ultimately, implies an individual right to control the dissemination of personal identity and communications. Here, an individual with an honest capacity for privacy has power, within reason, over the deceptive, genuine, careful and/or reckless distribution, circulation, and containment of self and self-generated information. Parker summarized, "Control over who can see us, hear us, touch us, smell us, and taste us, in sum, control over who can sense us, is the core of privacy" (p. 280).

Federal: A Right to Watch, Record, and Defend

Here, we describe a federal conception of privacy fostered and imposed by the U.S. government across the early 21st century. The U.S. government's conception of privacy in the new digital age is succinctly expressed through the implementation and continued interpretation of the "business records" provision of the PATRIOT Act, 50 USC section 1861. For example, interpreting this and similar provisions across the 21st century, the U.S. government enabled the National Security Agency (NSA) to collect the telephone records of millions of unsuspecting Verizon customers (Greenwald, 2013). Further, under an undisclosed program titled Prism, the NSA has direct daily access to Internet data (including search histories, email content, live chats, and more) available and mediated through Google, Facebook, Apple, and other Internet service providers (Greenwald & MacAskill, 2013).

These and similar surveillance methods have become common practice domestically in the United States. In 2013, justifying the NSA surveillance efforts after an internal leak disclosed their activities to a vast ignorant public, U.S. president Barack Obama declared, "it's important to recognize that you can't have 100% security and also then have 100% privacy and zero inconvenience" (Favole

& Nicholas, 2013). Indeed, the government often frames ongoing and intensifying domestic transparency as a necessary means for social and national security.

Senate Intelligence Committee Chair Senator Dianne Feinstein noted phone and Internet surveillance programs have been instrumental in thwarting terrorist attacks since, at least, 2009 (Delawala, 2013). A notable proportion of Americans appear to accept the government's proposed trade-off between investigating terrorism and protecting personal privacy. The Pew Research Center (2013) found a majority of Americans (56%) supported the NSA's undeclared annual program tracking the telephone records of millions of Americans as a means for investigating terrorism.

Alternative View: Devaluing the Power of Powerful Motives

The Atlantic *(2013) responded to the arguably exhausted antiterrorism defense for a perpetual federal "breach" of individual "liberty" in America (Friedersdorf, 2013). In an article titled, "The Irrationality of Giving Up This Much Liberty to Fight Terror,"* The Atlantic *proposed terrorism, while deserving significant attention, did not warrant the extreme, ongoing invasive practices we experience today (Friedersdorf, 2013). Friedersdorf (2013) highlighted that, over an interval of approximately 11 years, guns posed more than 100 times greater a threat to American safety, responsible for over 360,000 deaths. Drunk driving threatened American safety 50 times more than terrorism, responsible for roughly 150,000 deaths. Bailey (2011) further observed, between 2006 and 2011, an American's chance of being killed by a terrorist was approximately 1 in 20 million. These and other poignant figures prompted both Bailey and* The Atlantic *to question the use of terrorism threats to repeatedly justify surrendering more and more individual privacy and other traditional American liberties.*

Commercial: A Right to Harvest and Profit

The government and commercial conceptions of privacy often pursue, enact, and enable similar practices, promoting transparency as a necessary and benevolent social good. Indeed, many argue, Silicon Valley helped build the thriving surveillance state we experience daily (Hirsh, 2013). As proposed by one Google executive, for online organizations, "thriving data sets" are a "gift," "enabling [entitled organizations] to better respond to citizen and customer concerns, to precisely target specific demographics of the population, and, with the emergent field of **predictive analytics**, to predict what the future will hold" (Schmidt & Cohen, 2013, p. 57).

In contrast to the federal conception of privacy outlined above, commercial agencies predominantly seek and nurture increased transparency online for profit. Advertising, marketing, Internet convenience, and Internet browsing are all purportedly enhanced, and thus made more profitable, through routine **data mining** practices (explored in more detail below).

Radical: A Right to Know and Understand

The radical perspective is primarily concerned with the current obfuscation and inscrutable applications of data mining and other guarded operations conducted by hegemonic forces. For example, Solove (2013) argued information processing today is too often conducted in secret and data mining practices that forbid people's knowledge or involvement, are "a structural problem involving the way people are treated by government institutions" (Rosen). Indeed, the full breadths of commercial and federal information-gathering practices and applications are often imposed without consent and remain undisclosed for years. Americans experienced this reality in 2013 when they discovered that the NSA was for years secretly amassing massive data sets of personal information based on e-mail and mobile phone records. Further, these particular NSA data sets were completely inaccessible to the average individual (Rosen, 2013).

Agents embodying the radical perspective believe such practices expose a troublesome power imbalance between the institutions collecting data and the individuals generating data. In response to these imbalances, a resistance has emerged, seeking to combat this imbalance through nurturing and promoting political transparency and the liberation and de-commercialization of information (White, 2013). Agencies and agents representing this perspective include WikiLeaks, Anonymous, AdBusters, and whistleblowers such as Edward Snowden, William Binney, Bradley Manning, and Amber Lyon. In the next section, we transition toward a discussion of the trends occurring online diminishing traditional expectations of privacy, nourishing lucrative commercial and federal surveillance, and prompting radical agents and agencies to seek more commercial and federal transparency.

A Shift in Expectations: Entrepreneurs Embrace Unregulated Framework

As explicit across this text, the Internet overflows with endless human traffic and communications. From our comments to our clicks to our queries to the emails we write to the geographic locations we cross and the approximate times of our last browsing, all our online acts create recordable tracks. Most importantly here, each sugary digital bit we create is easily attainable fodder for harvesting and trading. When organized, archived, and analyzed, this data becomes infinitely revealing and immensely profitable.

Rauhofer (2008) observed, the digitization of "everyday tasks has resulted in a paradigm shift where vast amounts of information about individuals, their opinions and habits, is generated and stored in the databases of those providing online services" (p. 185). Indeed, in spring 2007, "Google conceded that until then it had stored every single search query entered by one of its users, and every single search result a user subsequently clicked on to access it" (Mayer-Schonberger, 2009, p. 6). Here, we see the beginnings of the individual losing control, initially unknowingly, over the containment and repurposing of their self-generated information online.

Since their genesis, online organizations such as Google have maintained relatively unlimited and unregulated access to any and all digital communications hosted on and through their servers. As a result, excited entrepreneurs quickly discovered and created lucrative benefits in the aggregation, archiving, and trading of any and all digital crumbs. Google, and similar "free" Internet services, evolved into big business from "a particular set of circumstances in which [they] had access to consumer information and the freedom to use it" (Harper, 2010, p. W2). Efficient advertising and search results fueled by efficient spying helped fuel Google's growth and popularity. In an article titled, "The Web's New Gold Mine: Your Secrets," *The Wall Street Journal* (2010) declared, "one of the fastest growing businesses on the Internet…is the business of spying" (Angwin, p. W1). In 2010, America's "50 top websites on average installed 64 pieces of tracking technology onto the computers of visitors, usually with no warning" (Angwin, p. W1).

In the physical world, if a room is tapped with surveillance technology, the tapper needs legal permission and the tapped need to be verifiably suspect of criminal activity. Online, new rules and procedures for surveillance developed. Self-regulation has been the encouraged default. Accordingly, in the digital world, software developers, commercial agencies, federal defense organizations, and corporate interests have developed their own rationale and rhetoric for the implementation of surveillance devices. Various ambiguous degrees of transparency and monitoring have quickly become the imposed price for most Internet services rendered. Possibility did not become an announcement, discussion, or negotiation, but a profitable and exciting opportunity.

Digital Portfolios and Data Mining: Big Data, Big Bucks

Data miners, or electronic voyeurs for financial gain, exploit various digital technologies to build extensive digital portfolios on Internet users. As a result, each of us has multiple digital counterparts brimming with eclectic insights into our daily electronic activities, from the favorite quotes we highlighted in an e-book in the morning to the scandalous videos we streamed from our bed in the evening. Our digital portfolios reflect both monumental and minute details on our endlessly evolving identities and behaviors. Accordingly, the Silicon Valley memory banks, and other similar archives amassing our digital paw prints, contain "details we have long forgotten, discarded from our minds as irrelevant, but which nevertheless shed light on our past: perhaps we once searched for an employment attorney when we considered legal action against a former employer, researched a mental health issue, looked for a steamy novel, or booked ourselves into a secluded motel room to meet a date while still in another relationship" (Mayer-Schonberger, 2009, p. 7).

With their depthless insights and obvious potential for marketing, digital portfolios have become as fashionable and coveted as petroleum and pork bellies (Rule, 2009). For example, **behavioral targeting**, sustained and made possible by digital portfolios, has become big business for commercial industries. In 2005, the Direct Marketing Association estimated its profits as exceeding $161 billion (Rule, 2009). In 2011, "the financial value companies derived from personal data in Europe was $72 billion" (Regalado & Leber, 2013). Data brokers amass, buy, and sell dossiers of our personal data like kids from the 1990s exchanged baseball cards. One such broker boasted about having portfolios collected on over 70% of active U.S. email addresses and responded to more than one billion requests for that information in one month (Stross, 2011).

Alternative View: Big Data Can Change the World

The appeal of big data for research and academic purposes is seemingly undeniable. Yet, some propose we approach research based on these prodigious data sets with judicious trepidation (Taleb, 2013). Taleb expressed one mounting concern, observing, "big data means anyone can find fake statistical relationships, since the spurious rises to the surface … in large data sets, large deviations are vastly more attributable to variance (or noise) than to information (or signal)." After finding ample amounts of data supporting a hypothesis, the researcher can persuasively present her findings, discounting deviations as noise, and circumventing further analysis that might discount her initial findings.

Tools Constructing Commercial Transparency

Over time, across cultures, disparate public and private scenes emerged in physical space. For instance, offline, closed doors, in the neighborhood or office or home, allow self-contained communicative transactions to occur, shared only by those invited or explicitly granted access. Solove argued private spaces of this nature protect individuals from disruption, shame, and anxiety while encouraging experimentation and social exploration (p. 93).

Online, in its current form, the Internet would appear to be a predominantly public scene with limited potential for closing doors or blinds to create private space for unique or controversial information exchanges. Indeed, various popular agencies typically promote and design transparency online as a means towards achieving social goods such as efficient consumer transactions and national (or international) security (Solove, 2008, p. 89). In this section, we describe specific digital technologies and practices encouraging and endorsing a more transparent and public scene online.

Email

Multiple email service providers, from Yahoo to Google, regularly sift through each of our online letters to add to our individual digital portfolios. Each and every user, knowingly or, more generally, unknowingly, must open up their online correspondences to digital opportunists. Email services typically defend their practices by highlighting a need to fuel more personal advertising (Vega, 2012). Additionally, outside of our email providers, police can also gain, without warrant or announcement, instant access to any email over 180 days old (Sengupta, 2012).

In mid-2011, Yahoo Mail officially opened their latest update to 284 million users worldwide (Walsh, 2011). Somewhat ironically, Yahoo promoted new features that limit email spam (Spam 1.0) while unveiling new scanning technology that will deliver more and more targeted advertisements (or Spam 2.0). The revamped Yahoo Mail championed an updated email reader that scans all emails being sent and received by Yahoo Mail. Yahoo further promised that their latest scanning machines would help upgrade Spam 2.0 across the Yahoo multiverse (Walsh, 2011). Not only would users be flooded with specialized advertisements inside of Yahoo Mail, but also users would now be followed by specialized advertisements (based on their email content) across Yahoo and third-party affiliates.

Alternative View: The Dastardly "Opt-Out" Approach

Spam 2.0 floods the web with accelerating speed and bravado. If the Internet continues down its current course, users appear destined to be surrounded by more and more personal advertisements. Unfortunately, building these targeted advertisements demands a significant diminishment of user control over self-generated information. Each personalized advertisement exists only because of scanning machines (or spies) that read through each of our intimate and professional online correspondences. For some, the price of the stamp begins to seem small when compared to the price of using market research firms as hosts for our conversations. In the digital world, our mail does not appear to be sealed in an envelope until received by the intended party. The precedence set in the physical world was deliberately redesigned when letters migrated online.

However, if a user becomes aware of, and uncomfortable with, outsiders reading her electronic letters, Yahoo Mail purportedly offers an off switch located somewhere within their website (Walsh, 2011). Here, the "opt-out" trend persists, continuing the commercial promotion of transparency as the preferred default. As has become common online, users are watched, recorded, and targeted unaware until they become aware and choose to "opt out." However, "opting out" is only possible if the user can discover and navigate the obscure "opt out" mechanisms for every single practitioner online planting tracking technology onto each and every consumer device. Arguably, the average consumer may be unequipped to keep pace. Indeed, Turow (2012) discovered only one in five Web users actually notice the "AdChoices" icon, which allows consumers to learn about, and "opt out," of behavioral targeting and other privacy-infringing practices (Davis, 2012).

Cookies

The drive to know and record our every digital juke and jive does not end when we exit a particular emailing service or search engine. For instance, online video streaming websites, such as Hulu and YouTube, autonomously plant surveillance devices called Flash cookies deep inside our computers (Vega, 2010). Cookies, of this nature and intent, are ubiquitous online. Cookies are small text files implanted on the consumer's hard drive by visited websites (or third parties affiliated with visited websites). Cookies can be configured to last for days, months, years, or indefinitely (Miyazaki, 2008).

These undercover tagalongs surreptitiously follow us from website to website; relaying juicy bits back to their mother ship (and any affiliated third parties). Depending on their programming, cookies can collect endless user data, ranging from any and all text entered on a webpage to the specific sites we visit and how long we visit them (Miyazaki, 2008). The data collected by these cookies inevitably become packaged and sellable to advertisers and other undisclosed third parties with buying power.

Alternative View: Cookie Backlash

Concerned privacy advocates observe, "if enough data is collected over time, [data collectors] can create detailed profiles of users including personally identifiable data" and these profiles can become a personal liability (Vega, 2010, p. B3; Franceschi-Bicchierai, 2013). In response to these and similar concerns, many contemporary web browsers now allow users to detect, reject, and remove cookies implanted on their hard drives. The average consumer, however, appears unaware or ill-equipped to regularly exploit these applications. Various studies show consumers are unaware of the full breadth of cookie practices, and, accordingly, do not employ removal services (Davis, 2012; Ha, Shaar, Inkpen, & Hdeib, 2006; Jensen, Potts, & Jensen, 2005; Milne, Rohm, & Bahl, 2004). Even aware consumers struggle to discern "which cookies to delete to enhance privacy protection and which to save to aid in online functional efficiency" (Miyazaki, 2008, p. 21). Further, more advanced cookies, such as the Flash cookie, have been programmed to avoid detection and removal.

Mobile Devices

A 2012 study conducted by mobile ad network InMobi declared that, of those surveyed, people were likely to use mobile devices while in bed (Walsh, 2012). Echoing these findings, in 2013, the Pew Research Center reported some 44% of American adults sleep with their cell phone, consulting the device throughout the night (Pew Research Center's Internet & American Life Project). Within the boundaries of that most intimate of hideaways, the bedroom, one might reasonably expect privacy when communicating. As noted above, before convergence and mass digitization, there existed an obvious line between private and public spaces. For instance, the books underneath our bedside lamps, the sweet nothings whispered to an intimate partner, and the magazines tucked under the bedsprings were known only to those invited to see and listen. Today, however, the line is muddled, the bedroom door creaked open.

Mobile devices and tablets appear to be the most vulnerable and fertile of all digital devices. The Pew Research Center found smartphone users are "twice as likely" to experience feelings of privacy invasion (Boyles, Smith, & Madden, 2012). When Americans bring e-readers, tablets, and cell phones into their bedrooms or restrooms or barrooms for reading or communicating or time-keeping, they are frequently inviting autonomous commercial agents carrying pens to follow close behind, scribbling extensive notes, for present and future use.

Cell Phones and Smartphones

Depending on the state and the court, police can gain seamless access to any and all cell phone data including location records and text message (Sengupta, 2012). Furthermore, police are often granted access without warrant and without notifying the consumer. Here, agents are granted unsolicited secondary use of self-generated information that many privacy advocates would define as "inherently private" (Sengupta, 2012). In 2011, cell phone carriers reported over 1.3 million demands from law enforcement agencies for consumer's cell phone data (Sengupta, 2012). Authorities defend these practices, citing "consumers have no privacy claim over signals transmitted from an individual mobile device to a phone company's communication tower, which they refer to as third-party data" (Sengupta, 2012).

Police are not the only agents with predominantly unregulated access to our physical movements through mobile, Internet-equipped devices. Multiple "apps" installed on tablets and smartphones have been found to pilfer location and contact information, without consent or announcement (Perlroth & Bilton, 2012). In 2011, researchers discovered Apple's iPhone, since an iOS update released as early as 2010, was tracking anywhere and everywhere users carried their phone (Arthur, 2011). All data would be saved on a file hidden on the device until the device could transmit that data back to Apple (Arthur, 2011). Matching their competition, Google's Android cell phones were caught doing the same, recording and relaying geographic data without user consent (Stross, 2011).

In response, contemporary consumers are increasingly uninstalling technologies exposed as invasive. The Pew Internet & American Life Project reported, "57 % of all app users have either uninstalled an app over concerns about having to share their personal information, or declined to install an app in the first place for similar reasons" (Boyles, Smith, & Madden, 2012).

E-Readers and Digital Books

Our emergent relationship with digital books further reflects a cultural shift towards a more transparent existence. As of 2012, there is no option to not share what books one buys and reads through Amazon (Alter, 2012). Reading was once largely a "solitary and private act, an intimate exchange between the reader and the words on the page," but the rise of digital books has transformed reading into something "measureable and quasi-public" (Alter, 2012). Tablets and e-readers give publishers and other agencies novel and salivating access to any and all reading habits, from how long a user reads a book to the sentences she highlights.

While publishers mine and celebrate these new, profitable insights into consumers, others worry readers will increasingly steer clear of sensitive content (Alter, 2012). Further, the precedence set for e-readers, tablets, and cell phones will inevitably transfer to other Internet-equipped devices. For instance, as our vehicles become "giant rolling smartphones" connected to the Internet, they too will likely invite multiple acts of routine surveillance with limited regulation and lucrative commercial applications (Timberg, 2013).

Tools Designing and Marketing Individual Control

As Internet-enabled devices inviting surveillance become more intimately entwined with daily life, consumers representing traditional conceptions of privacy have become increasingly sensitive and aware of invasive digital practices (Singer, 2012; Boyles et al., 2012). In response, various artists, culture jammers, and software developers can be seen encouraging increased user control online. Here, we present a list of digital technologies and tools promoting and producing control and awareness in the digital communications experience (see Table1 3.1). Collectively, these offerings represent a rising, active resistance against a lack of privacy and disclosure in digital communications.

Table 13.1. Tools Combating Surveillance Trends in Digital Communications.

Tool	Debut Year	Service
Hushmail	1999	Hushmail is an email provider that refuses to perform content analysis of emails sent and received through their servers. Digital communications hosted through Hushmail are sealed, opened, and scanned only by sender and receiver. Further, the service provides **encryption** services that limit third parties from accessing the user's professional and personal correspondences.

Tool	Debut Year	Service
The Onion Router (Tor)	2002	Tor is an alternative Internet network supporting and creating an anonymous web surfing experience. Tor reroutes a user's Internet data between "a series of random volunteer 'node' computers" making it nearly impossible to "trace the data back to the original user" (Fowler, 2012). Some believe Tor provides "lifesaving privacy and security for people who otherwise could face extreme reprisal from their governments" (Fowler, 2012). Others see it as a challenge to law enforcement (Fowler, 2012).
AdBlock Plus (ABP)	2005	ABP is a content-filtering, open-source extension for installation on various web browsers (i.e., Mozilla Firefox, Google Chrome, and Opera). According to adblockplus.org, ABP "blocks all annoying ads on the web by default," from pop-ups to flashy banners and beyond. ABP enables users to circumvent intrusive behavioral targeting and, as a result, manage the exposure of preferences and interests between those sharing the same Internet equipped device.
IxQuick	2006	In contrast to popular search engines, such as Bing, IxQuick enables users to search the Internet without capturing users' IP addresses and planting tracking cookies to record users' patterns, behaviors, and preferences.

Table 13.1. *Continued*

Tool	Debut Year	Service
Duck Duck Go	2008	Duck Duck Go is a search engine that explicitly avoids tracking user activity (in contrast to services such as Yahoo and Google). Further, the engine is void of discernible, targeted advertisements and does not personalize search results. As of October 2012, Duck Duck Go processes some 45 million searches monthly (Rosenwald, 2012).
iPredator	2009	iPredator is one of several virtual private networking (VPN) services available online. A VPN allows users to disguise and reassign their IP addresses while encrypting the user's Internet communications (making data accessed by external agents appear like "jumbles of nonsense") (Richmond, 2011).
Snapchat	2011	Snapchat enables users to send texts, pictures, and videos from cell phone to cell phone with an implicit self-destruct timer. In response to similar concerns voiced by Mayer-Schonberger (2011), Snapchat aims to reintroduce forgetting and deleting to the digital age. Using Snapchat, after the receiver views a delivered message on her phone, the message will disappear, forever. The hosts behind the service do not keep or mine the message and the receiver cannot hold on to the message for future (potential secondary) use. As declared by one founder, "There is real value in sharing moments that don't live forever" (Wortham, 2013, p. A1).

Table 13.1. *Continued*

Tool	Debut Year	Service
Disconnect	2011	As declared on their website, disconnect.me, "Your personal info should be your own. But today thousands of companies invisibly collect your data on the Internet, including the pages you go to and the searches you do. Often, this personal data is packaged and sold without your permission." In response, *Disconnect* enables users to evade ad trackers and data miners while also encrypting traffic over Wi-Fi networks. Further, the application provides a breakdown of all the various agencies trying to access information online and insight on why these agencies are attempting to access that information. Over one million people use the application each week (disconnect.me).
PrivacyChoice	2012	Privacy Choice, accessed via privacyscore.com, scores websites, on a scale of 0 to 100, according to how they use personal data. The tool shows users whether a website "shares personal user data with other sites, how long a site retains that data and whether the site confirms that data has been deleted" (Vega, 2004, p. B4).
Silent Circle	2012	Silent Circle enables users to send easily encrypted, surveillance-resistant messages (text, voice, and video) to fellow Silent Circle users, peer-to-peer. Silent Circle's encryption methods, purportedly, ensure each message can only be accessed and viewed by the sender and receiver (Gallagher, 2012).

Table 13.1. *Continued*

Tool	Debut Year	Service
Stealth Wear	2013 (beta)	Developed by Adam Harvey, Stealth Wear provides various physical garments for avoiding and prohibiting tracking technologies. For example, in 2013, Harvey debuted an "off-pocket" that zeroes out a consumer's phone signal when it's inside. The "off-pocket" would enable consumers to avoid location data collection practices covered previously in this chapter (under "Mobile Devices"). Harvey also showcased a hoodie, "engineered to conceal the body's thermal signature that can be picked up by the IR sensors built into many surveillance cameras and drones" (Dillow, 2013).

Promoting Transparency, Defending the Digital Panopticon

Convenience, efficiency, and security persuade and justify. Today, we see this across debates, analyses, and conversations circulating around privacy in the new digital age. On the web, routine monitoring and data mining are regularly defended and promoted as means toward a more effective, efficient, and secure relationship with the Internet. In this section, we introduce and define four popular justifications for online transparency.

1. **Behavioral Targeting**

   On Tuesday, a consumer sends a lengthy, personal email to a family member about being severely depressed and anxious about her second year in college. One week or several hours later, that same consumer, while searching for a puppy with her roommate, becomes blessed with various banner advertisements and pop-up windows singing and selling the praises of Zoloft, Lexapro, Prozac, and Paxil. The above would be an extreme example of effective behavioral targeting. (Also, the above example would be an instance where information about oneself becomes used by third parties and exposed to secondary parties, unintentionally.) Behavioral targeting and retargeting are the micro targeting of information and, most importantly, advertisements, based on behaviors and information amassed on particular individuals while they interact with the web.

Most websites on the current (2013) Internet depend on advertising to survive. Websites dependent on advertising revenue for sustenance fail to attract advertisers without efficient behavioral targeting practices. According to the Pew Research Center, multiple online websites directly suffer if they do not utilize targeted advertisements (Vega, 2012). Many individuals embrace this practice, trading personal information for material benefits and convenience, while others find it uncomfortable and intrusive (Rauhofer, 2008).

2. Personalization

Personalization takes many forms online. Personalization occurs whenever a search result or a shopping query tactfully delivers results reflecting previous online behaviors and purchases. Particular results, products, and services are fed to consumers according to information available in their particular digital portfolios. The industry promotes personalization as an efficient method for giving the consumer what they want today, based on what they wanted yesterday.

Alternate View: Filters Filter Knowledge

Responding to the question, "Is the Internet closing our minds politically?" Fuse (2012) argued the personalization of search results and other content was trapping consumers in "information bubbles" (NPR, 2012). Algorithms designed to shape searches to reflect searchers pre-established interests aid in narrowing one's perspective. Personalization continually feeds users similar perspectives, attitudes, voices, and thoughts. Accordingly, personalization can lead to less opportunity for discovering contrasting ideas and new ways of thinking.

3. Ethic Maintenance/Enhanced Accountability

When applied to institutions, advocates for radical conceptions of privacy argue a transparent network helps maintain good values and accountability (TED, 2012). A more transparent Internet can enable interested and conscious users to discover what's going on inside institutions, scrutinize their practices, inform others of those practices, and create collective responses (TED, 2012). If an institution is misbehaving or lacking integrity, routine transparency will ensure they are exposed and judged accordingly. Schmidt and Cohen (2013) predicted, in a more transparent future, "information blackouts, propaganda and "official" histories will fail to compete with the public's access to outside information, and cover-ups will backfire in the face of an informed and connected population" (p. 62).

4. **Safety From Harm**

Federal surveillance programs that diminish privacy and enable domestic transparency are often cited as necessary for national defense. For example, in September 2009, a potential terror plot by Najibullah Zazi aimed at bombing the New York subway system was foiled through phone and Internet surveillance programs (Delawala, 2013). Representative Mike Rogers, chair of the House Intelligence Committee, argued that, in the Zazi case, federal surveillance programs were "the key piece that allowed us to stop a bombing" (Delawala, 2013). Further, NSA Chief General Keith Alexander alleged, surveillance programs "have protected the U.S. and our allies from terrorist threats across the globe…helping prevent potential terrorists events over 50 times [since late 2001]" (Fitzpatrick, 2013). Despite these proclamations, in spring 2013, active NSA surveillance practices were unable to prevent the Boston Marathon bombing that killed three and injured over 200.

Alternative View: Profits in Surveillance

In America, money from the private industry often helps fund and drive political campaigns. When private money nurtures leaders and politicians, many question the weight and significance of ideals and value in informing public policy (Sirota, 2013). One active political contributor is Booz Allen Hamilton, a private contractor aiding national and international surveillance programs (and the former employer of whistleblower Edward Snowden). Notably, 99% of Booz Allen Hamilton's multibillion-dollar annual revenues come from the federal government (Riley, 2013). Further, Booz Allen Hamilton and its parent company, The Carlyle Group, make regular and sizable financial contributions to politicians such as Barack Obama, John McCain, and other surveillance defenders (Sirota, 2013). Connections of this nature lead journalists to observe; "There are huge corporate forces with a vested financial interest in making sure the debate over security is tilted toward the surveillance state and against critics of that surveillance state" (Sirota, 2013; Riley, 2013).

Promoting Privacy, Celebrating Identity Management

Tim Berners-Lee, inventor of the World Wide Web, cautioned,

> over the last two decades, the web has become an integral part of our lives.... A trace of our use of it can reveal very intimate personal things. A store of this information about each person is a huge liability. Whom would you trust to decide when to access it, or even to keep it secure? (Franceschi-Bicchierai, 2013)

Countless communication scholars voice concern over the increasingly vulnerable digital exchange of information. Many believe, in the digital age, traditional expectations of privacy are withering into obsolescence, to our advantage and disadvantage (Rauhofer, 2008; Rule, 2007).

In response, across the last decade, scholars from different disciplines have rallied to define the explicit merits of privacy while offering insights on the rippling effects of a more transparent social existence. Books addressing the topic from all angles abound, carrying audacious titles, including Solove's (2013) *Nothing to Hide: The False Tradeoff Between Privacy and Security*, Rule's (2007) *Privacy in Peril: How We Are Sacrificing a Fundamental Right in Exchange for Security and Convenience*, Mayer-Schonberger and Cukier's (2013) *Big Data: A Revolution That Will Transform How We Live, Work and Think*, and Nissenbaum's (2009) *Privacy in Context: Technology, Policy, and the Integrity of Social Life*. Here, several voices and concerns surrounding privacy and its merits are captured and summarized.

1. **Protection From Discrimination (or Risk Management)**

   Discrimination, of varying degrees and forms, persists across cultures. Knowing when, how, and why discrimination might occur, in the present and in the future, can be difficult or impossible to manage. Oftentimes, one party might condemn some preference or affiliation that another party views as harmless. In cases of social experimentation and beyond, privacy often acts as a blanket of protection against these types of discrimination, managing risk to oneself. Solove (2008) found that the shadows of privacy nurture cultural experimentation, spontaneous acts, and social indulgences while protecting participants from discrimination, disruption, and unnecessary anxiety (p. 93).

   Currently, an action, affiliation, or preference expressed in one's personal environment and shared online (with a specific and deliberate network of invited friends) too often invites unwanted external discrimination. Internet users continue to show an interest in sharing and an implicit expectation of protection from future harm. For example, Stacy Snyder, an aspiring teacher, was denied her teacher's license because of a party photograph posted online titled "Drunken Pirate" (Mayer-Schonberger, 2009). Here, transparency limited Snyder from effectively entering the workplace. She was unprotected from discrimination against something she likely perceived as harmless: jovially drinking alcohol in her social environment, at home with friends, a legal and popular activity seen across America.

Others have discovered an unknown lack of privacy or protection online leads to termination from well-established social and professional positions. Successful California polo coach Mitch Stein was fired in 2012 because of some photographs (of Stein posing with drag queens) that a parent found posted on his Facebook page (Edwards-Stout, 2012). Similarly, in 2013, Pennsylvania police chief Tom Keller was suspended after a single Facebook picture (of Keller and a young woman holding a pistol) became public fodder, without his apparent consent (Matyszcyk, 2013). Both Stein and Keller had a desire to share an impromptu and jovial portrait with their respective online communities, but believed they were sharing these glimpses within trusted and protected circles.

2. **Boundary Management**

Viewing privacy as in peril, Rule (2007) cautioned, "In countless ways, in every social setting, people stand to gain or lose by controlling what others know about them, and certainly by keeping certain 'personal' information to themselves" (p. 3). Snyder, Stein, and Keller stand as testament to this argument. To control what others know and manage perceptions, individuals often employ boundary management practices. Here, senders establish explicit or implicit privacy rules with receivers, in an attempt to limit the circulation of disclosed information (Venetis, Greene, Magsamen-Conrad, Banerjee, Checton, & Bagdasarov, 2012). After rules are understood and tested, information (of varying degrees of sensitivity) is revealed and concealed accordingly. Rules for boundary management can often be "ambiguous" and "not clearly articulated," but serve a tangible function in promoting healthy relationships and preventing "embarrassing," "stigmatizing," and/or "shameful" information from circulating too broadly (Venetis et al., 2012, pp. 346–347).

With some exceptions, currently, when communicating digitally, the server hosting the communications, not the receiver or the sender, typically determines the rules for how disclosed information will be contained and circulated to third parties. If the sender refuses to accept these rules, they are refused access without negotiation or conversation. For example, when a bill proposing users be able to have more control over information about themselves online was presented to the California Senate, Facebook, Google, Twitter, Yahoo, and Skype stepped into the ring to argue that greater control would somehow undermine the individuals' ability to "make informed choices about the use of their personal information" (AP, 2011). The power over boundary management, in their view, should reside in the host's hands, not the senders and/or receivers.

3. **Boundary Protection**

The discloser of information often perceives, implicitly, that receivers will respect and protect their information (Venetis et al., 2012). Online, however, to date, senders and receivers cannot expect established boundaries will be respected or protected by those hosting their communications. The vulnerability of digital portfolios is continually underscored in popular culture. In 2006, AOL accidentally leaked three months of data on over 650,000 of its users (Kawamoto & Ellis, 2006). In early 2011, a group of activists downloaded mass amounts of sensitive user data from the purportedly impenetrable Playstation Network, crippling its activities for months. In 2013, the National Human Genome Research Institute began posting purportedly anonymous genetic data online (Kolata, 2013). The subjects, however, were easily identifiable and not adequately protected with the anonymity the researchers promised. Indeed, casually combing the data, one researcher identified five randomly selected participants and their entire families (Kolata, 2013).

4. **Empowerment of the Present**

Our lack of control over information exchanged online can also limit our capacity to forget, forgive, and move forward in modern culture. As discussed, surveillance, digitization, and cheap storage have spawned detailed data warehouses built to preserve "inordinate amounts of information," from our business transactions to our emails to our mobile messages (Mayer-Schonberger, 2009, p. 68). Individually and as a society, escaping these detailed databases and online shadows becomes increasingly more difficult and costly as we spend more time communicating digitally without privacy. In contrast, accessing these databases has become increasingly easy.

For example, there are low-cost applications and websites allowing users to comb our digital trails, uncovering reports on criminal histories, address histories, approximate incomes, and summaries of social networking profiles (Rosen, 2010). In these and similar instances, digital artifacts potentially representing outdated versions of our selves can corrupt and obstruct the possibilities of the present. Today, the digital crumbs and portfolios we generate in the past have begun to follow us from scene to scene deep into the present, despite individual intent or personal growth.

5. **Self-Development (and the Chilling Effect)**
   Research indicates heightened awareness of surveillance practices can often interfere with self-development and self-discovery (Solove, 2008). Schmidt and Cohen (2013) observed, in certain political climates,

> just being in the background of a person's photo could matter if a government's facial-recognition software were to identify a known dissident in the picture…though this scenario is profoundly unfair, we worry that it will happen all too often, and could encourage self-censoring behaviors among the rest of society. (p. 62)

Indeed, continual surveillance, information permanence, and the passive potential for future discrimination lead many to alter behavior in the present.

Many see a chilling effect descending online, censoring what people explore or say in the present (see Kang, 1998; Swire, 1999; Mayer-Schonberger, 2009; Solove, 2008). The potential for surveillance at any moment without the individual knowing creates effects similar to those proposed by Jeremy Bentham's 1791 prison design, the **Panopticon**. Here, the feeling of "persistent gawking" promotes discomfort and inhibition, with a constant fear of rippling exposure (Solove, 2008, p. 108). Indeed, with the rise of information permanence, "if retrospective redemption is unattainable, what remains is prospective caution" (Mayer-Schonberger, 2009, p. 108).

When Ideologies, Interests, and Attitudes Collide

Occasionally, agents and agencies embodying differing perspectives unite for collaboration and compromise. For example, the National Day of Civic Hacking, a White House–supported event, asks citizens, hackers, and developers to access newly opened and vast federal databases to create data visualizations and apps aimed at making government practices more accessible to the public (Franceschi-Bicchierai, May 2013). In June 2013, the U.S. government teamed with the American Civil Liberties Union to make a collection of torture documents available for the National Day of Civic Hacking, hoping to give Americans a clearer picture of the Bush administration's "enhanced interrogation" practices. In this instance, federal and radical agencies united toward the betterment of society through transparency. More frequently, however, agents and agencies representing conflicting interests and expectations collide online and offline in sparks, blasts, and fizzles. In this section, we offer four snapshots of various public and legal collisions revolving around privacy and transparency practices and technologies.

Case Study 1: Hulu Employs "Supercookies" (Traditional vs. Commercial)

In summer 2011, researchers found Hulu, a popular online video delivery service, employing "supercookies" that were capable of regeneration after deletion and unauthorized tracking (Angwin, 2011). As consumers become more aware, techniques for mining data must evolve to circumvent that awareness and maintain profitability. Hulu and their advertisers feigned ignorance in response to the discovery of their "supercookies" but were still delivered a lawsuit for their actions (Davis, 2012a). The lawsuit claimed Hulu violated the federal video privacy law (VPPA) by "sharing data about users' video watching history with ad networks, as well as companies engaged in analytics and market research" (Davis, 2012a). In 2012, Hulu attempted and failed to dismiss the potentially expensive and blossoming lawsuit.

Case Study 2: Opportunists Capitalize on Software Settings (Traditional vs. Commercial)

In 2012, *The Wall Street Journal* reported Google was bypassing user privacy settings, exploiting a loophole in users' Safari browsers to install secret "cookies." The implanted "cookies" tracked users' online movements and behaviors, without consent. Further, these specific targeted users intentionally restricted monitoring on their devices and believed all "cookies" of this nature were being blocked. Despite user intent, 22 of 100 tested websites installed the Google tracking code on a test computer while advertisements on 23 different websites installed the tracking code on a tested iPhone browser (Angwin & Valentino-Devries, 2012). After discovery, in response, Google was charged with surreptitiously bypassing the privacy settings of millions of Apple Inc. users (Angwin, 2012). Later, the U.S. Federal Trade Commission (FTC) forced Google to pay $22.5 million to settle the issue. The case was a victory for advocates for traditional conceptions of privacy, marking the largest penalty ever levied on a single company by the FTC (Angwin, 2012). However, from a macro perspective, a single rock hitting the face of the ocean produces a very limited ripple. Google is the world's largest Internet service, worth over $250 billion.

Case Study 3: WikiLeaks Imposes Military Transparency (Radical vs. Federal)

Organizations of power often employ secrecy to disguise actions the public would not support. Accordingly, Julian Assange, a representative of the radical perspective (outlined above) and a public martyr for WikiLeaks, argued,

> because different actors will always try to destroy or otherwise cover up parts of [our] shared history out of self-interest, it should be the goal of everyone who seeks and values truth to get as much as possible into the record, to prevent deletions from it, and then to make this record as accessible and searchable as possible for people everywhere. (Schmidt & Cohen, 2013, p. 40)

Living by these ideals, WikiLeaks employs technology to post and disseminate confidential government documents online for public global consumption. Across 2010, WikiLeaks made secret military archives and troves of confidential military documents available to the Internet and the popular press worldwide. These documents provided insights into various undisclosed American military operations, regional alliances and controversial collaborations occurring across both Afghanistan and Iraq conflicts (Manzzetti, Perlez, Schmitt, & Lehren, 2010; *The New York Times*, 2010).

Some contend the ultimate goal behind Assange and WikiLeaks is not political transparency, but to "hobble the U.S. government" (Crovitz, 2010). Indeed, government agencies may balkanize to limit reoccurring information leaks, leading to less efficient operations. Further, leaked documents pose a threat to present and future conflicts and relationships. The Pentagon responded to the activities of WikiLeaks in 2010, warning,

> We know terrorist organizations have been mining the leaked Afghan documents for information to use against us and this Iraq leak is more than four times as large. By disclosing such sensitive information, WikiLeaks continues to put at risk the lives of our troops, their coalition partners and those Iraqis and Afghans working with us. (*The New York Times*, 2010)

Case Study 4: The Shelved "Right to Know" Bill (Federal and Radical vs. Commercial)

In February 2013, legislators in California teamed with the Electronic Frontier Foundation (EFF) to endorse and introduce the "Right to Know" data access bill in America. The intent of the proposed bill was to increase transparency between those that collect information online and those that produce that collected information. The bill proposed to allow Americans to access, upon request, a complete view of their various digital portfolios, including names and contact information for all third parties with which these portfolios have been shared (Regalado & Leber, 2013). The European Union already has similar legislation in place allowing its citizens to submit an individual request to acquire data profiles. The proposed legislation in America, however, was quickly opposed by Silicon Valley as financially damaging to Internet firms and "unworkable" (Regalado & Leber, 2013). As a result, the "Right to Know" bill was immediately shelved.

Conclusion and Future Possibilities

As we continue to live inside and through virtual systems, we believe education on issues of online privacy, transparency, security, and information distribution online will become increasingly vital. Indeed, Schmidt and Cohen (2013) proposed, "as children live significantly faster lives online than their physical maturity allows, most parents will realize that the most valuable way to help their child is to have the privacy-and-security talk even before the sex talk" (p. 37). An increase in awareness will better inform action and policy while helping individual users better manage perceived risks and rewards.

Private and Public Sectors Solidify Online

Further, in the brewing future, we anticipate a distinct line will be demanded, designed, and drawn in the virtual sand. As evident in this chapter, across our modern culture, from our media to our academic journals to our courts, expectations of transparency and privacy online are being debated and reconceptualized daily. In some areas, however, neither side appears ready for compromise. In response, we believe two definitive and disparate Internet experiences will emerge and solidify in the future virtual mainstream. This virtual separation will mirror the separation of private and public spaces in our physical scene. The Internet experience created by a distinct splitting of virtual space would support and solidify opting-in over opting-out.

On one side will be the "Transparent Network," where hosts employ and exploit behavioral tracking, location recording, digital portfolios, personalization, and other transparency-favoring practices fostered by routine surveillance and data mining. The "Transparent Network" will be explicitly designed, promoted, and treated as a public space. Commercial agencies, such as Amazon.com and Google.com, might elect to only be available via the "Transparent Network."

On the other side will be the "Private Network," or the "Anonymous Network," where users and hosts gain access to an anonymous (if desired), untracked, and predominantly private Internet experience. Tor, currently adopted in niche pockets in the United States and abroad, represents, in our minds, the beginnings of, or inspiration for, a massively adopted and disseminated "Private Network" (Fowler, 2012). On the "Private Network," there will be no "cookies,""supercookies," behavioral targeting, or other monitoring technologies. Window blinds, closed doors, and temporality will be reintroduced to communications conducted on the "Private Network." In essence, the consumer will be given more control over the distribution and circulation of her self-generated information. Virtual social environments, such as Facebook and/or its offspring, might elect to only be available via the "Private Network." Further, services available on the "Private Network" might implement subscription charges for profit instead of relying on advertising for sustenance.

Classroom Exercises

Debate It:

Resolved: Installing tracking technology onto the average consumer's Internet equipped device should require legal permission or notification.

Resolved: Installing tracking technology onto the average consumer's Internet equipped device should not require legal permission or notification.

Many websites, organizations, and governments have adopted the practice of automatically installing tracking technology onto the unsuspecting consumer's Internet-equipped device, from laptops to cell phones. Divide students into two opposing groups and debate whether such practices should require consent, notification, and/or legal permission. Consider social context, political tensions, and the various potential motivations behind the nonconsensual and warrantless implementation of tracking technologies.

Current Issues: In this chapter, we have discussed and outlined the emergence of various new technologies being designed and promoted as options for reclaiming privacy in a transparent age. For this exercise, instruct students to search for information on software being built to promote and reinstate semblances of privacy in digital communications. Are these types of software threatening digital transparency? Are they reaching the mainstream or residing inside niche pockets? Has privacy become a niche interest?

Everyday Impact: Behavioral targeting has become common practice across the online multiverse. For this exercise, encourage students to dissect their own Internet navigation experiences and how behavioral targeting might be impacting those experiences. Have students ever noticed behavioral targeting while surfing the Web? If "Yes," when and how? Did it make the students uncomfortable or did they enjoy a more personalized commercial experience? Did it lead any students to purchase a product or service?

Everyday Impact: New risks surround the publication of personal data online. At the time of posting their pictures, for Snyder, Stein, and Keller, each discussed above, the benefits outweighed any risks, if risks were considered. For this exercise, instruct students to describe, interpret, and evaluate how and what types of personal information they do and do not share online. When posting pictures online, do students consider future potential risks? Have students ever not shared something they wanted to share because of a perceived risk? Should personal photographs available online influence employment decisions? If "Yes," under what circumstances?

Key Terms

Risk-Benefit Ratio

Big Data

Digital Portfolios

Cookies

Information Bubbles

Chilling Effect

Encryption

Data Mining

Behavioral Targeting

Personalization

Panopticon

Predicative Analysis

14

International, Corporate, and Radical Politics

The Internet is an international and multilingual communications device diffused across boundaries, across the human species. Over two billion people from around the globe use the Internet, accessing it from private homes and cybercafés, from landlines and satellites, laptops and Internet-enabled phones (U.S. Census Bureau, 2012). Indeed, 356 million users in China alone access the web via Internet-enabled phones (CINIC, 2012).

Google maintains offices in more than 60 countries and offers its search interface in more than 130 languages (google.com). Popular social network, Twitter, is available in 28 different languages with over half of its users operating outside its American home (twitter.com). Indeed, across 2012, an astronaut tweeted from outside Earth. Users can "tweet" left to right and right to left in Spanish, Arabic, Farsi, Hebrew, Urdu, and more. Etsy.com, called the "world's most vibrant handmade marketplace," allows connected users to buy and sell hand-crafted art from Russia, the United States, and the United Kingdom. Here, users exchange eclectic goods seamlessly across borders, languages, and cultures.

The supranational quality of the Internet, however, has limits. The processes guiding its global diffusion and applications are social phenomena grounded in particular political economies (John, 2011, p. 328). Various ideologies embedded and supported online often conflict with local beliefs and practices supported offline. Local forces frequently impinge and redesign processes of globalization (such as the Internet's universal diffusion) (Sassen, 2007). As a result, the Internet retains universal characteristics when viewed across regions, but also produces a relative and regional experience.

Preview: A Regional Experience With Global Implications

Oftentimes, the relationships between digital communication technologies and populations vary significantly based on specific political climates. In this chapter, we use the description of specific events, popular social movements, and government actions to highlight the Internet as a dynamic and relative media system with significant, evolving international power. First, we begin with a discussion of government control of the Internet. While Americans experience limited censorship online (as of 2013), other regions routinely experience censorship online. Here, we describe the Internet in China; arguably one of the largest and most accomplished censored networks on planet Earth.

Second, we shift to a discussion of radical politics in the United States and abroad. At the dawn of the 21st century, social networks and other digital communication technologies, typically exploited for entertainment, identity management, marketing, and social interaction, evolved into invaluable tools for collective action against authoritarian agencies. Third and finally, we explore corporate and government relationships and battles with and over the Internet. Here, we describe the rise of cyber warfare, Twitter's global struggle for freedom of human expression online, and contrast Google's antitrust investigations in Europe and America.

Questions Previewing This Chapter

1. *How do applications of the Internet differ across countries, governments, and corporations?*

2. *What steps have local authorities taken to control their particular region's Internet experience?*

3. *How have Internet communication technologies aided radical politics?*

4. *How have online systems of power collided with offline systems of power in the 21st century?*

Government Control of the Internet

In the United States of America, users experience and navigate a predominant-ly unregulated and uncensored Internet. Currently (2013), North Americans can browse, stream, download, and upload an eclectic, near endless range of unrestrained, manic content. In contrast, regions and governments outside of North America limit and constrain the Internet experience available to their populace. North Africa, for instance, experiences highly repressed Internet activity and "egregious forms of censorship" in Sudan, Libya, Morocco, and Eritrea (Warf & Vincent, 2007; Warf, 2010). In these censored regions, partic-ular information and websites are restricted, blocked, and removed, while cyber activists and online journalists are systematically imprisoned (Warf, 2010). In-deed, some governments control and monitor the flow of information online like trusty faucets control the flow of hot and cold water offline. Motivations for censorship vary, but typically stem from notions of national security, needs for protecting public morality, and prohibiting blasphemy (Powell, 2001; Baghat, 2004; Warf, 2010).

Alternative View: Internet Censorship for Profit

While Americans generally enjoy an uncensored Internet experience, Amer-ican corporations have been repeatedly suspect of supporting censorship efforts abroad. The American-based corporation, Blue Coat, reportedly transfers software updates regularly to servers in both Syria and Iran (Hop-kins, 2013). The Blue Coat software and its regular updates are designed to help each respective country filter and spy on Internet activities. These updates, or goods shipped digitally, are considered violations of the U.S. embargo law, which has placed trade sanctions on both Syria and Iran. Syria is not new to Internet censorship. During escalating internal conflicts in spring 2013, Syria's Internet experienced two separate blackouts (Franceschi-Bicchierai, 2013).

In this section, we isolate and discuss one country, China, attempting to cen-sor, manipulate, and hack the flow of information online, nationally and interna-tionally. Here, a conversation on censorship abroad serves four notable functions. First, it shows the Internet as a dynamic communication device with unique re-lationships forming and occurring inside particular political environments. Sec-ond, it highlights the perceived media power attributed to information made available online. Third, it highlights the tools and methods used to manipulate and manage the flow of digital data. Fourth, it exemplifies a cross-cultural belief in the power of media to help shape, inform, and manage "reality."

The Function and Implementation of Media Manipulation Online: The Case of China

Since its beginnings in China, political and economic modes of power continually encroach upon the media power of the Internet (Lei, 2011). In 1996, two years after the Internet's integration into Chinese society, the government formally introduced regulations (Zhao, 2008). Today, as many as 12 different government-sanctioned agencies monitor and censor information exchanged online, including the General Administration of Press and Publication (GAPP), Ministry of Information Industry (MII), Central Propaganda Department (CPD), Ministry of Public Security (MPS), and State Secrecy Bureau (SSB) (Zhao, 2008). In part, these agencies help fuel and embody the "**Great Firewall**" of China and the "**Golden Shield Project**" (McDonald, March 2012). The "Great Firewall" aids in restricting access to undesirable foreign websites such as YouTube, Facebook, and Twitter, while the "Golden Shield Project" describes China's vast digital surveillance network designed to monitor the online behaviors and discourse of Internet users.

Media are invaluable tools for exercising power and controlling populations. In democratic societies, profit maximization and market demand often lead to "invisible" censorship and other forms of surreptitious media manipulation (Bourdieu, 2001; Lei, 2011). In China's authoritarian environment, the media is manipulated explicitly to maintain the rule of the Chinese Communist Party (CCP). Censorship and propaganda online help manage China's capitalist economy while combating the dissemination of rumors, defamation, and harmful information that might incite action against (or alter perceptions of) China's government and socialist system (Zhao, 2008; Lei, 2011).

Harmonization: The Purging of Suggestive and Critical Content

The Chinese Internet experience is often portrayed as "a giant cage" (The Economist, 2013). In China, users producing or disseminating anti-China information online are regularly targeted as cyber criminals or cyber dissidents. Posted comments or articles questioning or critiquing China are regularly "harmonized" (censored or deleted), within minutes of their posting. Terms and words such as "Tibet,""immolation," the "Dalai Lama,""democracy movement,""Sheng Xue" (a dissident writer), "Ai Weiwei" (an outspoken artist), and "Playboy" are commonly flagged for investigation or restrictions (McDonald, March 2012).

Analyzing 70 million messages created in one summer, one study found over 16% of digital messages sent in China were being systematically deleted (McDonald, March 2012). In addition to these instances of removal, China also pays thousands of trained commentators, coined the "50-Cent Party," to post and promote pro-government rhetoric across the web, steering opinion and criticism (Shane, 2011, p. WK1).

Virtual Loopholes

Before the Internet, mass media in China was only accessible to an elite class of journalists, politicians, experts, and intellectuals (Habermas, 2006). Today, the diffusion of the Internet has significantly increased the number of citizens that can access, produce, and disseminate information (Lei, 2011). China has the largest Internet population in the world with some 513 million growing Internet users (CINIC, 2012).[1] As more Chinese become entrenched in a mass media environment, awareness of media manipulation and censorship grows (The Economist, 2013). In response, some Chinese Internet users choose to adopt and employ methods for circumventing censorship. Savvy natives, international squatters, and other media conscious "netizens" avoid online censorship through various digital applications and coded discourse (see McDonald, March 2012).

Tools and applications continue to emerge that enable users to access restricted and blocked content. One significant and popular tool is the VPN (or Virtual Private Network; discussed in detail in Chapter 13), which enables users to navigate the Internet as if they were operating outside of their current physical location. Through VPNs, Chinese users effectively access content only available in specific regions of their choosing, from Europe to North America and beyond. As a result, content blocked in China, but available in other countries, becomes readily accessible under the cloak of a VPN.

Maintaining Control

China offers citizens an array of virtual clones (or Chinese doppelgängers) to redirect interest and illegal traffic towards globally entrenched websites such as Facebook, Twitter, and YouTube (services often accessed in China through VPNs). China provides their own government-controlled and monitored versions of these popular web services and social networks (i.e., Sina Weibo and Renren). Oftentimes, the design and aesthetic even matches the exact design and aesthetic of the non-Chinese counterpart.

Further, in late 2012, China explicitly responded to citizens employing VPNs to sidestep their information-management efforts. The government tweaked and upgraded China's "great firewall" and "golden shield," effectively limiting various popular illegal activities. Indeed, several VPN providers (i.e., Astrill, WiTopia, and StrongVPN) became immediately unavailable inside China after the 2012 upgrade (McDonald, December 2012). However, observing the diffusion of Internet in China, Lei (2011) proposed, "even though the political forces controlling the media system remain strong in China, a more decentralized media system enabled by technological advancements has weakened the power of the authoritarian state in monopolizing the production and dissemination of information and meaning" (p. 311).

Radical Politics and the Internet

Across history, communication technologies, from the telegraph to the printing press, have been creatively repurposed for political disruptions and culture jamming. Entering the early 21st century, Web 2.0 evolved similarly into an invaluable asset for radical actions. Various oppressed collectives across the world employed, and continue to employ, Web 2.0 to broadcast discontent globally, organize dissidence locally, and disrupt hegemonic forces nationally. Indeed, protests across 2011 in North America, Egypt, Spain, London, Tunisia, and Russia amassed varying degrees of success with and through the support of Internet technologies. In December 2011, the success of these radical movements led *Time* magazine to declare the protestor, "Person of the Year" (Andersen, 2011).

In this section, we discuss two radical acts of political resistance, the 2011 uprising in Egypt resulting in the dethroning of President Mubarak, and the North American "Occupy Wall Street" movement (OWS). Our discussion does not interpret effectiveness or evaluate outcomes, but seeks to describe how digital communication technologies have, and can, influence radical political movements. Today, as witnessed in Tunisia, Egypt, and North America, resistance against authoritative structures appears more public and more accessible than ever before, in part due to the exploitation, use, and reach of the Internet.

The 2011 Egypt Uprising: A Case Study

On January 28, 2011, the government in Egypt cut nearly all internal access to the Internet, causing a 90% drop in data traffic to and from Egypt (Richtel, 2011). Egypt, in essence, went offline. *The New York Times* declared the shutdown, "unprecedented in scope and scale" and "a concern for the global community" (Richtel, 2011, p. A13). Leading up to the shutdown, the subordinates of Egypt had been exploiting social networks online to inform, direct, and empower protests against the hegemony and, more specifically, the current president, Hosni Mubarak (Shane, 2011).

In Egypt, as with Tunisia before it, Facebook, Twitter, and blogging sites became transcendent, self-multiplying megaphones exposing government weaknesses and atrocities across the country. The citizens of Egypt, dissatisfied with government dealings, inspired action with organizational tweets, video uploads of authoritarian brutality, and photographs from the streets posted directly online in near real time. The connected globe quickly became witness to what Scott (1990) called, an "electrifying moment" where the hidden transcripts of subordinate classes are "spoken directly and publicly in the teeth of power" (p. xiii).

Dethroning Mubarak

President Mubarak's suppression of his country's Internet for five consecutive days underlined the Internet's capacity to influence and shake authoritarian governments from the ground level. Mubarak silenced the Internet because it was being used as a weapon against his regime, but, tellingly, the Internet blackout drove more people to the streets to protest his regime (Goodman, 2011). By February 2011, the public protests (nurtured online) resulted in the forced early departure of President Mubarak from office and the resurrection of the Internet in Egypt (Goodman, 2011).

Alternative View: A Vulnerable Agency for Resistance

While some celebrate the ability of the Internet to combat oppression, others cite the Web's simultaneous ability to thwart and routinely monitor dissidence and potential resistance (Morozov, 2011; Shane, 2011). Already, online surveillance is effectively used in China and Russia, with Facebook posts quoted during interrogations (Shane, 2011). Discussed at length in Chapter 13, transparency abounds online. Our social networks and digital footprints create easily cracked windows into our thoughts and lives. Today, police and governments can seamlessly access cell phone records, location data, and social network activities without the pre-Net need for street surveillance, wiretapping, and legal paperwork (see Cohen, 2011; Helft, 2011; Morozov, 2011; Perlroth, 2012; Sengupta, 2012; Shane, 2011). A dissident's social network and Twitter feeds are "a handy guide to his political views, his career, his personal habits and his network of like-thinking allies, friends, and family" (Shane, 2011 p. WK1).

The North American Occupy Wall Street Movement (OWS): A Case Study

In North America in 2011, the malcontent spent months repurposing public spaces into encampments for resistance against disproportionate economic and political power structures. They operated collectively under the ominous banner, "Occupy Wall Street." Online, their battle cry was easy to find, "We are the 99% that will no longer tolerate the greed and corruption of the 1%" (occupywallst.org). Online banners and tweets called for general assemblies in "every backyard, on every street corner because we don't need Wall Street and we don't need politicians to build a better society" (occupywallst.org). Participants hoped the occupation of public spaces and the public dissemination of their transcripts online would empower "real people to create real change from the bottom up" (occupywallst.org).

Across fall 2011, the founders, New York City protestors and their Internet mouthpiece, occupywallst.org, witnessed an explosive global adoption of their cause. OWS was embraced nationally and globally, from California to Indiana to Hong Kong to London to Ireland. In this section, we focus on the digital tools and strategies employed by the OWS movement to disseminate information on their movements, ideas, and political repression.

Circumventing Traditional Media

In late 2011, Web 2.0 allowed OWS participants to circumvent traditional media outlets to disseminate news and updates on their cause. OWS participants used the Internet to instantly deliver incessant and interactive coverage of OWS to an interested and engaged online population. Average citizens and amateur journalists generated their own raw news content. In two weeks (from November 1, 2011 to November 15, 2011), less than one month after the launch of their public occupation, occupywallst.org garnered over 14,000 individual personalized comments across 39 published articles and 15 uploaded videos (approximately 1,000 new comments every day or 40 new comments every hour). At two months old, OWS had 323,286 likes on Facebook.com, 21,357 subscribers on Reddit.com, and 121,170 followers on Twitter. Indeed, Twitter was a frequent tool for live OWS updates from inside and around each encampment, public march, and street clash with police. During one OWS march, in one hour, from 1:00 PM to 2:00 PM on November 17, 2011, there were over 1,500 unique tweets (approximately 25 new tweets every minute) mentioning #OWS (a digital tag indicating the message is about OWS) from over 1,000 twitter accounts.[2]

Live Streaming and Telepresence

November 17, 2011 marked the two-month anniversary of OWS's physical presence in Lower Manhattan. To celebrate, participants organized an all-day march on Foley Square and Brooklyn Bridge. Using wireless Internet and a smartphone, amateur journalist Tim Pool live-streamed the march on www. ustream.tv/theother99. Here, persons from around the globe were granted open, unending access to the public march. Indeed, over 30,000 users marched virtually alongside Pool across the day and into the night. If Pool mentioned his batteries were low on the live feed, within minutes, a battery would appear to sustain the stream. If online users used a text-based chat feature to request Pool engage or interview certain protestors on the ground in New York City, then Pool eagerly obliged (as observed live on www.ustream.tv/theother99 on November 17, 2011). Pool, wireless Internet, and his mobile phone enabled a unique form of telepresence. Through Pool, persons not able to physically march across New York City were able to feel present among the agitated crowds and rhythmic chants, engaging physical protestors with virtually launched questions.

Video Uploads

The Pew Research Center reported that over half (56%) of all adults now own a smartphone (Pew Internet & American Life Project). Across OWS encampments and marches, these smartphones were endlessly visible, snapping pictures and recording video, without reprieve. Complementing live-streaming efforts, OWS members regularly uploaded content to YouTube.com and Vimeo.com. Online videos became an attempt for OWS members to gather support and ensure accountability during tense cultural clashes.

When police turned violent, an OWS member, or a group of OWS members would eagerly record and upload videos of the incidences online. For example, during OWS protests in Oakland, California, American police were recorded aggressively raiding public encampments with tear gas, riot gear, and rubber bullets. During the Oakland raids, police severely wounded one unarmed and nonviolent American protestor, Scott Olsen. Videos of the Scott Olsen assault spread quickly online, garnering hundreds of thousands of views and touting ominous slogans such as, "We are all Scott Olsen" (youtube.com). The videos helped amass public support across Oakland and the nation, culminating in a physical march that momentarily shutdown various high-profile, commercial ports on the west coast (Chea, Left, & Collins, 2011).

> ### *Alternative View: Who Watches the Watchmen?*
>
> *Social theorist Michel Foucault posited that systems of overt or covert observation implicitly carry disciplinary power (Foucault, 1975, pp. 170–177). Visibility, or the threat of visibility, can influence the way we act in certain contexts. Architectures (or networks) of surveillance can and do control our behaviors (see Foucault, 1975). We see this in prisons, churches, and schools. As seen across 2011, 2012, and 2013, protestors around the globe, equipped with smartphones, wireless Internet, digital cameras, Twitter accounts, and Facebook pages, created mobile networks of surveillance that directed a disciplinary gaze at the disciplinarians. Across the globe, we witnessed protestors generating what Foucault (1975) called, "supervisors, perpetually supervised" (p. 177).*

Corporations and the Internet, Part 1: Connected and Exposed

Digital warfare is rising online, across the globe, threatening any and all plugged-in corporations and governments. In Russia, Kaspersky Lab, an online security firm, uncovered a global malware campaign actively stealing research from some 40 different countries. Hackers used a tool called NetTraveler to steal data on space exploration, nanotechnologies, energy production, medicine, and communications from government institutions, embassies, research centers, military contractors, and energy industries (Franceschi-Bicchierai, 2013). In 2013, using his State of the Union address, United States President Barack Obama publicly conceded, "We know foreign countries and companies swipe our corporate secrets [online]" (Sanger, Barboza, & Perlroth, 2013, p. A9). Further in 2013, in an unprecedented declaration, the U.S. Pentagon explicitly alleged the Chinese government actively engages in cyber warfare against the United States.

Today, governments and hackers exploit the Internet for cyber espionage with limited and sensitive national responses. Indeed, corporations and governments often hesitate to participate in investigations or acknowledge attacks as it can expose weakness (Gladstone, 2013). In this section, we discuss some of the known victims, perpetrators, and various popular methods used in digital warfare online.

Digital Warfare: Hacking Networks, Plundering Secrets

Economies across the globe increasingly rely on the Internet for a plethora of day-to-day operations, from communications to data storage. Currently, however, online agencies continue to be vulnerable to "cyberwarriors" on the prowl for corporate and government secrets (Sanger et al., 2013). In 2011, one study found malicious attacks accounted for some 37% of recorded data breaches (Ponemon Institute, 2012). Attackers employ various methods to penetrate protected networks, including rogue insiders, phishing, the theft of data-bearing devices, and electronic agents (i.e., malware, viruses, worms, and trojans).

The victims of cyber attacks are numerous, including major corporations from across the globe. For instance, Coca-Cola, *The New York Times*, *The Wall Street Journal*, the United States military, and Canada's Telvent, a company specializing in energy, oil, and gas management systems, have each been hacked by electronic agents (Sanger et al., 2013; Gladstone, 2013).

Spearphishing and PLA Unit 61398

Many believe China's People's Liberation Army (PLA) Unit 61398, operating inside Shanghai, routinely infiltrates and pillages corporate infrastructures across the globe (Mandiant, 2013; Sanger et al., 2013). According to *The New York Times*, PLA Unit 61398 routinely executes **spearphishing** techniques to enter protected networks, remaining undetected for months or years (Sanger et al., 2013). Spearphishing places surreptitious malware onto a particular computer when a user opens an infected e-mail attachment sent from a deceptive source. Malware implanted through spearphishing enabled international agents to leisurely scan and hide inside targeted networks, including *The New York Times* network (Gladstone, 2013). One study found 141 explicit attacks using electronic agents originated from the same neighborhood in Shanghai, the alleged home of PLA Unit 61398 (Mandiant, 2013).

Alternative View: Defending Data Breaches

Corporate and government defenses are necessarily responding to the increase in cyber attacks. In 2011, the organizational costs loss for data breaches declined for the first time in seven years, dropping from $7.2 million to $5.5 million (Ponemon Institute, 2012). The decline in costs suggests organizations have improved their methods for preparing and responding to a data breach. However, internal negligence continues to enable information theft (Singer, 2013). In 39% of cases, research found negligent employees or contractors inadvertently aided data breaches in their organization (Ponemon Institute, 2012; Singer, 2013).

Twitter Account Hijacking

For journalists and other prominent information peddlers, Twitter is a convenient and efficient means for disseminating urgent news and live updates on prolific and spontaneous happenings. Unfortunately, as a result, hacking Twitter has become an emergent means for digital warfare. Activists, jammers, and global dissidents repurpose established and refutable Twitter accounts domestically and abroad to disseminate damaging misinformation. Hijacked Twitter accounts can disrupt economies, governments, and the flow of valid and reliable information. For example, in spring 2013, a U.S. Associated Press Twitter account was hacked to deviously announce, "Breaking: Two Explosions in the White House and Barack Obama is injured" (Sternstein, 2013). The U.S. stock market flinched in response before the White House Press Secretary Jay Carney made a public announcement discrediting the Tweet.

Corporations and the Internet, Part 2: Free and Restricted

Internet agencies are often permitted to design their own rules and ideologies for their company's specific operations and practices online. These rules and ideologies often reflect the nation from which the company originates and resides. As a result, as these companies spread internationally, they frequently conflict with political interests resulting in government investigations, legal battles, and the dissolving of affiliations. Here, the freedoms of the Internet as a global entity and the companies operating inside it are exposed and tested. In this section, we describe two instances where companies operating online clash with, and adjust to, outside political forces.

Twitter and the Struggle for Unrestricted Content Abroad

Twitter was founded in the liberal glove of San Francisco, CA, in the United States in 2006. As of 2012, Twitter now has over 100 million users with well over half of all accounts registered outside the United States (Sengupta, 2012; twitter. com). With their rabid international diffusion, Twitter faces increased pressure to censor content regionally.

Regional requests to restrict Twitter content conflict with the company's idealistic belief that the freedom of expression is a universal human right (twitter. com). However, different regions have particular definitions and parameters for freedom of expression. As a result, Twitter has agreed to censor particular tweets within particular regions when explicitly requested by government officials. User tweets will still remain visible in uncensored countries outside of the particular countries requesting censorship. Further, toward transparency, Twitter will notify users when and if information is withheld (twitter.com).

Google and Antitrust Allegations: USA Versus Europe

As a search engine, Google accounts for nearly 80% of all web searches in the United States (Harbour, 2012). Like McDonald's before it, Google now offers the best fast food in town. Google's increased exposure and widespread adoption has brought increased and ongoing scrutiny. For instance, in 2012, both the United States and Europe uncovered potentially harmful practices occurring throughout and across Google search results (Kanter & Lohr, 2012). The Google search engine appeared to favor the company's commerce and other services in search queries, deceiving customers and limiting the potential for unbiased discovery (Kantner et al., 2012, p. B1). Journalists, lawyers, and academics argue practices such as **search bias** inhibit competition and consumer welfare (Harbour, 2012; Kanter et al., 2012).

Accordingly, both Europe and the United States developed separate antitrust investigations centered on Google's practices inside their respective countries. The current (2013) aftermath of these investigations is further indication the Internet, like water, takes different forms across different environments. In the United States, the Federal Trade Commission (FTC) backpedaled and softened their aggressive investigation into Google. In response, *The New York Times* warned, search bias online is a "severe setback for Internet users" (Harbour, 2012). The acceptance and proliferation of search bias online can endanger consumer choice and the free flow of information, now and into the future.

In Europe, Google remained under investigation due to ongoing concerns of search bias, **screen scraping**, and market obstruction. In contrast to the United States, Europe's investigation of the Google antitrust allegations reflected a belief that online competition should be shielded by the government, to a degree, against the potential abuses of a dominant company.

Conclusion and Future Predictions

While the Internet radically reshapes and impacts cultures around the globe, a lack of access still persists among various demographics and regions. Africa, for example, has struggled to provide Internet access across the region's 54 countries. Warf (2010) observed,

> whereas 26.8 percent of the world's people used the internet at the end of 2009, in Africa the internet penetration rate was only 8.7 percent; home to almost 15 percent of the world's people, Africa has less than five percent of its Internet users. (p. 43)

Today, however, the evolution of low-cost Internet-enabled phones and wireless networks now challenges the once persistent digital divide manifest across various geographies (Chen, 2004; Rice & Katz, 2003; Schmidt & Cohen, 2013).

A Diminishing Digital Divide: An Emerging Skill Divide

Akiyoshi and Ono (2008) suggested that mobile phones offer users with limited technological readiness a more accessible and affordable means for navigating the Internet. For instance, Internet users once marginalized in Japan, women, the less-educated and the less-affluent, have successfully adopted Internet-enabled phones in significant numbers (Akiyoshi & Ono, 2008). Here, gender, language skills, keyboard skills, telecommunication costs and other barriers preventing Internet access were circumvented with the diffusion of highly ergonomic and low-cost, Internet-enabled phones.

Outside of Japan, in Africa, the increase in mobile phones, wireless networks, and cybercafés also shows encouraging signs of diffusing Internet to marginalized audiences (Warf, 2010). Africa now (2013) boasts more than 650 million mobile-phone users, while Asia reports over three billion mobile-phone users (Schmidt & Cohen, 2013). Across rural provinces in China, users connecting via mobile devices already outnumber those connecting from computers (The Economist, 2013). Indeed, we believe the continued evolution and dispersal of Internet-enabled phones will continue to reduce the digital divide while nurturing enhanced global connectivity. Accordingly, more immediate concerns will shift focus from a lack of access to a lack of advanced hardware-software skills and repressed networks (Ynalvez & Shrum, 2006).

Yet, despite this emergence of a skill divide and ongoing network repression online, we maintain the next stage of Internet diffusion will only continue to radically revolutionize communities across the globe: improving and increasing communications, narrowing education gaps, challenging power structures, and connecting producers, consumers, and the disenfranchised operating on the peripheries of popular culture.

Classroom Exercise

Everyday Impact: The Internet wants to be free, but across the globe it faces resistance. Purposeful and radical censorship abounds across the globe. For this exercise, encourage students to evaluate their own experiences navigating the Internet. Have students ever experienced content restrictions online? How and why was this content restricted? Where was this content restricted? Considering age, social context, and political tensions, should content ever be restricted online (if yes, under what circumstances)?

Debate It:

Resolved: Governments should be permitted to shut down the Internet.

Resolved: Governments should not be permitted to shut down the Internet. Divide students into two opposing groups to debate when, where, how, and why governments should be, or should not be, permitted to shut down the Internet. While exploring this topic, consider social revolutions and political tensions seen nationally and abroad.

Current Issues: Misinformation is growing online with dangerous lurking and realized consequences. For this exercise, instruct students to search for instances or reports of misinformation being disseminated online. How was this information distributed? How did it impact society? Is the dissemination of misinformation an ongoing threat of significant concern? How can misinformation be combated?

Key Terms

Harmonization

Golden Shield Project

Spearphishing

Search Bias

The Great Firewall

Screen Scraping

Endnotes

1. Net citizens in China spend approximately 18.7 hours surfing the web per week (CINIC, 2012).
2. Comparatively, there were 2,175 individual tweets from 6:00 PM to 7:00 PM on November 17, 2011 discussing #OWS (approximately 36 new tweets every minute) and 920 individual tweets from 12:30 PM to 1:30 PM on November 20, 2011 discussing #OWS (approximately 15 new tweets every minute). #OWS is not the only digital tag associated with conversations about the OWS movement. However, #OWS is arguably the most popular tag associated with the movement. Alternative digital tags generate fewer tweets. For instance, there were 300 tweets from 1:00 PM to 2:00 PM on November 20, 2011 mentioning #occupyoakland (approximately five new tweets every minute) and 618 tweets from 1:45 PM to 2:45 PM on November 20, 2011 mentioning #occupy (approximately 10 new tweets every minute).

15

The Future of the Internet and Web Reconsidered: Back to the Future With Tim Berners-Lee

We want to end this volume on a speculative note. We think it is healthy and desirable—from a cognitive perspective—to just think about what the future of digital technologies and the Internet might become.

For some reason, thinking and writing about digital technologies and the Internet necessarily forces your mind—not only to think about where society and cultures have been and are—but where they are going. In other words, contemplating digital technologies and the Internet creates a powerful sense of where we are going. During the years we have contemplated this volume, a host of predictions have crossed our minds. We have kept track of these predictions. We share 29 of them here with you in the hope they stimulate your future thinking processes. The first seven of these predictions are derived from Anderson and Rainie (2006) and predictions 8 through 12 are based on Pew Internet & American Life Project's *The Future of the Internet III* (Anderson & Rainie, 2008). We continue to believe that such predictions require trend extrapolation data as well as the results of diverse Delphi procedures (see Chapter 4) if we are to feel confident about such predictions, we do think these 29 predictions reveal a host of potential developments worth considering.

Questions Previewing This Chapter

1. Based upon the published literature, what 29 predictions set the stage for future thinking?
2. How might the Internet foster and create a new reality that competes with our everyday face-to-face reality?
3. What five consequences and future projects might emerge from a virtual reality?
4. How has one of the founders of the Internet, Tim Berners-Lee, conceived of and anticipated the future of the Internet?

Beyond these 29 predictions, we want to add a 30th (a prediction that requires separate treatment because of the distinctions we draw between everyday and virtual realities) and then add five more after that. We end this chapter by considering the future orientation of Tim Berners-Lee. In all, then, we will consider some 36 predictions about digital technologies and the Internet. We begin, then, with our first 29 predictions.

These first 29 predictions include:

1. Global, low-cost networks exist, providing mobile wireless communications available to anyone anywhere at any time at an extremely low cost.
2. English will *not* displace other world languages but become one of many multiple language options.
3. The potential for surveillance, security, and tracking systems getting out of control will require intensive efforts to revise the movement toward autonomous technology.
4. Privacy will be redefined, but efforts to preserve the personal/public distinction will be under way.
5. Virtual reality on the Internet will come to enable more productivity from most people in technologically savvy communities than working in the real world.
6. Human organizations will blur current national boundaries as they are replaced by city-state, corporation-based cultural groupings and/or other geographically diverse and reconfigured human organizations tied together by global networks.
7. A new cultural group of technology "refuseniks" (the "New Luddites"), who separate themselves from "modern" society, will commit terrorist acts.

8. Next-generation research will be used to improve the current Internet; it won't replace it.
9. The mobile phone is the primary connection tool for most people in the world.
10. Talk and touch are common technology interfaces.
11. Few lines divide professional time from personal time, and that's OK.
12. Many lives are touched by the use of augmented reality or spent interacting in artificial spaces.
13. By 2020, Internet usage in the United States will reach 90%.
14. By 2015, computer memory capacity will increase by at least 10 times the current level, transforming personal computer databases, ultimately enabling individuals to merge or converge as many computer systems into one as they prefer.
15. A new generation of Internet usages—perhaps under the label of *Web 2.0*—will dominate computer use by 2020.
16. The democratization of the Internet will continue to dominate and overwhelm professional gate keepers, critics, and experts, although more effective search and category systems will make Web 2.0 more useful.
17. The Internet will become increasingly globalized, reflecting—not diminishing—the diversity of languages, cultures, and politics of the world.
18. Mobile communication systems will increasingly dominate networked digital communication systems by 2020.
19. Television and the Internet will merge, increasingly blending over-the-air and broadband connections, creating and sharing a merger of content through Internet technologies (e.g., laptop computers and smartphones) and television, creating a "hybrid" technology.
20. There will be a demographic shift in the focus of digital marketing. The young and old become important when there is a shift in content, strategy, and advertising focus from the Millennial Generation (born 1982 through 2003 or 2004) to generations both younger and older than the Millennial Generations.
21. By 2030, a transition toward a more seamless distinction between virtual reality and "real life" (everyday reality) will begin to occur, and the transition will seem increasingly appropriate and necessary for a significant group of computer users.
22. By 2030, sensor-based computing will dominate human–machine interactions.

23. By 2030, because of miniaturization and implants, the computer will "disappear"—links or access devices to Internet users and massive data bases will be smaller and smaller, with computers built into more and more devices.

24. By 2025, low-level computer chips will be in virtually all everyday objects thereby enabling virtually all everyday objects to be linked to and to process information and talk to the web.

25. By 2020, professions will regularly utilize digital technology to create such tangible and physical objects as parts and tools. By 2025, this practice will become widespread in homes.

26. Warfare will increasingly take place through digital technologies, more so than on a battlefield. Combatants on battlefields will be entirely connected through integrated digital technology. These advancements in physical warfare will speed the development of organic-digital hybrids.

27. By 2035, it will become more common for someone in a professional, managerial, or administrative position to work from home or remotely than to work in an office.

28. Economies of many nations will continue to merge as the Internet links the commerce of nations and as national economies become more based on the Internet.

29. Personal relationships will continue to influence and be influenced by the Internet. In this regard, a series of predictions concerning the Internet's influence on personal relationship is appropriate.

First, romantic relationships will become increasingly dependent on the Internet. The Internet will soon become a predominant method of finding romantic partners for all groups, not just those groups with lean dating markets.

Second, relational interaction and maintenance will increasingly take place on the Internet. In many cases, friends, associates, family members, and romantic partners will interact more through the Internet than through any other means.

Third, although not widespread in the near future, it will nevertheless become common for romantic relationships to take place entirely through digital means. In many cases, participants may never physically meet. People will meet through the Internet, interact and maintain their relationship through the Internet, get married/committed through the Internet, make love through the Internet, and divorce or separate through the Internet.

Fourth, as just alluded, engaging in sex will take place digitally, initially relying on teledildonics and then moving toward neural sensors and interfaces connected to the brain and nervous system. This trend will result in a decrease in sexually transmitted diseases and in a decrease in birthrates worldwide.

Fifth, and finally, human–human relationships will continue but human–digital relationships will become widespread. Both friendships and romantic relationships will develop among human users and digital entities. These digital entities will likely take the form of avatars, but in many cases will be developed into the physical form of a robot (see Levy, 2007, for example). In the case of romantic relationships, people will fall in love with these digital entities in whatever form is taken, and human–digital relationships will develop with same patterns and emotions as human–human relationships.

A 30th Prediction

Additionally, especially as we witness the power that dating sites have had in the lives of people as well as how powerful video and online games have been in directing lifestyles, we increasingly believe that the distinction between everyday reality and virtual reality can and is becoming blurred. Table 15.1 provides some of the distinctions that exist between these two reality systems. The differences seem worthy of consideration.

In our view, virtual realities are becoming more participatory, immersive, vivid, and manipulative. From mainstream culture to counterculture, many optimistically posit we will soon meld sensory experiences flawlessly with virtual reality (Kurzweil, 2005; Wahlmann, 2009). On the popular counterculture pages of *adbusters*, Wahlmann (2009) projected, "It just seems inevitable that people are going to continue to live more and more through technology. I think the gene-based, corporeal life we are familiar with is just the incipient stage of an evolutionary development of universal intelligence." Wahlmann, like fellow technological determinists, believes humans are fecundating virtual systems and artificial intelligences like bees fecundate plants.[1]

Mass crowds will migrate to these advanced, autonomous virtual environments, preferring to spend more time living online than offline, interacting seamlessly with both artificial intelligence and augmented, virtual humans. The precedence we set today for privacy online and digital security becomes exponentially more meaningful if virtual migration is to occur safely in the future. Currently, for instance, there are few private virtual bedrooms without integrated digital voyeurs.

Table 15.1. Exploring the Differences Between Everyday Reality[2] and Virtual Reality: A Set of Heuristic Propositions.

Dimensions of Any System Cast as a Reality	Everyday Reality	Virtual Reality
Time	External, uniform, and standardized agreement of time in internationally recognized units (e.g., seconds, minutes, hours, etc.)[3]	Sense of time is determined by the sequence and activity involved within the process. "Fast" and "slow" in a video game, for example, depends on the conflict levels, number of players, and range of actions available to players.[4]
Space	Sense of perspective and sense of three-dimensional figure-background perception	Two-dimensional "flatland" can exist and be manipulated to introduce figure-ground experience by increasing the size of the figure and reducing the size of background items as well as through recent visualization technologies.[5]
Feedback	Synchronistic and immediate	Asynchronistic and delayed
People	Image manipulation within the context established by physiology of the individual human body	Avatars can reflect physiology of the individual human body or dramatically reconstruct the image, sense of the self, and become a reflection of identity rather than physiology
Agency	Agencies used are cast and treated as tools or extensions of the human senses, reflecting and designed to extend human goals and purposes	Agencies used are cast and treated as technologies that can process activities in ways that humans do not, and they can produce processes and outcomes that are unexpected by human designers.
Ambient Sound	Soundscapes exist and are inherently created by non-human factors and human activities unrelated to immediate human interactions in every human environment	Sound tracks must be consciously and explicitly created and consciously and explicitly adjusted as "background" as ongoing human interactions change

Table 15.1. *Continued*

Dimensions of Any System Cast as a Reality	Everyday Reality	Virtual Reality
Smell	One of the basic human senses used to imposed social interpretations and meanings upon human interactions	Not clearly a feature of virtual realities although "guesses" about smell might be inferred from background and actions of central human actors
Touch	One of the basic human senses used to impose social interpretations and meanings upon human interactions	Touch screen and computer keyboard entries can associate actions within virtual realities with the sense of human control through touch although the sense of touch among avatar or avatars and avatars' environment is currently not that explicit.
Purpose	Competing human philosophies are an ongoing source of human intrigue and conflict, especially as they impact political systems and decisions	Specific purposes and objectives govern different types or genres of virtual realities, although interactions among interactive parallel processing is seldom a feature of ongoing virtual realities
Coherence of the System[6]	Closed system in which external inputs are integrated into existing patterns	Open system in which external inputs determine and control the internal dynamics of the procedures of a virtual reality[7]
Type of Rhetoric or Storytelling[8]	Its rhetorical framework casts everyday reality as the most basic and fundamental sense of what is	Cast as a supplement to existing everyday reality systems in much the same way that Facebook maintains that it exists to reflect previously established friendships

This 30th prediction gives rise to an additional five predictions about the Internet that seem worthy of consideration to us.

The 31ˢᵗ, 32ⁿᵈ, 33ʳᵈ, 34ᵗʰ, and 35ᵗʰ Predictions

31. Internet Retreats

We believe media detoxification (or deliberate, perhaps forced vacations, short or extended, from digital technologies) will become increasingly popular with various groups and governments. Abroad, South Korea already imposes digital curfews and plans to launch education for minors on electronic addiction (Souppouris, 2012). Many will view these educational seminars and retreats from the digital as essential for healthy relationships with, and between, online communication technologies and the natural world.

Ongoing research suggests an extended separation from technology has the potential to restore and boost various cognitive functions underutilized or ignored in technology-saturated environments (Atchley, Strayer, & Atchley, 2012). In their study comparing control groups before and after a four-day, technology-void hiking trip into the wilderness, Atchley et al. (2012) found a significant increase (a 50% boost) in various cognitive functions after immersion in a natural setting free of technology.

32. Efficient and Cognitive Machines

IBM predicts within five years (from 2012) machines will be able to feel and smell (Dillow, 2012). Every month of every year, on stages, in laboratories, and on television, our programs and software systems showcase increasingly more accurate feats of speech and pattern recognition. From chess to *Jeopardy!* to identifying images in and across vast databases, our computer programs are outperforming our human experts, our organic minds (Markoff, 2012). In 2012, IBM began training Watson, the computer that thoroughly defeated two human champions in *Jeopardy!* in 2011, to assist doctors with diagnosis in hospitals (Hafner, 2012; Markoff, 2011). Efficient and cognizant machines are entering professional scenes with a snowballing potential to excel in data farming, civilian surveillance, and drug development (Markoff, 2012). Cognitive machines are rising.

33. Organic-Digital Hybrids

As our digital technologies continue to integrate themselves into society while out-processing the human mind without bias or exhaustion, populations will likely seek means to enhance their organic minds (self-perceived and media-promoted as slow or inferior) through a physical union with digital systems. Today, nearly half of all Americans (45%) own a smartphone device (Pew Internet & American Life Project, September 17, 2012). These mobile, wirelessly connected devices accompany their users everywhere, from the dining table to the theater to the bathroom, leading some to declare, we are all already cyborgs (Snider, 2012). Tapping these devices, users can be seen consulting the hive mind frequently. Users replace memory recall or face-to-face queries with a quick web search, a virtual consultation.

The synthetic boundaries between our intellectual technologies and organic intelligences are converging and primed to continue converging. Toward this end, both Apple and Google are developing commercial eyeglasses that will provide a constant connection to the Internet while digitally augmenting physical environments and interactions. The appeal and ease of merging the organic brain completely with digital technologies will further intensify with the continued evolution of artificial intelligence and the commercialization of wearable, constantly viewable, digital interfaces. Artificial neural networks (or "neural nets") already increasingly resemble the neural connections in the organic brain (Markoff, 2012).

Many believe an organic-digital fusion will improve on our current "suboptimal," "irrational," and "inconsistent" cognitive tendencies (Bailey, 2012; Kurzweil, 2012; Wickelgren, 2012). Promoting his text, "How to Create a Mind: The Secret of Human Thought Revealed," the provocative futurist Ray Kurzweil (2012) posited a merger between artificial intelligence and human intelligence will exponentially expand intelligence, quickly solving minor problems such as war, death, and material scarcity (Bailey, 2012).

34. Radical Media Separatism

The overconsumption of misinformation and catered, selectively targeted information will continue to encourage information intolerance and social separatism. Prior to the 2012 North American election, American "values and basic beliefs" were "more polarized along partisan lines than at any point in the past 25 years" (Pew Research Center for the People & the Press, June 4, 2012). The polarization, in part, must be attributed to media production and media consumption patterns. If current trends continue, social groups and governments will continue to fragment based on arguments and ideas supported and fueled, in part, by the consumption of selective and biased media experiences and self-perpetuating personalized information delivered online.

35. Nonlinear Normalization

Carr (2010) observed, "calm, focused, undistracted, the linear mind is being pushed aside by a new kind of mind that wants and needs to take in and dole out information in short, disjointed, often overlapping bursts—the faster the better" (p.10). The suppression of the linear thought process will continue as the mind becomes increasingly habituated to the nonlinear consumption patterns favored online. Expedient and diverse multitasking will become the preferred method of media consumption, as "single-tasking" and linear consumption becomes perceived as rudimentary, inefficient, sluggish, and restrictive.

The 36th Prediction: Berners-Lee's Future Orientations

We conclude this volume by mentioning the future orientation of Tim Berners-Lee. In such powerful ways, Berners-Lee was responsible for the birth of the Internet when he proposed a code for unifying the diverse voices that we so readily understand to be the World Wide Web. As we noted in Chapter 3 when we discussed the founders of the Internet, Berners-Lee's website provided this historical note about Berners-Lee's early efforts to create the World Wide Web:

> In 1989, he proposed a global hypertext project, to be known as the World Wide Web. Based on the earlier "Enquire" work, it was designed to allow people to work together by combining their knowledge in a web of hypertext documents. He wrote the first World Wide Web server, "*httpd*," and the first client, "*WorldWideWeb*" a what-you-see-is-what-you-get hypertext browser/editor which ran in the NeXTStep environment. This work was started in October 1990, and the program "WorldWideWeb" first made available within CERN in December, and on the Internet at large in the summer of 1991.

We intentionally end this volume by considering some of the forecasts and predictions that this early founder has contemplated.

Two of Berners-Lee's explorations have captured our attention here. First, we explore his notion of the semantic web or "Web 3.0." Second, we examine some of the basic conceptions of what a Web Science curriculum at the PhD level might involve.

The Semantic Web

Proposed in 2001, Berners-Lee, Hendler, and Lassila suggested that a "Semantic Web will bring structure to the meaningful content of web pages, creating an environment where software agents roaming from page to page can readily carry out sophisticated tasks for users" (*Scientific American*, May 2001). As Berners-Lee and his colleagues continued in greater detail:

> The Semantic Web is not a separate Web but an extension of the current one, in which information is given well-defined meaning, better enabling computers and people to work in cooperation. The first steps in weaving the Semantic Web into the structure of the existing Web are already under way. In the near future, these developments will usher in significantly new functionality as machines become much better able to process and "understand" the data that they merely display at present.

> The essential property of the World Wide Web is its universality. The power of a hypertext link is that "anything can link to anything." Web technology, therefore, must not discriminate between the scribbled draft and polished performance, between commercial and academic information, or among cultures, languages, media and so on. Information varies along many axes. At one end of the scale we have everything from the five-second TV commercial to poetry. At the other end we have databases, programs and sensor output. To date, the Web has developed most rapidly as a medium for people rather than for data and information that can be processed automatically. The Semantic Web aims to make up for this.

The Semantic Web, then, would process all data and sensory inputs based upon the meanings implied explicitly and implicitly of data and sensory inputs as human beings know and understand them. In this sense, Berners-Lee and his colleagues predict that machines will process information along the meaning dimensions that human beings now employ.

While cast by some as a form of artificial intelligence, others suggested that the Semantic Web required that earlier developments of the World Wide Web be reconsidered. In this regard, the notion of a "Web 1.0" was created to provide a more coherent description of how the anticipated development of the Semantic Web or Web 3.0 might emerge within the context of existing technologies. Within this context, Web 1.0 constituted "very loosely defined boundaries" of "strictly one-way published media" (Wikipedia, "Web 1.0," April 19, 2008). In this regard, Web 1.0 websites are analogous to the conception of *sources* of information in which the source is believed to solely create and is responsible alone for the transmission of messages to a relatively passive group of receivers.

Likewise, Web 2.0 websites were cast as environments in which users dominate and determine the messages and interactions among users (Wikipedia, "Web 2.0," April 19, 2008). In this regard, Web 2.0 employed technologies that aimed to "facilitate creativity, information sharing, and, most notably, collaboration among users" (Wikipedia, "Web 2.0," April 19, 2008). Berners-Lee himself has questioned whether one can use Web 2.0 in a meaningful way since many of the technological components of "Web 2.0" existed in the early days of the Web ("Work Interview" with Berners-Lee in 2006). The usefulness of Web 2.0 sites generated a host of reactions in 2006 through 2008, most of them tremendously positive.[9]

In all, the contrast between Web 1.0 and Web 2.0 was vividly characterized by Alex Wright (2008, p. 32) when he noted that Web 1.0 sites are "nondemocratic and top-down (a hierarchy)" or "peer-to-peer and open."

Yet, the Semantic Web or Web 3.0 would be profoundly different than these "earlier" versions of the Semantic Web. While Berners-Lee and his colleagues understood that mechanisms exist that might assist the development of a Semantic Web, they also understood that a "new mix of mathematical and engineering decisions" still had to be developed to make the Semantic Web a reality (*Scientific American*, May 2001). In their view, the

> real power of the Semantic Web will be realized when people create many programs that collect Web content from diverse sources, process the information and exchange the results with other programs. The effectiveness of such software agents will increase exponentially as more machine-readable Web content and automated services (including other agents) become available. The Semantic Web promotes this synergy: even agents that were not expressly designed to work together can transfer data among themselves when the data come with semantics.

As conceived by Berners-Lee and his colleagues, a revolutionary transformation in what knowledge is and how it becomes integrated would be essential. As Berners-Lee and his colleagues concluded: "The semantic web is not 'merely' the tool for conducting individual tasks we have discussed so far. In addition, if properly designed, the Semantic Web can assist in the evolution of human knowledge as a whole" (*Scientific American*, May 2001).

Beyond his conception of what a Semantic Web might be and do, Berners-Lee has also been involved in an effort to develop a PhD program in "Web Science." The description of this program is over 125 pages as published in *Foundations and Trends in Web Science* (see also Shadbolt & Berners-Lee, October 2008; Hendler, Shadbolt, Hall, Berners-Lee, & Weitzner, July 2008). Moreover, adding its complexity, Berners-Lee and his five other colleagues initially noted that "'Web Science' is a deliberately ambiguous phrase" (p. 3) for it merges two discrete paradigms and it "needs to be *studied* and understood, and it needs to be *engineered*"

(p. 3). A full description of this program goes beyond what we can provide here, and actually goes beyond our intention in this chapter at this point for we are only seeking to illustrate what Berners-Lee foresees as part of the future of the Internet. In terms of our objectives in this chapter, then, we can illustrate Berners-Lee's conception of the future by underscoring what he and his colleagues identify as the "new directions" for "web engineering" (p. 41). In this regard, five new directions (pp. 42–51) are specified, and they include:

1. Web services
2. Distributed approaches: Pervasive computing, P2P, and grids
3. Personalization
4. Multimedia
5. Natural language processing

While specific applications of these new directions in computer engineering will depend on the breakthroughs of particular candidates in the program, we would expect that those breakthroughs that are consistent with the program will extend the range and kinds of services offered on the Internet, increase the diversity in how these services are offered, personalize the services as well as offering them through an increasing range of devices such as smartphones and laptop computers. Perhaps most significantly, the scientific, engineer, and technical nature of these developments will be minimized because everyday language will become the way in which these developments are explained. If these engineering directions do begin to occur and reach the popular culture, we expect that the Internet revolution we have so far experienced will be understood to be only at its inception.

Conclusion

In many respects, predictions about the future are self-fulfilling prophecies. We focus on some future conceptions, devote time and attention to them, believe they are and will exert influence on us, and they do. The outcome is, in many ways, ironically predictable. In this chapter, we invite you to explore thoughts you have about the future of digital technologies and the Internet. But, also ask yourself which of these predictions can be said to be a product of long-term trends and to what degree do those involved with these trends perceive of themselves the way you do.

Classroom Exercise

Future Focus: Considering all of the topics and different futures each of these topics might generate, select one topic that you find particularly involving, and write a two- or three-page summary of how you think this topic will involve in the next 20 years. Try to make your ideas expressed in your topic as concise and operational as possible (think about the change in behavior you are suggesting), determine if you think there is a trend that you might extend that would lend credibility to your prediction, and assess the value of the change. Be prepared to provide an oral summary of your prediction as a presentation in your class and consider whether or not other predictions are consistent with your projection.

Endnotes

1. From this perspective, we are but mere mother hens with our technologies tucked warmly beneath our wings, growing, evolving, preparing to explode from their shell and flourish on their own; beside us and above us.

2. In this scheme, a definition of the concept *everyday reality* emerges from the characteristics attributed to it in contrast to the characteristic given to *virtual reality*. Indeed, conceptual definitions of *everyday reality* are not common. For example, Angeles has ignored the concept of *everyday reality* as a philosophical term; see Angeles, 1981). However, in 1981, Chesebro and Klenk posited the following definition of *everyday communication,* which actually may be more useful within this context:

   While the conception of everyday communication is seldom treated as a critical concept, we have conceived of the study of everyday communication to be the examination of (1) particular intentions rather than intentionality as an epistemological issue, (2) settings in which agents assume that a single reality exists dependent of perception; an intersubjective reality is unconsciously presumed, (3) agents who believe that imperative actions are required rather than dialective exchanges, (4) imminent actions and face-to-face interactions possessing the full scope of all verbal and nonverbal stimuli in which immediate, flexible, continuous and prereflective symbolic exchanges occur, (5) exchanges in which members of a communication system assume there is a correspondence among their meanings, and (6) agents engaged in continuity in their interactions in terms of geography, time, and social relations. Thus, everyday communication is the study of autobiographical meanings or the pragmatics and self-serving understandings of an inner circle or symbolic enclave in which there is a high degree of dependency, interest, and intimacy. We would not, therefore, perceive formal speeches, academic debates, most written essays, or highly ceremonial occasions typically to be "everyday communication."

For more extended discussions of these notions, see Berger and Luckmann, (1966; reprinted: Anchor Books, 1967), and Goffman, (1959, reprinted: Doubleday, 1959) pp. 327–328.

See also Chesebro and Klenk (1981) pp. 87–103 and 325–330.

3. For example, referring to everyday uses of time, Tullis and Albert (2008) argued that time is typically understood to exist in terms of a metric in which, by "agreement," a "consistent and reliable" set of intervals are used to designate days, hours, minutes, and seconds. As Tullis and Albert put it, "a second lasts for the same amount of time no matter what the time-keeping device is" (p. 7).

4. See Carney, (January 5, 2010). Variations in time are also discussed by Shulevitz; see: Judith Shulevitz (2010).

5. For example, see Markoff (February 23, 2010).

6. Every reality must establish that it is systematic or organized, with boundaries and subsystems that can be employed as different functions are relevant.

7. A Web 2.0 type of virtual reality is presumed here, much as social networks adjust to the desires of their users.

8. Every reality must have a set of persuasive "god terms" that invoke attention, human involvement, and human responses if it is to function as a relevant and significant reality in which humans believe they must or should participate.

9. See Madden & Fox, 2006; Hirschorn, April 2007; Padilla, New Creative Zen, September 15, 2007; eMarketer, September 25, 2007; Klaassen, October 1, 2007; Cowley, October 18, 2007; Madden, February 26, 2008; Elliott, March 28, 2008; Cromity, August 2008; Schenker, November 21, 2007; Pontin, August 2008; Pontin, August 2008; Widman, April 9, 2008; and, Anderson, May 4, 2010.

References

Abrams, L. (2012, December 11). Where ER doctors work entirely via webcam. *The Atlantic*. Retrieved from http://www.theatlantic.com/health/archive/2012/12/where-er-doctors-work-entirely-via-webcam/265935/

Ackerman, A. (2013, July 17). Cyberattacks target finance hubs. *The Wall Street Journal*, p. C2.

Akhlaq, A. A., & Ahmed, E. (2013). The effect of motivation on trust in the acceptance of internet banking in a low income country. *International Journal of Bank Marketing, 31*, 115–125.

Akhlaq, A. A., & Shah, A. (2011). Internet banking in Pakistan: Finding complexities. *Journal of Internet Banking and Commerce, 16*, 1–14.

Akiyoshi, M., & Ono, H. (2008). The diffusion of mobile Internet in Japan. *Information Society, 24*, 292–303.

Alan Turing's 100th birthday: Google doodles a Turing machine. (2009). Retrieved from http://ibnlive.in.com/news/alan-turings-100th-birthday-google-doodles-a-turing-machine/267430-11.html

Albrecht, K., & McIntyre, L. (2005). *Spychips: How major corporations and government plan to track your every move with RFID*. Nashville, TN: Nelson Current.

Alexander, S. (2008). Computer games. *The Encyclopedia Britannica 2008* (p. 226). New York, NY: Black Dog & Leventhal.

Allen, I. E., & Seaman, J. (2013). *Changing course: Ten years of tracking online education in the United States*. Babson Park, MA: Babson Survey Research Group.

Alsajjan, B., & Dennis, C. (2010). Internet banking acceptance model: Cross-market examination. *Journal of Business Research, 63*, 957–963.

Alter, A. (2011, December 9). How I became a best-selling author. *The Wall Street Journal*, pp. D1, D2.

Alter, A. (2012, July 19). Your e-book is reading you. *The Wall Street Journal*. Retrieved from http://online.wsj.com/news/articles/SB10001424405270230487030457749095005143 8304

Alterovitz, S. S., & Mendelsohn, G. A. (2013). Relationship goals of middle-aged, young-old, and old-old internet daters: An analysis of online personal ads. *Journal of Aging Studies, 27*, 159–165.

Anderson, C. (2006). *The long tail: Why the future of business is selling less of more*. New York, NY: Hyperion.

Andersen, K. (2011, December 14). The protester. *Time*. Retrieved from http://content.time.com/time/person-of-the-year/2011/

Anderson, J. Q., & Rainie, L. (2006). *The future of the Internet II*. Washington, DC: Pew Research Center.

Anderson, J. Q., & Rainie, L. (2008). *The future of the Internet III*. Washington, DC: Pew Research Center.

Anderson, J. Q., & Rainie, L. (2010). *The future of the Internet IV*. Washington, DC: Pew Research Center.

Anderson, J. Q., & Rainie, L. (2010). *The fate of the semantic Web*. Washington, DC: Pew Research Center.

Angeles, P. A. (1981). *Dictionary of philosophy*. New York, NY: Barnes & Noble Books.

Angwin, J. (2010, July 30). The web's new gold mine: Your secrets. *The Wall Street Journal*. Retrieved from http://online.wsj.com/news/articles/SB10001424052748703940904575395073512989404

Angwin, J. (2011, August 19). Latest in web tracking: Stealthy "supercookies." *The Wall Street Journal*. Retrieved from http://online.wsj.com/news/articles/SB10001424053111903480904576508382675931492

Angwin, J. (2012, July 10). Google, FTC near settlement on privacy. *The Wall Street Journal*. Retrieved from http://online.wsj.com/news/articles/SB10001424052702303567704577517081178553046

Angwin, J., & McGinty, T. (2010, July 30). Sites feed personal details to new tracking industry. *The Wall Street Journal*. Retrieved from http://online.wsj.com/news/articles/SB10001424052748703977004575393173432219064

Angwin, J., & Valentino-Devries, J. (2012, February 17). Google's iPhone tracking. *The Wall Street Journal*. Retrieved from http://online.wsj.com/news/articles/SB10001424052970204880404577225380456599176

Anstead, N., & Chadwick, A. (2009). Parties, election campaigning, and the Internet: Toward a comparative institutional approach. In A. Chadwick & P. N. Howard (Eds.), *Routledge handbook of Internet politics* (pp. 56–71). New York, NY: Routledge.

Arango, T., & Carter, B. (2009, November 21). An unsteady future for broadcast. *The New York Times*, pp. B1, B2.

Armstrong, J., Brooker, C. (Writers), & Welsh, B. (Director). (2011). The entire history of you [Television series episode]. In C. Brooker & A. Jones (Executive producers), *Black Mirror*. London, England: Channel 4.

Arndt, J. (1967). The role of product-related conversations in the diffusion of a new product. *Journal of Marketing Research, 4*, 291–295.

Arthur, C. (2011, April 20). iPhone keeps record of everywhere you go. *The Guardian*. Retrieved from http://www.theguardian.com/technology/2011/apr/20/iphone-tracking-prompts-privacy-fears

Associated Press. (2009, October 1). U.S. moves to lessen its oversight of Internet. Retrieved from http://www.nytimes.com/2009/10/01/technology/internet/01icann.html?_r=0

Associated Press. (2011). Calif. rejects social network privacy bill. Retrieved from http://abclocal.go.com/kabc/story?section=news/state&id=8158115

Atchley, R. A., Strayer, D. L., & Atchley, P. (2012). Creativity in the wild: Improving creative reasoning through immersion in natural settings. *PLoS ONE, 7*(12).

Aubrey, J. S., & Rill, L. (2013). Investigating relations between Facebook use and social capital among college undergraduates. *Communication Quarterly, 61*, 479–496.

Ayenson, M., Wambach, D. J., Soltani, A., Good, N., & Hoofnagle, C. J. (2011). Flash cookies and privacy II: Now with HTML5 and ETag respawning. *Social Science Research Network*. Retrieved from http://papers.ssrn.com/sol3/papers.cfm?abstract_id=1898390

Aylward, A. (2013, July 3). Prosecutors wrap up case in Manning trial. *The New York Times*, p. A4.

Ayoko, O. B., Konrad, A. M., & Boyle, M. V. (2012). Online work: Managing conflict and emotions for performance in virtual teams. *European Management Journal, 30*, 156–174.

Back, M. D., Schukle, S. C., & Egloff, B. (2008). How extraverted is honey.bunny77@hotmail.de? Inferring personality from e-mail addresses. *Journal of Research in Personality, 42*, 1116–1122.

Baghat, H. (2004). Egypt's virtual protection of morality. *Middle East Report, 230*, 22–25.

Bailey, R. (2011, September 6). How scared of terrorism should you be? *Reason: Free Minds and Free Markets*. Retrieved from http://reason.com/archives/2011/09/06/how-scared-of-terrorism-should

Bailey, R. (2012, November 16). Book review: Beyond human nature, How to create a mind. *The Wall Street Journal*. Retrieved from http://online.wsj.com/news/articles/SB10001424127887324556304578121040520084804

Banchero, S., & Simon, S. (2011, November 12). My teacher is an app. *The Wall Street Journal*. Retrieved from http://online.wsj.com/news/articles/SB10001424052970204358004577030600066250144

Banerjee, N. (2006, September, 1). Intimate confessions pour out on church's web site. *The New York Times*, p. A11.

Banker, D., & Corbeil, M. (2013, March). Not your mother's teacher ed course: Using immersive online worlds and online technologies to prepare teachers for next century learners. *Society for Information Technology & Teacher Education International Conference, 1*, 198–205.

Barack Obama's get-out-the vote machine is bigger, faster and smarter. (2008, October 23). *The Economist*. Retrieved from http://www.economist.com/node/12470573

Bargh, J. A., McKenna, K. Y. A., & Fitzsimons, G. M. (2002). Can you see the real me? Activation and expression of the "true self" on the Internet. *Journal of Social Issues, 58*, 33–48.

Barna Research Group. (2000). Teenagers embrace religion but are not excited about Christianity. Retrieved from http://www.youth-ministry.info/articles.php5?type=2&cat=20&art_id=38 bin/PagePressRelease.asp?PressReleaseID=57&Reference=B

Barna Research Group. (2006). State of the church. Retrieved from http://www.barna.org/FlexPage.aspx?Page=ResourceArea&ResourceAreaID=22

Barnes, B. (2010, September 26). In this war, movie studios are siding with your couch. *The New York Times*, pp. BU1, BU6.

Barrett, B. (2012, January 17). What is SOPA? Retrieved from http://gizmodo.com/5877000/what-is-sopa

Barry, C., Markey, R., Almquist, E., & Brahm, C. (2011). Putting social media to work. Retrieved from http://www.bain.com/Images/BAIN_BRIEF_Putting_social_media_to_work.pdf

Bauerlein, M. (2011). *The digital divide: Arguments for and against Facebook, Google, texting, and the age of social networking*. New York, NY: Jeremy P. Tarcher/Penguin.

Baumeister, R. F., & Wotman, S. R. (1992). *Breaking hearts: Two sides of unrequited love*. New York, NY: Guilford Press.

Baumeister, R. F., Wotman, S. R., & Stillwell, A. M. (1993). Unrequited love: On heartbreak, anger, guilt, scriptlessness, and humiliation. *Journal of Personality and Social Psychology, 64*, 377–394.

Baxter, M. (2011). Brain health and online gaming. *Generations, 35*, 107–109.

Baym, N. K., & Ledbetter, A. (2009). Tunes that bind? Predicting friendship strength in a music-based social network. *Information, Communication & Society, 12*, 408–427.

BBC News. (2010, February 12). Net neutrality: Are all bits created equal? Retrieved from http://news.bbc.co.uk/1/hi/programmes/click_online/8512843.stm

Beckett, L. (2012, October 12). How companies have assembled political profiles for millions of Internet users. *ProPublica*. http://www.propublica.org/article/how-companies-have-assembled-political-profiles-for-millions-of-internet-us

Belkin, D., & Korn, M. (2013, May 30). Web courses woo professors. *The Wall Street Journal*. Retrieved from http://online.wsj.com/news/articles/SB10001424127887324682204578513541557842934

Bell, G., & Gray, J. N. (1999). The revolution yet to happen. In P. J. Denning & R. M. Metcalfe (Eds.), *Technology workers in the United States* (pp. 60–61). Washington, DC: Computing Research Association.

Bennett, J., & White, J. B. (2013, June 6). Car apps that help you find a parking spot, and more. *The Wall Street Journal*. Retrieved from http://online.wsj.com/news/articles/SB10001424127887323844804578529081561343810

Berke, R. L. (1995, March 1). Tennessean brings down-home bid for Republican presidential nomination. *The New York Times*. Retrieved from http://www.nytimes.com/1995/03/01/us/tennessean-begins-down-home-bid-for-republican-presidential-nomination.html?src=pm

Berlin, L. (2005). *The man behind the microchip: Robert Noyce and the invention of Silicon Valley*. New York, NY: Oxford University Press.

Berners-Lee, T. (1999). *Weaving the web: The original design and ultimate destiny of the World Wide Web*. San Francisco, CA: Harper San Francisco.

Berners-Lee, T., Hall, W., Hendler, J. A., O'Hara, K., Shadbolt, N., & Weitzner, D. J. (2006). A framework for Web Science. *Foundations and Trends in Web Science, 1*, 1–139.

Berners-Lee, T., Hendler, J., & Lassila, O. (2001, May 17). The semantic web. *Scientific American*. Retrieved from http://www.scientificamerican.com/article.cfm?id=the-semantic-web

Berners-Lee, Tim. (n.d.). In *Wikipedia*. Retrieved from http://www.en.wikipedia.org/wiki/Tim_ Berners-Lee

Berners-Lee, Tim. (2003/2005). Foreword. In D. Fensel, J. Hendler, H. Lieberman, & W. Wahlster (Eds.), *Spinning the Semantic Web: Bringing the World Wide Web to its full potential* (pp. xi–xxiii). Cambridge, MA: The MIT Press.

Bezos, J. (2008, June 9). The way we read. *The Wall Street Journal*, pp. R3, R10.

Bloomberg News (2011, June 29). Apple TV sets may be sold in 2012. *The San Francisco Chronicle*, p. D–3.

Bialik, C. (2009, July 29). Marriage-maker claims are tied in knots. *The Wall Street Journal*. Retrieved from http://online.wsj.com/news/articles/SB124879877347487253

Bilton, N. (2013, August 27). Facebook releases report on government requests. *New York Times*, p. A5.

Bimber, B., & Davis, R. (2003). *Campaigning online: The Internet in U.S. elections*. New York, NY: Oxford University Press.

Blackstone, S., & Groth, A. (2012). The 20 most popular TED Talks ever. *Business Insider*. Retrieved from http://www.businessinsider.com/the-20-most-popular-ted-talks-of-all-time-2012-8

Blascovich, J., & Bailenson, J. (2011). *Infinite reality: Avatars, eternal life, new world, and the dawn of virtual revolution*. New York, NY: HarperCollins.

Bloom, N., Liang, J., Roberts, J., & Ying, Z. J. (2013). *Does working from home work? Evidence from a Chinese experiment* (No. w18871). Cambridge, MA: National Bureau of Economic Research.

Boase, J., Horrigan, J. B., Wellman, B., & Rainie, L. (2006). *The strength of Internet ties: The Internet and e-mail aid users in maintaining their social networks and provide pathways to help when people face big decisions*. Washington, DC: Pew Research Center.

Bogost, I. (2007). *Persuasive games: The expressive power of videogames*. Cambridge, MA: The MIT Press.

Bohn, R. E., & Short, J. E. (2009). *How much information? 2009 report on American consumers*. San Diego, CA: University of California San Diego Global Information Industry Center.

Book, C., Anderson, J. Q., & Rainie, L. (2008). *Whither the Internet?* Washington, DC: Pew Research Center.

Bosman, J. (2012, January 29). The bookstore's last stand. *The New York Times*, BU1.

Bourdieu, P. (2001). Television. *European Review*, 9, 245–256.

Bowen, H. (2005). Can videogames make you cry? Retrieved from http://www.bowenresearch.com/studies.php?id=3

Boyles, J. L., Smith, A., & Madden, M. (2012). *Privacy and data management on mobile devices*. Washington, DC: Pew Research Center.

Brand, R. J., Bonatsos, A., D'Orazio, R., & DeShong, H. (2011). What is beautiful is good, even online: Correlations between photo attractiveness and text attractiveness in men's online dating profiles. *Computers in Human Behavior, 28*, 166–170.

Brembeck, W. L., & Howell, W. S. (1952). *Persuasion: A means of social influence*. Englewood Cliffs, NJ: Prentice-Hall.

Brembeck, W. L., & Howell, W. S. (1976). *Persuasion: A means of social influence*. Englewood Cliffs, NJ: Prentice-Hall.

Broderick, D. (Ed.). (2008). *Year million: Science at the far edge of knowledge*. New York, NY: Atlas & Company.

Brooks, D. (2007, October 26). The outsourced brain. *The New York Times*. Retrieved from http://www.nytimes.com/2007/10/26/opinion/26brooks.html?_r=0

Brunton, F. (2013). *Spam: A shadow history of the Internet*. Cambridge, MA: The MIT Press.

Brundidge, J., & Rice, R. E. (2009). Political engagement online: Do the information rich get richer and the like-minded more similar? In A. Chadwick & P. N. Howard (Eds.), *Routledge handbook of Internet politics* (pp. 144–156). New York, NY: Routledge.

Buchuk, D. (2013, July 27). UK online porn ban: Web traffic analysis of Britain's porn affair. *SimilarWeb*. Retrieved from http://blog.similarweb.com/uk-online-porn-ban-web-traffic-analysis-of-britains-porn-affair/

Bumiller, E. (2010, June 8). Army leak suspect is turned in, by ex-hacker. *The New York Times*, pp. A1, A10.

Business Wire. (2013). PayPal global study spells doom for wallet. *Business Wire*. Retrieved from http://www.businesswire.com/news/home/20130521006662/en/PayPal-Global-Study-Spells-Doom-Wallet

Buskirk, E. V. (2007, October 19). Estimates: Radiohead made up to $10 million on initial album sales. Retrieved from http://www.wired.com/listening_post/2007/10/estimates-radio/

Cabage, N. (2013, July 12). SEO is dead. Now what? *Inc.* Retrieved from http://www.inc.com/neal-cabage/seo-is-dead-now-what.html

Cai, X. (2004). Is the computer a functional alternative to traditional media? *Communication Research Reports, 21*, 26–38.

Campbell, H. (2004). Challenges created by online religious networks. *Journal of Media & Religion, 3*, 81–99.

Campbell, H. (2005). *Exploring religious community online: We are one in the network*. New York, NY: Peter Lang.

Campbell, K. A., Coulson, N. S., & Buchanan, H. (2013). Empowering processes within prostate cancer online support groups. *International Journal of Web Based Communities, 9*, 51–66.

Campbell-Kelly, M. (2009, September). Origins of computing. *Scientific American, 301*, 3, 62–69.

Canfield, J. (2012). Digital trends: Consumer behavior in key financial sectors. Retrieved from https://blog.compete.com/2012/05/25/free-report-digital-trends-consumer-behavior-in-key-financial-sectors/

Canon, S. (2006, September 16). The library's niche in a wired world. *Kansas City Star*. Retrieved from http://www.kansascity.com

Cantoni, L., & Tardini, S. (2006). *Internet*. London, England; New York, NY: Routledge/Taylor & Francis Group.

Caramanica, J. (2011, November 19). For some, free music is an investment that pays off. *The New York Times*, pp. C1, C5.

Carpenter, C. J. (2012). Narcissism on Facebook: Self-promotional and anti-social behavior. *Personality and Individual Differences, 52*, 482–486.

Carpenter, E., & McLuhan, M. (Eds.). (1960). *Explorations in communication, an anthology*. Boston, MA: Beacon Press.

Carr, D. (2008, November 10). How Obama tapped into social networks' power. *The New York Times*. Retrieved from http://www.nytimes.com/2008/11/10/business/media/10carr.html?_r=0

Carr, N. (2008, July 1). Is Google making us stupid? What the Internet is doing to our brains. *The Atlantic*. Retrieved from http://www.theatlantic.com/magazine/archive/2008/07/is-google-making-us-stupid/306868/

Carr, N. (2011). *The shallows: What the Internet is doing to our brains*. New York, NY: W. W. Norton.

Carroll, J. (2005, December 23). Americans inventory their gadgets. *The Gallup Poll*. Retrieved from http://www.gallup.com/poll/20593/americans-inventory-their-gadgets.aspx

Carvajal, D. (2007, June 18). Focus of video games shifts to lure more casual players. *The New York Times*, p. C5.

Castle, S. (2013, July 18). British agency is cleared of illegal data gathering. *The New York Times*, p. A14.

Castle, S., & Schmitt, E. (2013, July 1). Europeans voice anger over reports of spying by U.S. on its allies. *The New York Times*, p. A9.

Cellular-News. (2011, May 12). Mobile banking surges as emerging markets embrace mobile finance. Retrieved from http://www.cellular-news.com/story/49148.php

Chaulk, K., & Jones, T. (2011). Online obsessive relational intrusion: Further concerns about Facebook. *Journal of Family Violence, 26*, 245–254.

Chea, T., Leff, L., & Collins, T. (2011, November 2). Thousands of Occupy protesters disrupt busy port. Retrieved from http://www.syracuse.com/news/index.ssf/2011/11/thousands_of_occupy_oakland_pr.html

Chen, W., & Wellman, B. (2004). The global digital divide—Within and between countries. *IT and Society, 1*(7), 39–45.

Chesebro, J. W. (1993). Communication and computability: The case of Alan Mathison Turing. *Communication Quarterly, 41*, 9–121.

Chesebro, J. W. (2000). Communication technologies as symbolic form: Cognitive transformations generated by the Internet. *Qualitative Research Reports in Communication, 1*, 8–13.

Chesebro, J. W. (2003). Communication, values, and popular television series—A twenty-five-year assessment and final conclusions. *Communication Quarterly, 51*, 367–418.

Chesebro, J. W. (2007, March 29). *A Transformation in the basic media configuration: From the traditional mass communication scheme to the new digital configuration*. Paper presented at the annual meeting of the Central States Communication Association, Minneapolis, MN.

Chesebro, J. W. (2009). *The research agenda for the discipline of communication in the year 2030*. Paper presented at the annual meeting of the Eastern Communication Association, Philadelphia, PA.

Chesebro, J. W., & Bonsall, D. G. (1989). *Computer-mediated communication: Human relationships in a computerized world*. Tuscaloosa: University of Alabama Press.

Child, J. T., & Petronio, S. (2011). Unpacking the paradoxes of privacy in CMC relationships: The challenges of blogging and relational communication on the Internet. In K. B. Wright & L. M. Webb (Eds.), *Computer-mediated communication in personal relationships* (pp. 21–40). New York, NY: Peter Lang.

China Internet Network Information Center (CINIC). (2012). Statistical report on Internet development in China. Retrieved from http://www1.cnnic.cn

China's Internet: A giant cage. (2013, April 6). Retrieved from http://www.economist.com/news/special-report/21574628-internet-was-expected-help-democratise-china-instead-it-has-enabled

Chorney, D. B., & Morris, T. L. (2008). The changing face of dating anxiety: Issues in assessment with special populations. *Clinical Psychology: Science and Practice, 15*, 224–338.

Chozick, A. (2013, June 18). Longing to stay wanted, MTV turns its attention to younger viewers. *The New York Times*, p. B4.

Chung, J. E. (2013a). Social networking in online support groups for health: How online social networking benefits patients. *Journal of Health Communication* (ahead-of-print), 1–21.

Chung, J. E. (2013b). Social interaction in online support groups: Preference for online social interaction over offline social interaction. *Computers in Human Behavior, 29*, 1408–1414.

Chung, J. E. (2013a). Social networking in online support groups for health: How online social networking benefits patients. *Journal of Health Communication*, (ahead-of-print), 1–21.

Cisco. (2009). Cisco study finds telecommuting significantly increases employee productivity, work-life flexibility and job satisfaction. Retrieved from http://newsroom.cisco.com/dlls/2009/prod_062609.html

Cisco. (2012). Cisco visual networking index: Forecast and methodology, 2011–2016. Retrieved from http://www.cisco.com

Cisco. (2013). Cisco visual networking index: Global mobile data traffic forecast update, 2012–2017. Retrieved from http://www.cisco.com/en/US/solutions/collateral/ns341/ns525/ns537/ns705/ns827/white_paper_c11-520862.html

Cisco. (2013a). Cisco study reveals 74 percent of consumers open to virtual doctor visit. Retrieved from http://newsroom.cisco.com/press-release-content?articleId=1148539

Cisco. (2013b). Consumers want a more seamless and personalized customer experience from their bank. Retrieved from http://newsroom.cisco.com/release/1174098

Clark, R. E. (1985). Evidence for confounding in computer-based instruction studies: Analyzing the meta-analysis. *Education Communication & Technology Journal, 33*, 249–262.

Clarke, R. A. (2010). *Cyber war: The next threat to national security and what to do about it.* New York, NY: HarperCollins.

Clemes, M. D., Gan, C., & Du, J. (2012). The factors impacting on customers' decisions to adopt Internet banking. *Bank and Bank Systems, 7*, 1–18.

Clifford, S. (2010, February 8). Magazines' newsstand sales fall 9.1 percent. *The New York Times.* Retrieved from http://mediadecoder.blogs.nytimes.com/2010/02/08/magazines-newsstand-sales-fall-91-percent/

Cohen, J., & Metzger, M. (1998). Social affiliation and the achievement on ontological security through interpersonal and mass communication. *Critical Studies in Mass Communication, 15*, 41–60.

Cohen, N. (2011, March 26). It's tracking your every move and you may not even know. *The New York Times*, pp. A1, A3.

comScore. (2009, August 27). U.S. online video market soars in July as summer vacation drives pickup in entertainment and leisure activities online. Retrieved from http://www.comscore.com/Insights/Press_Releases/2009/8/U.S._Online_Video_Market_Soars_in_July_as_Summer_Vacation_Drives_Pickup_in_Entertainment_and_Leisure_Activities_Online

Conway, F., & Siegelman, J. (2005). *Dark hero of the information age: In search of Norbert Wiener, the father of cybernetics.* New York, NY: Basic Books.

Cooper, A. (1998). Sexuality and the Internet: Surfing into the new millennium. *CyberPsychology & Behavior, 1,* 187–193.

Coopersmith, J. (2006). Does your mother know what you really do? The changing nature and image of computer-based pornography. *History and Technology, 22,* 1–25.

Cortada, J. W. (2002). *Making the information society: Experience, consequences, and possibilities.* Upper Saddle River, NJ: Prentice-Hall.

Coulson, N. S., & Shaw, R. L. (2013). Nurturing health-related online support groups: Exploring the experiences of patient moderators. *Computers in Human Behavior, 29,* 1695–1701.

Cowley, S. (2007, October 18). Web 2.0: IT revolution or Pandora's box? Retrieved from http://www.crn.com/news/applications-os/202404411/web-2-0-it-revolution-or-pandoras-box.htm

Craig, E., & Wright, K. B. (2012). Computer-mediated relational development and maintenance on Facebook. *Communication Research Reports, 29,* 119–129.

Creating an effective online environment. (2013). *Community College Research Center, Teachers College, Columbia University.* Retrieved from http://ccrc.tc.columbia.edu

Crovitz, L. G. (2010, December 6). Julian Assange, information anarchist. *The Wall Street Journal.* Retrieved from http://online.wsj.com/news/articles/SB10001424052748703989004575653113548361870

Crovitz, L. G. (2012, November 4). How campaigns hypertarget voters online. *The Wall Street Journal.* Retrieved from http://online.wsj.com/news/articles/SB10001424052970204846304578096982338148870

Crovitz, L. G. (2012, November 26). The U.N.'s Internet sneak attack. *The Wall Street Journal,* p. A15.

D'Alessio, D. (2000). Adoption of the World Wide Web by American political candidates, 1996–1998. *Journal of Broadcasting & Electronic Media, 44,* 556–568.

Darvell, M. J., Walsh, S. P., & White, K. M. (2011). Facebook tells me so: Applying the theory of planned behavior to understand partner-monitoring behavior on Facebook. *Cyberpsychology, Behavior, and Social Networking, 14,* 717–722.

Davies, D. R. (2006). *The postwar decline of American newspapers, 1945–1965.* Westport, CT: Praeger.

Davis, W. (2009, November 2). Net neutrality proponents warn FCC of loopholes in regulations. Retrieved from http://www.mediapost.com/publications/article/116653/

Davis, W. (2010, April 6). Fed court rules FCC has no authority in net-neutrality case. Retrieved from http://www.mediapost.com

Davis, W. (2011, January 6). Study: Do Americans really not support FCC neutrality rules? Retrieved from http://www.mediapost.com/publications/article/142516/

Davis, W. (2011, February 11). EFF warns that open Internet rules could backfire. Retrieved from http://www.mediapost.com/publications/article/144393/

Davis, W. (2012). 20% of web users notice behavioral-advertising icons. Retrieved from http://www.mediapost.com/publications/article/187186/20-of-web-users-notice-behavioral-advertising-ico.html

Davis, W. (2012a). Hulu loses bid to dismiss privacy lawsuits. Retrieved from http://www.media post.com/publications/article/180703/hulu-loses-bid-to-dismiss-privacy-lawsuit.html

Dawsey, J. (2012, December 18). YouTube's top ten viral videos of 2012. *The Wall Street Journal.* Retrieved from http://blogs.wsj.com/speakeasy/2012/12/18/youtubes-top-ten-viral-videos-of-2012/

Deahl, R., & Milliot, J. (2011, December 15). Is Amazon pushing publishers to brink on terms, co-op? Retrieved from http://www.publishersweekly.com/pw/by-topic/industry-news/book selling/article/49874-is-amazon-pushing-publishers-to-brink-on-terms-co-op.html

DeAndrea, D. C., Tong, S. T., Liang, Y. J., Levine, T. R., & Walther, J. B. (2012). When do people misrepresent themselves to others? The effects of social desirability, ground truth, and account-ability on deceptive self-presentations. *Journal of Communication, 62,* 400–417.

Deheane, S. (2009). *Reading in the brain: The science and evolution of a human invention.* New York, NY: Viking.

Delawala, I. (2013, June 9). Intelligence committee leaders defend NSA surveillance. Retrieved from http://abcnews.go.com/blogs/politics/2013/06/intelligence-committee-leaders-defend-nsa-surveillance/

Deutschman, A. (2011, September 5). Exit the king. *Newsweek,* pp. 30–35.

DeVoss, D. (2002). Women's porn sites—Spaces of fissure and eruption or "I'm a little bit of every-thing." *Sexuality & Culture, 6,* 75–94.

Dillow, C. (2012, December 17). IBM predicts: Cognitive computers that feel and smell, within the next five years. Retrieved from http://www.popsci.com/technology/article/2012-12/ibms-pre dicts-cognitive-computers-enhanced-sensory-perception-next-five-years

Dillow, C. (2013, January 16). Anti-surveillance hoodie and scarf prevent drones from tracking you. Retrieved from http://www.popsci.com/technology/article/2013-01/drone-proof-your-wardrobe-artist-unveils-surveillance-thwarting-hoodie-and-scarf

Dochterman, M. A. (2010). Part 1: The determination of web credibility: A thematic analysis of a web user's judgments. *Qualitative Research Reports in Communication, 11,* 37–43.

Dokoupil, T. (2012, July 8). Is the Internet making us crazy? What the new research says. Retrieved from http://www.newsweek.com/internet-making-us-crazy-what-new-research-says-65593

Duck, S. W. (1998). *Meaningful relationships* (3rd ed.). London, England: Sage.

Duck, S. W., & McMahan, D. T. (2012). *The Basics of communication: A relational perspective* (2nd ed.). Thousand Oaks, CA: Sage.

Duck, S. W., & McMahan, D. T. (2015). *Communication in everyday life* (2nd ed.). Thousand Oaks, CA: Sage.

Duck, S. W., Rutt, D. J., Hurst, M. H., & Strejc, H. (1991). Some evident truths about conversa-tions in everyday relationships: All communications are not created equal. *Human Communi-cation Research, 18,* 228–267.

Dyson, E. (2008). Reflections on privacy 2.0. *Scientific American, 299*(3), 50–56.

Dyson, G. (2012). *Turing's cathedral: The origins of the digital universe.* New York, NY: Pantheon Books.

Eberhart, G. M. (2008). Redefining the library in the digital age. In *Encyclopedia Britannica 2008 book of the year* (pp. 188–189). Chicago, IL: Encyclopedia Britannica.

Edidin, P. (2005, August 28). Confounding machines: How the future looked. *The New York Times*, p. 12WK.

Edwards-Stout, K. (2012, June 20). Fired for Facebook photos, a gay high school coach speaks out. Retrieved from http://www.huffingtonpost.com/kergan-edwardsstout/gay-families-school-teachers_b_1610182.html

Efrati, A. (2013, June 25). FTC warns search engines to better identify ads. *The Wall Street Journal*. Retrieved from http://blogs.wsj.com/digits/2013/06/25/ftc-warns-search-engines-to-better-identify-ads/

Election spending. (2008). Retrieved from http://www.cnn.com/ELECTION/2008/map/ad.spending

Elliott, S. (2008, March 28). Helping the help desk, the Intel way. *The New York Times*, p. C5.

Ellison, N., Hancock, J. T., & Toma, C. L. (2012). Profile as promise: A framework for conceptualizing veracity in online dating self-presentations. *New Media & Society, 14*, 45–62.

Ellison, N., Heino, R., & Gibbs, J. (2006). Managing impressions online: Self-presentation processes in the online dating environment. *Journal of Computer-Mediated Communication, 11*, 415–441.

Ellison, N. B., Steinfield, C., & Lampe, C. (2007). The benefits of Facebook "friends": Social capital and college students' use of online social network sites. *Journal of Computer-Mediated Communication, 12*, 1143–1168.

Ellison, N. B., Steinfield, C., & Lampe, C. (2011). Connection strategies: Social capital implications of Facebook-enabled communication practices. *New Media & Society, 13*, 873–892.

eMarketer. (2007). Web 2.0 users go shopping. *eMarketer*. Retrieved from http://www.emarketer.com

eMarketer. (2012). Advertising and marketing. Retrieved from http://www.emarketer.com

eMarketer. (2013). Digital to account for one in five ad dollars. Retrieved from http://www.emarketer.com/Article/Digital-Account-One-Five-Ad-Dollars/1009592

ENIAC. (n.d.). In *Wikipedia*. Retrieved from http://www.en.wikipedia.org/wiki/Eniac

Entertainment Software Association. (2012). Essential facts about the computer and video game industry. Retrieved from http://www.theesa.com/facts/pdfs/esa_ef_2012.pdf

Entous, A., & Gorman, S. (2013, October 30). Europeans shared spy data with U.S. *The New York Times*, pp. A1, A12.

Erlanger, S. (2013, July 2). Outrage in Europe grows over spying disclosures. *The New York Times*, pp. A4, A6.

Erlanger, S. (2013, July 5). France, too, is sweeping up data, newspaper reveals. *The New York Times*, p. A3.

Etherington, D. (2013, June 5). eBay to make true window shopping a reality with new NYC virtual retail stores. Retrieved from http://techcrunch.com/2013/06/05/ebay-to-make-true-window-shopping-a-reality-with-new-nyc-virtual-retail-stores/

Evans, C. (1979). *The micro millennium*. New York, NY: The Viking Press.

Exit polls. (2008). Retrieved from http://elections.nytimes.com/2008/results/president/exit-polls.html

Family Safe Media. (2013). Pornography statistics. Retrieved from http://www.familysafemedia.com/pornography_statistics.html

Favole, J. A., & Nicholas, P. (2013, June 7). Obama: "Nobody is listening to your telephone calls." *The Wall Street Journal.* Retrieved from http://online.wsj.com/news/articles/SB100014241278 873238448045785313433379996824

Fay, M. J., & Kline, S. L. (2011). Coworker relationships and informal communication in high-intensity telecommuting. *Journal of Applied Communication Research, 39,* 144–163.

Fay, M. J., & Kline, S. L. (2012). The influence of informal communication on organizational identification and commitment in the context of high-intensity telecommuting. *Southern Communication Journal, 77,* 61–76.

Federal Communications Commission. (2010, December 23). *Report and order.* Washington, DC: Federal Communications Commission (FCC 10-201).

Federal Reserve. (2012). *Consumers and mobile financial services.* Washington, DC: Board of Governors of the Federal Reserve System.

Feigenbaum, E. A., & McCorduck, P. (1983). *The fifth generation: Artificial intelligence and Japan's computer challenge to the world.* Reading, MA: Addison-Wesley.

Fiegerman, S. (2013, May 29). More than 500 million photos are shared every day. Retrieved from http://mashable.com/2013/05/29/mary-meeker-internet-trends-2013/

Feingold, A. (1990). Gender differences in effects of physical attractiveness on romantic attraction: A comparison across five research paradigms. *Journal of Personality and Social Psychology, 59,* 981–993.

Finkel, E. J., Eastwick, P. W., Karney, B. R., Reis, H. T., & Sprecher, S. (2012). Online dating: A critical analysis from the perspective of psychological science. *Psychological Science in the Public Interest, 13,* 3–66.

Fiore, A. T., Taylor, L. S., Mendelsohn, G. A., & Hearst, M. (2008). Assessing attractiveness in online dating profiles. *Proceedings of Computer-Human Interaction* (pp. 797–806). New York, NY: ACM Press.

Fitzpatrick, A. (2012, December 21). Quick donate transformed political fundraising in 2012. Retrieved from http://mashable.com/2012/12/21/obama-quick-donate/

Fitzpatrick, A. (2013, June 18). NSA chief: Surveillance programs have foiled 50 terrorist plots. Retrieved from http://mashable.com/2013/06/18/nsa-surveillance-plots/

Foehr, U.G. (2006, December). *Media multitasking among American youth: Prevalence, predictors and pairings.* Menlo Park, CA: The Henry J. Kaiser Family Foundation.

Fogg, B. J., Soohoo, C., Danielson, D. R., Marable, L., Stanford, J., & Tauber, E. R. (2003). How do users evaluate the credibility of web sites? A study with over 2,500 participants. *Proceedings of the 2003 Conference on Designing for User Experiences,* pp. 1–15.

Foot, K. A., & Schneider, S. M. (2006). *Web campaigning.* Cambridge, MA: The MIT Press.

Forester, T. (Ed). (1985). *The information technology revolution.* Cambridge, MA: The MIT Press.

Fost, D. (2009, October 1). Internet governance goes global as U.S. loosens grip on ICANN. *The Los Angeles Times.* Retrieved from http://articles.latimes.com/2009/oct/01/business/fi-internet1

Foucault, M. (1977). *Discipline and punish: The birth of the prison.* New York, NY: Pantheon Books.

Fowler, G. A. (2012, December 17). Tor: An anonymous, and controversial, way to web-surf. *The Wall Street Journal.* Retrieved from http://online.wsj.com/news/articles/SB100014241278873 246772045781853823771144280

Fox, S. (2011). *Peer-to-peer healthcare*. Washington, DC: Pew Research Center.

Fox, S., Anderson, J. Q., & Rainie, L. (2005). *The future of the Internet*. Washington, DC: Pew Research Center.

Fox, S., & Duggan, M. (2013). *Health online 2013*. Washington, DC: Pew Research Center.

Franceschi-Bicchierai, L. (2013, May 13). Syria suffers yet another Internet blackout. Retrieved from http://mashable.com/2013/05/15/syria-internet-outage/

Franceschi-Bicchierai, L. (2013, May 29). ACLU opens torture database for National Day of Civic Hacking. Retrieved from http://mashable.com/2013/05/29/aclu-torture-data/

Franceschi-Bicchierai, L. (2013, June 4). Global cyberattack hits 350 victims, 40 countries. Retrieved from http://mashable.com/2013/06/04/nettraveler-malware-cyberespionage-campaign/

Franceschi-Bicchierai, L. (2013, June 11). Tim Berners-Lee: NSA surveillance an "intrusion on basic human rights." Retrieved from http://mashable.com/2013/06/10/tim-berners-lee-nsa-surveillance/

Fraser, M., & Dutta, S. (2008, November 19). Barack Obama and the Facebook election. *US News and World Report*. Retrieved from http://www.usnews.com/opinion/articles/2008/11/19/barack-obama-and-the-facebook-election

Freeth, T. (2009, December). Decoding an ancient computer. *Scientific American, 301*(6), 76–83.

Friedersdorf, C. (2013, June 10). The irrationality of giving up this much liberty to fight terror. Retrieved from http://www.theatlantic.com/politics/archive/2013/06/the-irrationality-of-giving-up-this-much-liberty-to-fight-terror/276695/

Friedman, W. (2013, April 4). "Game of Thrones," "Dexter" piracy cost cablers multimillions. Retrieved from http://www.mediapost.com/publications/article/197344/game-of-thrones-dexter-piracy-cost-cablers-mu.html

Fritz, B., & James, M. (2012, February 21). Comcast and Netflix escalate fight for viewers. Retrieved from http://articles.latimes.com/2012/feb/21/business/la-fi-ct-comcast-vod-20120222

Gallagher, R. (2012, October 16). New "surveillance-proof" app to secure communications has governments nervous. Retrieved from http://www.slate.com/articles/technology/future_tense/2012/10/silent_circle_mike_janke_s_iphone_app_makes_encryption_easy_governments.html

Gantz, J., & Reinsel, D. (2011). Extracting value from chaos. Retrieved from http://www.emc.com/collateral/analyst-reports/idc-extracting-value-from-chaos-ar.pdf

Gantz, J., & Reinsel, D. (2012). The digital universe in 2020: Big data, bigger digital shadows, and biggest growth in the Far East. Retrieved from http://www.emc.com/collateral/analyst-reports/idc-the-digital-universe-in-2020.pdf

Garrett, R. K. (2009). Politically motivated reinforcement seeking: Reframing the selective exposure debate. *Journal of Communication, 59*, 676–699.

Ghonim, W. (2012). *Revolution 2.0: The power of the people is greater than the people in power. A memoir*. New York, NY: Houghton Mifflin Harcourt.

Gibbs, J. L., Ellison, N. B., & Heino, R. D. (2006). Self-presentation in online personals: The role of anticipated future interaction, self-disclosure, and perceived success in Internet dating. *Communication Research, 33*, 152–177.

Gibbs, J. L., Ellison, N. B., & C.-H., Lai. (2011). First comes love, then comes Google: An investigation of uncertainty reduction strategies and self-disclosure in online dating. *Communication Research, 38*, 70–100.

Giles, J. (2005). Internet encyclopedias go head to head. *Nature, 438*, 900–901.

Gladstone, B. (Producer). (2012, January 31). *Hacking the New York Times, tweeting revolutions, and more* [Podcast]. Retrieved from http://itunes.apple.com

Gladstone, B. (Producer). (2013, May 9). *Who's gonna pay for this stuff?* [Podcast]. Retrieved from http://itunes.apple.com

Global Internet usage. (n.d.). In *Wikipedia*. Retrieved from http://en.wikipedia.org/wiki/Global-Internet-usage

Goffman, E. (1962). *Behavior in public places*. New York, NY: Free Press.

Golan, G. J., & Zaidner, L. (2008). Creative strategies in viral advertising: An application of Taylor's six-segment message strategy wheel. *Journal of Computer-Mediated Communication, 13*, 959–972.

Goodman, D. J. (2011, January 11). World leaders cheer but remain wary. *The New York Times*. Retrieved from http://www.nytimes.com/2011/02/12/world/middleeast/12global.html

Google. (2012). Mobile banking trends 2012. Retrieved from http://www.google.com/think/research-studies/mobile-banking-trends-2012.html

Google/Compete. (2011a). Search to rx: Depression. Retrieved from http://www.google.com/think/research-studies/search-to-rx-depression.html

Google/Compete. (2011b). Search to rx: Cancer. Retrieved from http://www.google.com/think/research-studies/search-to-rx-cancer.html

Google/Compete. (2011c). OTC pain relief: The e-patient's path to treatment. Retrieved from http://www.google.com/think/research-studies/otc-pain-relief-the-e-patients-path-to-treatment.html

Google/Compete. (2012a). The digital journey to wellness: Hospital selection. Retrieved from http://www.google.com/think/research-studies/the-digital-journey-to-wellness-hospital-selection.html

Google/Compete. (2012b). The digital journey to recovery. Retrieved from http://www.google.com/think/research-studies/the-digital-journey-to-recovery-treatment.html

Google/Manhattan Research. (2012). Screen to script: The doctor's digital path to treatment. Retrieved from http://www.google.com/think/research-studies/the-doctors-digital-path-to-treatment.html

Google/Shopper Sciences. (2011). The zero moment of truth finance study—Banking. Retrieved from http://www.google.com/think/research-studies/zero-of-truth-for-bank-deposits-study.html

Google/TNS & Tru. (2012). Digital and the new college experience. Retrieved from http://www.google.com/think/research-studies/digital-and-the-new-college-experience.html

Gorman, S. (2008, October 17). Obama buys first video game campaign ads. Retrieved from http://www.reuters.com/article/2008/10/17/us-usa-politics-videogames-idUSTRE49EAGL20081017

Gorman, S. (2013, October 29). U.S. spying to undergo senate review. *The New York Times*, pp. A1, A8.

Gorman, S., Barrett, D., & Valentino-DeVries, J. (2013, August 22). Secret spy court raps NSA snooping. *The Wall Street Journal*, pp. A1, A6.

Gorman, S., & Entous, A. (2013, October 28). Obama unaware as U.S. spied on world leaders: Aides. *The Wall Street Journal*, pp. A1, A8.

Gorman, S., Lee, C. E., & Hook, J. (2013, August 13). Obama proposes surveillance-policy overhaul. *The Wall Street Journal*, p. A4.

Gorman, S., & Valentino-DeVries, J. (2013, August 21). NSA's reach into U.S. net is deep, wide. *The Wall Street Journal*, pp. A1, A8.

Greenberg, A. (2012). *This machine kills secrets: How Wikileakers, hacktivists, and cyberpunks aim to free the world's information.* New York, NY: Random House.

Greenwald, G. (2013, June 5). NSA collecting phone records of millions of Verizon customers daily. Retrieved from http://www.theguardian.com/world/2013/jun/06/nsa-phone-records-verizon-court-order

Greenwald, G., & MacAskill, E. (2013, June 6). NSA Prism program taps in to user data of Apple, Google and others. Retrieved from http://www.theguardian.com/world/2013/jun/06/us-tech-giants-nsa-data

Grimes W. (2008, May 23). Volumes to go before you die. *The New York Times*, pp. B21, B23.

Grodal, T. (2000). Video games and the pleasures of control. In D. Zillman & P. Vorderer (Eds.), *Media entertainment the psychology of its appeal* (pp. 197–213). Mahwah, NJ: Lawrence Erlbaum.

Gross, A. (1996). *The rhetoric of science.* Cambridge, MA: Harvard University Press.

Gross, D. (2011, February 9). New app helps Catholics confess on the go. Retrieved from http://edition.cnn.com/2011/TECH/mobile/02/09/confession/

Grossman, L. (2011, February 21). 2045: The year man becomes immortal. *Time*, 40–46.

Grossman, L. (2012, January 30). The beast with a billion eyes. Retrieved from http://content.time.com/time/magazine/article/0,9171,2104815,00.html

Gudkova, D. (2013). Kaspersky security bulletin: Spam evolution 2012. Retrieved from http://www.securelist.com/en/analysis/204792276

Ha, V., Shaar, F. A., Inkpen, K., & Hdeib, L. (2006). An examination of user perception and misconception of Internet cookies. *Proceedings of CHI '06 Extended Abstracts on Human Factors in Computing Systems* (pp. 833–838). New York, NY: ACM. Retrieved from http://dl.acm.org/citation.cfm?id=1125615

Habermas, J. (2006). Political communication in media society: Does democracy still enjoy an epistemic dimension? The impact of normative theory on empirical research. *Communication Theory, 16*, 411–426.

Hafner, K. (2012, December 4). Could a computer outthink this doctor? *The New York Times*, pp. D1, D6.

Hall, J. A., Park, N., Song, H., & Cody, M. J. (2010). Strategic misrepresentation in online dating: The effects of gender, self-monitoring, and personality traits. *Journal of Social and Personal Relationships, 27*, 117–135.

Hampton, K. N., Goulet, L. S., Raine, L., & Purcell, K. (2011). *Social networking sites and our lives.* Washington, DC: Pew Research Center.

Hampton, K. N., Sessions, L. F., Her, E. J., & Rainie, L. (2009). *Social isolation and new technology.* Washington, DC: Pew Research Center.

Hancock, J. T., & Toma, C. L. (2009). Putting your best face forward: The accuracy of online dating photographs. *Journal of Communication, 59*, 367–386.

Harbour, P. J. (2012, December 19). The emperor of all identities. *The New York Times*, p. A31.

Hardy, B. W., Jamieson, K. H., & Winneg, K. (2009). The role of the Internet in identifying deception during the U.S. presidential campaign. In A. Chadwick & P. N. Howard (Eds.), *Routledge handbook of Internet politics* (pp. 131–143). New York, NY: Routledge.

Harper, J. (2010, August 8). The great privacy debate. *The Wall Street Journal*, pp. W1–W2.

Hart, M. (2000). *The American Internet advantage: Global themes and implications.* Lanham, MD: University Press of America.

Haskins, C. H. (1957). *The rise of universities.* Ithaca, NY: Cornell Paperbacks /Cornell University Press. Originally given as Colver Lectures in 1923 at Brown University, with the first edition of this volume published in 1923 by Henry Holt and Company.

Haugney, C. (2012, August, 8). Magazine sales decline on newsstands by 10%. *The New York Times*, p. B3.

Hauser, M. L., Fleuriet, C., & Estrada, D. (2012). The cyber factors: An analysis of relational maintenance through the use of computer-mediated communication. *Communication Research Reports, 28*, 34–43.

Helft, M. (2011, April 26). Apple and Google use phone data to map the world. *The New York Times*, pp. B1, B2.

Hellman, H. (1969). *Communications in the world of the future.* Philadelphia, PA: Lippincott.

Hendler, J., Shadbolt, N., Hall, W., Berners-Lee, T., &Weitzner, D. (2008). Web science: An interdisciplinary approach to understand the Web. *Communications of the ACM, 51*(7), 60–69.

Hendy, N. (2013, April 4). Research shows teleworking has revenue benefits. Retrieved from http://www.smh.com.au/small-business/research-shows-teleworking-has-revenue-benefits-20130404-2h7uh.html

Hernandez, B. A. (2013, June 7). "Red Wedding" is the most social episode of any HBO show ever. Retrieved from http://mashable.com/2013/06/07/red-wedding-game-of-thrones-social/

Hewitt, S. (1998, September 10). Can there be a REAL internet church? *Christian Computing Magazine*. Retrieved from http://www.gospelcom.net/ccmag/articles/cov0998.shtml. No longer available online.

Hirschorn, M. (2007, April). The Web 2.0 bubble. *The Atlantic*, pp. 134–138

Hirsh, M. (2013, June 10). Silicon Valley doesn't just help the surveillance state—It built it. Retrieved from http://www.theatlantic.com/national/archive/2013/06/silicon-valley-doesnt-just-help-the-surveillance-state-it-built-it/276700/

Hite, S. (1981). *The Hite report on male sexuality.* New York, NY: Ballantine Books.

Hofstadter, D. R. (1983, November 13). Mind, body, and machine [Review of the book *Alan Turing: The Enigma*, by A. Hodges]. *The New York Book Review*, p. 1.

Hollander, S. (2012, September 25). Online holdouts no more. Retrieved from http://online.wsj.com/news/articles/SB10000872396390044371370457760324100600159

Hoofnagle, C. J., & Good, N. (2012, October). The web privacy census. Retrieved from http://www.law.berkeley.edu/privacycensus.htm

Hoover, S., Clark, L. S., & Rainie, L. (2004). *Faith online.* Washington, DC: Pew Research Center.

Hope, D. A., & Heimberg, R. G. (1990). Dating anxiety. In H. Leitenberg (Ed.), *Handbook of social and evaluation anxiety* (pp. 217–246). New York, NY: Plenum.

Hopkins, C. (2013, May 29). U.S. company allegedly caught aiding Syria and Iran in censorship efforts. Retrieved from http://www.dailydot.com/news/blue-coat-syria-iran-spying-software/

Houser, M. L., Fleuriet, C., & Estrada, D. (2012). The cyber factor: An analysis of relational maintenance through the use of computer-mediated communication. *Communication Research Reports, 29,* 34-43.

Howe, N., & Strauss, W. (2000). *Millennials rising: The next great generation.* New York, NY: Vintage Books.

Howe, N., & Strauss, W. (2007). *Millennials go to college.* Great Falls, VA: Life Course Associates.

Huberman, B. A. (2001). *The laws of the Web: Patterns in the ecology of information.* Cambridge, MA: The MIT Press.

IAB, InMobi, & Viggle. (2013). Mobile and money: Consumer awareness and adoption of smartbased financial application. Retrieved from http://www.iab.net

ICEF Monitor. (2013). 8 countries leading the way in online education. Retrieved from http://monitor.icef.com/2012/06/8-countries-leading-the-way-in-online-education/

IDC Financial Insights. (2012). New IDC Financial Insights consumer payments survey reveals over a third of all U.S. residents use mobile payments. Retrieved from http://www.idc.com/getdoc.jsp?containerId=prUS23585112

Imperva. (2012). *Imperva's hacker intelligence summary report: The anatomy of an anonymous attack.* Redwood Shores, CA: Imperva.

Internet World Statistics. (2013, May 10). 2013 Q1 Internet world statistics. Retrieved from http://tek-tips.nethawk.net/2013-q1-internet-world-statistics

Ipsos. (2012a). Interconnected world: Shopping and personal finance. Retrieved from http://www.ipsos-na.com/download/pr.aspx?id=11513

Ipsos. (2012b). Telecommuting: Citizens in 24 countries assess working remotely for a total global perspective. Retrieved from http://www.ipsos-na.com/download/pr.aspx?id=11327

Iraq archive: The strands of a war. (2010, October 23). *The New York Times,* p. A1.

Isaac, M. (2012, December 22). Facebook pushes "gifts" hard in time for the holidays. Retrieved from http://allthingsd.com/20121222/facebook-pushes-gifts-hard-in-time-for-the-holidays/

Isaacson, W. (2012). *Steve Jobs.* New York, NY: Simon & Schuster.

Jacobs, A. J. (2013, April 20). Two cheers for web U! Retrieved from http://www.nytimes.com/2013/04/21/opinion/sunday/grading-the-mooc-university.html?_r=0

Jacobs, B. (2012, November 11). Orca failed; but so did Obama's 2008 version of the same. Retrieved from http://www.theatlantic.com/politics/archive/2012/11/orca-failed-but-so-did-obamas-2008-version-of-the-same/265077/

Jacobsen, W. C., & Forste, R. (2011). The wired generation: Academic and social outcomes of electronic media use among university students. *Cyberpsychology, Behavior, and Social Networking, 14,* 275–280.

Jaggars, S. S. (2013). Choosing between online and face-to-face courses. *The American Journal of Distance Education, 28*(1). Retrieved from http://ccrc.tc.columbia.edu/publications/online-demand-student-voices.html

Jaggars, S. S., & Xu, D. (2013). *Predicting online student outcomes from a measure of course quality.* Community College Research Center, Teachers College, Columbia University. Retrieved from http://ccrc.tc.columbia.edu/publications/predicting-online-student-outcomes-and-course-quality.html

Jansen, J. (2011). *The civic and community engagement of religiously active Americans.* Washington, DC: Pew Research Center.

Jaret, P. (2013, January 14). Mining electronic records for revealing health data. Retrieved from http://www.nytimes.com/2013/01/15/health/mining-electronic-records-for-revealing-health-data.html

Jarvis, B. (2013, March 4). Twitter becomes a tool for tracking flu epidemics and other public health issues. Retrieved from http://articles.washingtonpost.com/2013-03-04/national/37429814_1_twitter-data-tweets-mark-dredze

Jensen, C., Potts, C., & Jensen, C. (2005). Privacy practices of Internet users: Self-reports versus observed behavior. *International Journal of Human-Computer Studies, 63,* 203–227.

John, N. A. (2011). The diffusion of the Internet to Israel: The first 10 years. *Israel Affairs, 17,* 327–340.

Johnson, S. (2005). *Everything bad is good for you: How today's popular culture is actually making us smarter.* New York, NY: Riverhead Books.

Jones, S., & Fox, S. (2009, January 28). *Generations online in 2009.* Washington, DC: Pew Research Center.

Jonson, H., & Siverskog, A. (2012). Turning vinegar into wine: Humourous self-presentations among older GLBTQ online daters. *Journal of Aging Studies, 26,* 55–64.

Jurgensen, J. (2008, June 28). Studies in videogames. *The Wall Street Journal,* p. W3.

Justice, G. (2004, November 6). Kerry kept money coming in with the Internet as his ATM. Retrieved from http://www.nytimes.com/2004/11/06/politics/campaign/06internet.html

Kafka, P. (2006, January 13). Musicland's sad song. Retrieved from http://www.forbes.com/2006/01/13/musicland-itunes-retailing_cx_pak_0113musicland.html

Kaiser Family Foundation. (2010, January). *Generation M²: Media in the lives of 8-to-18- year-olds.* Menlo Park, CA: Henry J. Kaiser Family Foundation.

Kane, Z. (2012, July 20). Steve Jobs: Apple was 90 days from going bankrupt. Retrieved from http://thenextweb.com/apple/2010/06/02/steve-jobs-90-days/#!pQs12

Kang, J. (1998). Information privacy in cyberspace transactions. *Stanford Law Review, 1193,* 1212–1220.

Kanter, J., & Lohr, S. (2012, December 18). Google deal on antitrust seen in U.S. *The New York Times,* pp. B1, B2.

Kawamoto, D., & Mills, E. (2006). AOL apologizes for release of user search data. Retrieved from http://news.cnet.com/AOL-apologizes-for-release-of-user-search-data/2100-1030_3-6102793.html

Keen, A. (2006, February 15). The second generation of the Internet has arrived. It's worse than you think. Retrieved from http://www.weeklystandard.com/Content/Public/Articles/000/000/006/714fjczq.asp

Keen, A. (2007). *The cult of the amateur: How today's Internet is killing our culture.* New York, NY: Doubleday/Currency.

Keeter, S. (2006, May 30). *Politics and the "dotnet" generation.* Washington, DC: Pew Research Center.

Keller, B. (2011, July 17). Let's ban books, or at least stop writing them. *The New York Times Magazine*, p. MM9.

Kelly, L., & Duran, R. (2012). Narcissism or openness: College students' use of Facebook and Twitter. *Communication Research Reports, 29*, 108–118.

Kempers, M. (2002). *Community matters: An exploration of theory and practice.* Chicago, IL: Burnham.

Kennedy-Lightsey, C. D., Martin, M. M., Thompson, M., Himes, K. L., & Clingerman, B. Z. (2012). Communication Privacy Management Theory: Exploring coordination and ownership between friends. *Communication Quarterly, 60*, 665–680.

Kim, M., Kwon, K.-N., & Lee, M. (2009). Psychological characteristics of Internet dating service users: The effect of self-esteem, involvement, and sociability on the use of Internet dating services. *CyberPsychology & Behavior, 12*, 445–449.

Klaassen, A. (2007, October 1). Do home pages have a place in Web 2.0's future? Retrieved from http://adage.com/article/digital/home-pages-a-place-web-2-0-s-future/120789/

Koetsier, J. (2013, March 5). Facebook: 15 million businesses, companies, and organizations now have a Facebook page. Retrieved from http://venturebeat.com/2013/03/05/facebook-15-million-businesses-companies-and-organizations-now-have-a-facebook-page/

Kohut, A. (2000). *Youth vote influenced by online information.* Washington, DC: Pew Research Center.

Kohut, A. (2012). *Social networking popular across the globe.* Washington, DC: Pew Research Center.

Kolata, G. (2013, January 18). Web hunt for DNA sequences leaves privacy compromised. *The New York Times*, pp. A1, A15.

Korn, M. (2013). Coursera makes case for MOOCs. Retrieved from http://online.wsj.com/news/articles/SB10001424127887324715704578483570761525766

Korn, M. (2013, February 21). Online-education provider Coursera signs 29 more schools. Retrieved from http://online.wsj.com/news/articles/SB100014241278873238643045783165305449240 00

Kujath, C. (2011). Facebook and MySpace: Complement of substitute for face-to-face interaction? *Cyberpsychology, Behavior, and Social Networking, 14*, 75–78.

Kulish, N., & Cieply, M. (2011, December 6). Around the world in one movie: Film financing's global future. *The New York Times*, p. A1.

Kurzweil, R. (2005). *The singularity is near: When humans transcend biology.* New York, NY: Viking.

Kurzweil, R. (2012). *How to create a mind: The secret of human thought revealed.* New York, NY: Viking.

Kurzweil, R., & Grossman, T. (2004). *Fantastic voyage: Live long enough to live forever.* Emmaus, PA: Rodale.

Kutner, L., & Olson, C. K. (2008). *Grand theft childhood: The surprising truth about violent video games and what parents can do.* New York, NY: Simon & Schuster.

LaFraniere, S. (2013, July 21). Crackdowns on leaks prompted by 153 cases that led nowhere. *The New York Times*, pp. 1, 18.

Landler, M. (2011, February 15). U.S. policy to address Internet freedom. *The New York Times*, p. A10.

Landler, M., & Sanger, D. E. (2013, October 29). Obama may ban spying on heads of allied states. *The New York Times*, pp. A1, A9.

LaPointe, P. (2011, January 4). Predictions for social media metrics: 2011. Retrieved from http://www.mediapost.com/publications/article/142366/

Lavallee, A. (2009, September 18). Net neutrality in the spotlight. *The Wall Street Journal*, p. A4.

Lee, K. M. (2004). Presence, explicated. *Communication Theory, 14*, 27–50.

Lee, M. C. (2009). Factors influencing the adoption of internet banking: An integration of TAM and TPB with perceived risk and perceived benefit. *Electronic Commerce Research and Applications, 8*, 130–141.

Lei, Y. (2011). The political consequences of the rise of the Internet: Political beliefs and practices of Chinese netizens. *Political Communication, 28*, 291–322.

Lenhart, A. (2000). *Who's not online: 57% of those without Internet access say they do not plan to log on*. Washington, DC: Pew Research Center.

Lenhart, A. (2011). *"How do [they] even do that?" Myths & facts about the impact of technology on the lives of American teens*. Washington, DC: Pew Research Center.

Lenhart, A., Kahne, J., Middaugh, E., & Evans, C. (2008). *Teens, video games, and civics*. Washington, DC: Pew Research Center.

Lenhart, A., Madden, M., Smith, A., Purcell, K., Zickuhr, K., & Raine, L. (2011). *Teens, kindness and cruelty on social networking sites: How American teens navigate the new world of digital citizenship*. Washington, DC: Pew Research Center.

Levine, D. (2000). Virtual attraction: What rocks your boat? *CyberPsychology & Behavior, 3*, 565–573.

Levitz, J. (2011, October 7). Post office's rescue plan: Junk mail. Retrieved from http://online.wsj.com/news/articles/SB10001424052970204612504576606743516301586

Levy, D. (2007). *Love and sex with robots: The evolution of human-robot relationships*. New York, NY: Harper.

Levy, P. (2001). *Cyberculture*. Minneapolis: University of Minnesota Press.

Levy , S. (2010). *Hackers: Heroes of the computer revolution*. Cambridge, MA: O'Reilly Media.

Lewin, T. (2013a, May 10). Universities team with online course provider. Retrieved from http://www.nytimes.com/2013/05/30/education/universities-team-with-online-course-provider.html

Lewin, T. (2013b, April 29). Colleges adapt online courses to ease burden. Retrieved from http://www.nytimes.com/2013/04/30/education/colleges-adapt-online-courses-to-ease-burden.html

Lichtblau, E. (2013, July 7). In secret, court vastly broadens powers of N.S.A. *The New York Times*, pp. 1, 4.

Lichtblau, E., & Schmidt, M. E. (2013, August 4). Other agencies clamor for data N.S.A. compiles. *The New York Times*, pp. 1, 16.

Lieberman, D. (2010, April 12). Cable wins ruling on Internet traffic. *The Star Press* [Muncie, IN], p. 8A.

Lindgren, S. (2010). Widening the glory hole: The discourse of online porn fandom. In F. Attwood (Ed.), *Porn.com: Making sense of online pornography* (pp. 171–185). New York, NY: Peter Lang.

Lo, S. K., Hsieh, A. Y., & Chiu, Y. P. (2013). Contradictory deceptive behavior in online dating. *Computers in Human Behavior, 29*, 1755–1762.

Loane, S. S., & D'Alessandro, S. (2013). Communication that changes lives: Social support within an online health community for ALS. *Communication Quarterly, 61*, 236–251.

Loechner, J. (2011, December 2). Almost 120 million people worldwide are social game players. Retrieved from http://www.mediapost.com/publications/article/163040/

Long, J. (2012). *Anti-porn: The resurgence of anti-pornography feminism*. London, England: Zed Books.

Lovink, G. (2002). *Dark fiber: Tracking critical Internet culture*. Cambridge, MA: The MIT Press.

Luo, M., & Zeleny, J. (2008, June 20). Obama, in shift, says he'll reject public financing. Retrieved from http://www.nytimes.com/2008/06/20/us/politics/20obama.html?pagewanted=all&_r=0&gwh=1B9AE8889F73078A4BB812D9ADD32F76&gwt=pay

Luo, S., & Zhang, G. (2009). What leads to romantic attraction: Similarity, reciprocity, security, or beauty? Evidence from a speed-dating study. *Journal of Personality, 77*, 933–964.

Luo, X., Lee, C. P., Mattila, M., & Liu, L. (2012). An exploratory study of mobile banking services resistance. *International Journal of Mobile Communications, 10*, 366–385.

Lutz, M. (2009). *The social pulpit: Barack Obama's social media toolkit*. New York, NY; Chicago, IL: Edelman.

Madden, M., & Fox, S. (2006, October 5). *Riding the waves of "Web 2.0": More than a buzzword, but still not easily defined*. Washington, DC: Pew Research Center.

Madden, M., Lenhart, A., Cortesi, S., Gasser, U., Duggan, M., & Smith, A. (2013). *Teens, social media, and privacy*. Washington, DC: Pew Research Center.

Madden, N. (2008, February 26). Portals in a Web 2.0 world. Retrieved from http://www.technewsworld.com/story/61820.html

Madrigal, A. C. (2012, November 16). When the nerds go marching in. Retrieved from http://www.theatlantic.com/technology/archive/2012/11/when-the-nerds-go-marching-in/265325/

Manago, A. M., Graham, M. B., Greenfield, P. M., & Salimkhan, G. (2008). Self-presentation and gender on MySpace. *Journal of Applied Developmental Psychology, 29*, 446–458.

Mandiant. (2013). APT1: Exposing one of China's cyber espionage units. *Mandiant*. Retrieved from http://intelreport.mandiant.com/Mandiant_APT1_Report.pdf

Manguel, A. (1996). *A history of reading*. New York, NY: Penguin Books.

Marinetti, F. T. (1983). *Manifesto of futurism*. New Haven, CT: Yale Library Associates. Published for the first time on February 5, 1909 in *La gazzettadell' Emilia;* an article then reproduced in the French Daily newspaper *Le Figaro* on February 20, 1909.

Markoff, J. (2011, February 17). On "Jeopardy!" Watson win is all but trivial. *The New York Times*, p. A1.

Markoff, J. (2011, November 8). It started digital wheels turning. *The New York Times*, pp. D1, D4.

Markoff, J. (2011, December 21). Jacob Goldman, a father of computing, dies at 90. *The New York Times*, p. A25.

Markoff, J. (2012, November 24). Learning curve: No longer just a human trait. *The New York Times*, pp. A1, A3.

Markoff, J. (2013, May 27). At high speed, on the road to a driverless future. Retrieved from http://www.nytimes.com/2013/05/28/science/on-the-road-in-mobileyes-self-driving-car. html?_r=0

Martino, J. P. (1972). *Technological forecasting for decision making*. New York, NY: American Elsevier.

Mathews, A. W. (2012, December 12). Doctors move to webcams. Retrieved from http://online. wsj.com/news/articles/SB10001424127887324731304578189461164849962

Matyszcyk, C. (2013, March 1). Police chief suspended for Facebook pic with gun-toting woman. Retrieved from http://news.cnet.com/8301-17852_3-57572063-71/police-chief-suspended-for-facebook-pic-with-gun-toting-woman/

Mayer-Schönberger, V. (2009). Delete: The virtue of forgetting in the digital age. Princeton, NJ: Princeton University Press.

Mayer-Schönberger, V., & Cukier, K. (2013). *Big data: A revolution that will transform how we live, work, and think*. New York, NY: Houghton Mifflin Harcourt.

Mazzetti, M., Perlez, J., Schmitt, E., & Lehren, A. W. (2010, April 26). Pakistan aids insurgency in Afghanistan, reports assert. *The New York Times*, p. A1.

Mazzetti, M., & Shane, S. (2013, August 10). Balancing act put to the test. *The New York Times*, pp. A1, A11.

McDonald, M. (2012, March 13). Watch your language! (In China, they really do). Retrieved from http://rendezvous.blogs.nytimes.com/2012/03/13/watch-your-language-and-in-china-they-do/

McDonald, M. (2012, December 23). Adding more bricks to the great firewall of China. Retrieved from http://rendezvous.blogs.nytimes.com/2012/12/23/adding-more-bricks-to-the-great-fire wall-of-china/

McGillion, C. (2000, December 23). Web of disbelief: Religion has staked a big claim in cyberspace, but has it done a Faustian deal? *Sydney Morning Herald*.

McGonigal, J. (2011). *Reality is broken: Why games make us better and how they can change the world*. New York, NY: Penguin Press.

McHale, J. (1973). *World facts and trends*. New York, NY: Collier Macmillan.

McKenna, K. Y. A. (2008). MySpace or your place: Relationship initiation and development in the wired and wireless world. In S. Sprecher, A. Wnezel, & J. Harvey (Eds.), *Handbook of relationship initiation* (pp. 235–247). New York, NY: Psychology Press.

McKinney, M. S., & Banwart, M. C. (2011). The election of a lifetime. In M. S. McKinney & M. C. Banwart (Eds.), *Communication in the 2008 U.S. election: Digital natives elect a president* (pp. 1–9). New York, NY: Peter Lang.

McLuhan, M. (1951). *The mechanical bride: Folklore of the industrial man*. New York, NY: Vanguard Press.

McLuhan, M. (1964). *Understanding media: The extensions of man*. New York, NY: McGraw-Hill.

McLuhan, M. (1970). *Culture is our business*. New York, NY: McGraw-Hill.

McLuhan, M. (2006/1943). *The classical trivium: The place of Thomas Nashe in the learning of his time*. Corte Masdera, CA: Gingko Press. This work is an edition (edited by W. Terrence Gordon) of McLuhan's Cambridge University doctoral dissertation submitted and approved in 1943.

McLuhan, M., & Fiore, Q. (1967). *The medium is the message*. New York, NY: Bantam Books.

McLuhan, M., & Fiore, Q. (1968). *War and peace in the global village: An inventory of some of the current spastic situations that could be eliminated by more feedforward.*

McLuhan, M., & McLuhan, E. (1988). *Laws of media: The new science.* Toronto, Canada: University of Toronto Press.

McLuhan, M., & Watson, W. (1970). *From cliché to archetype.* New York, NY: Viking Press.

Mehrabian, A. (2001). Characteristics attributed to individuals on the basis of their first names. *Genetic, Social, and General Psychology Monographs, 127,* 59–88.

Mehta, M. D., & Plaza, D. (1997). Content analysis of pornographic images available in the Internet. *Information Society, 13,* 153–161.

Messaris, P., & Humphreys, L. (2006). *Digital media: Transformations in human communication.* New York, NY: Peter Lang.

Miller, C. C. (2008, November 7). How Obama's Internet campaign changed politics. Retrieved from http://bits.blogs.nytimes.com/2008/11/07/how-obamas-internet-campaign-changed-politics/

Miller, C. C. (2011, October 28). With new device, Google tries again on Internet TV. Retrieved from http://bits.blogs.nytimes.com/2013/07/24/with-new-device-google-tries-again-on-internet-tv/

Miller, C. C., & Bosman, J. (2011, May 20). E-books outsell print books at Amazon. *The New York Times,* pp. B2.

Milne, G. R., Bahl, S., & Rohm, A. (2008). Toward a framework for assessing covert marketing practices. *Journal of Public Policy & Marketing, 27,* 57–62.

Mindlin, A. (2009, June 29). Settling in for longer online videos. *The New York Times,* p. B3.

Mindlin, A. (2010, September, 6). Life without a TV set? Not impossible. *The New York Times,* pp. B2.

Miyazaki, A. D. (2008). Online privacy and the disclosure of cookie use: Effects on consumer trust and anticipated patronage. *American Marketing Association, 27,* 19–33.

Moore, K. (2011). *71% of online adults now use video-sharing sites.* Washington, DC: Pew Research Center.

Moreno, M. A., Swanson, M. J., Royer, H., & Roberts, L. J. (2011). Sexpectations: Male college students' views about displayed sexual references of female's social networking web sites. *Journal of Pediatric Adolescence Gynecology, 24,* 85–89.

Morozov, E. (2011). *The net delusion: The dark side of Internet freedom.* New York, NY: PublicAffairs.

Morozov, E. (2013, February 23). Is smart making us dumb? Retrieved from http://online.wsj.com/news/articles/SB10001424127887324503204578318462215991802

Morse, A. (2012, July 4). Online education for boomers. Retrieved from http://online.wsj.com/news/articles/SB10001424052702304708604577505222718048642

Mosk, M. (2008, March 28). Obama rewriting rules for raising campaign money online. Retrieved from http://www.washingtonpost.com/wpdyn/content/article/2008/03/27/AR2008032702968.html

MTV Press Releases. (2013, June 18). Young Millennials will keep calm & carry on. Retrieved from http://mtvpress.com/press/release/young_millennials_will_keep_calm

Mueller, M. L. (2002). *Ruling the root: Internet governance and the taming of cyberspace.* Cambridge, MA: The MIT Press.

Muise, A., Christofides, E., & Desmarais, S. (2009). More information than you ever wanted: Does Facebook bring out the green-eyed monster of jealousy? *CyberPsychology & Behavior, 12,* 441–444.

Muntinga, D. G., Moorman, M., & Smit, E. G. (2011). Introducing COBRAs: Exploring motivations for brand-related social media use. *International Journal of Advertising, 30,* 13–46.

Murphy, S. (2013, March 19). Apple sets sights on augmented reality for iOS devices. Retrieved from http://mashable.com/2013/03/19/apple-augmented-reality-app/

Murphy, S. (2013, June 3). Zynga lays off 18% of its workforce. Retrieved from http://mashable.com/2013/06/03/zynga-cuts-workforce/

Naik, G. (2013, January 24). Storing digital data in DNA. Retrieved from http://online.wsj.com/news/articles/SB10001424127887324539304578259883507543150

Naisbitt, J. (1982). *Megatrends: Ten new directions transforming our lives.* New York, NY: Warner Books.

Naisbitt, J., & Aburdene, P. (1990). *Megatrends 2000: Ten new directions for the 1990s.* New York, NY: William Morrow.

National Endowment for the Arts. (2007). *To read or not to read: A question of national consequence.* Washington, DC: National Endowment for the Arts.

National Telecommunications and Information Administration. (1998). *Memorandum of understanding between the U.S. Department of Commerce and Internet Corporation for Assigned Names and Numbers.* Retrieved from http://www.ntia.doc.gov/page/1998/memorandum-understanding-between-us-department-commerce-and-internet-corporation-assigned-

Needleman, S. E., & Ante, S. E. (2013, May 8). Bitcoin primer: What you need to know about the new virtual currency. Retrieved from http://blogs.wsj.com/digits/2013/05/08/a-bitcoin-primer-what-you-need-to-know-about-the-new-virtual-currency/

Nielsen. (2012a). *State of the media: The social media report 2012.* Retrieved from http://www.targetspot.com/wp-content/uploads/2012/12/The-Social-Media-Report-2012.pdf

Nielsen. (2012b). *Global trust in advertising and brand messages.* Retrieved from http://www.nielsen.com/us/en/reports/2012/global-trust-in-advertising-and-brand-messages.html

Nixon, R. (2013, February 7). Trying to stem losses, Post Office seeks to end Saturday letter delivery. *The New York Times,* pp. A15.

Nobel, C. (2013, June 18). Advertising symbiosis: The key to viral videos. *Working Knowledge.* Retrieved from http://hbswk.hbs.edu/item/7267.html

NPR. (2012, April 23). *Is the Internet closing our minds politically?* [Audio podcast]. Retrieved from http://npr.org/2012/03/23/151037080/is-the-internet-closing-our-minds-politically

O'Brien, K. J. (2010, November 28). Mobile banking in the emerging world. Retrieved from http://www.nytimes.com/2010/11/29/business/global/29iht-mobilebanks29.html?

Ogas, I., & Gaddam, S. (2012). *A billion wicked thoughts: What the Internet tells us about sexual relationships.* New York, NY: Plume.

Ong, W. J. (1982). *Orality and literacy: The technologizing of the word.* New York, NY: Methuen & Company.

Open Secrets. (2008a). Obama fundraising and spending. *Open Secrets.* Retrieved from http://www.opensecrets.org/pres08/summary.php?cid=n00009638

Open Secrets. (2008b). McCain fundraising and spending. *Open Secrets*. Retrieved from http://www.opensecrets.org/pres08/summary.php?cycle=2008&cid=N00006424

Oracle. (2013). *Consumer views of live help online 2012: A global perspective*. Retrieved from http://www.oracle.com/us/products/applications/commerce/live-help-on-demand/oracle-live-help-wp-aamf-1624138.pdf

Our view: Is phone program worth price? Burden of proof rest with NSA. (2013, October 21). *USA Today*, p. 8A.

Ovide, S., & Rusli, E. M. (2013, May 28). Apple's Cook hints at wearable devices. Retrieved from http://online.wsj.com/news/articles/SB10001424127887323855804578512042192416604

Paasonen, S. (2011a). Online pornography: Ubiquitous and effaced. In M. Consalvo & C. Ess (Eds.), *The handbook of Internet studies* (pp. 424–439). Malden, MA: Wiley-Blackwell.

Passonen, S. (2011b). *Carnal resonance: Affect and online pornography*. Cambridge, MA: The MIT Press.

Padilla, R. (2007, September 15). Getting to the bottom of a buzzword: What is Web 2.0? Retrieved from http://www.techrepublic.com/blog/tech-decision-maker/getting-to-the-bottom-of-a-buzzword-what-is-web-20/

Parker, R. B. (1974). A definition of privacy. *Rutgers Law Review, 27*, 280.

Pask, G., & Curran, S. (1982). *Microman: Living and growing with computers*. London, England: Century.

Passy, C. (2013, March 18). Online-only banks start to yield more than virtual advantage. Retrieved from http://online.wsj.com/news/articles/SB10001424127887323869604578368342555203164

Patsiotis, A. G., Hughes, T., & Webber, D. J. (2012). Adopters and non-adopters of Internet banking: A segmentation study. *International Journal of Bank Marketing, 30*, 20–42.

Pease, R. (2012, June 26). Alan Turing: Inquest's suicide verdict "not supportable." Retrieved from http://www.bbc.co.uk/news/science-environment-18561092

Pellman, P. (2013, March 14). How and why we track: Confessions of an ad "tracking" company. Retrieved from http://allthingsd.com/20130314/how-and-why-we-track-confessions-of-an-ad-tracking-company/

Pempek, T. A., Yermolayeva, Y. A., & Calvert, S. L. (2009). College students' social networking experiences on Facebook. *Journal of Applied Developmental Psychology, 30*, 227–238.

Perez, S. (2013, May 20). Tumblr's adult fare accounts for 11.4% of site's top 200k domains, Adult sites are leading category of referrals. Retrieved from http://techcrunch.com/2013/05/20/tumblrs-adult-fare-accounts-for-11-4-of-sites-top-200k-domains-tumblrs-adult-fare-accounts-for-11-4-of-sites-top-200k-domains-adults-sites-are-leading-category-of-referrals/

Perez-Pena, R. (2008, June 9). Uncertainty as *Tribune* prepares to retrench. *The New York Times*, pp. C1, C2.

Perlroth, N. (2012, August 31). Software meant to fight crime is used to spy on dissidents. *The New York Times*, pp. A1, B2.

Perlroth, N., & Bilton, N. (2012, February 15). Mobile apps take data without permission. Retrieved from http://bits.blogs.nytimes.com/2012/02/15/google-and-mobile-apps-take-data-books-without-permission/?

Perlroth, N., & Hardy, Q. (2013). Bank hacking was the work of Iranians, officials say. Retrieved from http://www.nytimes.com/2013/01/09/technology/online-banking-attacks-were-work-of-iran-us-officials-say.html?

Peters, J. (2011, November 20). At 154, a digital milestone. Retrieved from http://mediadecoder.blogs.nytimes.com/2011/11/20/at-154-a-digital-milestone/?

Peters, J. (2012, February 12). A newspaper, and a legacy, reordered. *The New York Times*, p. BU1.

Petrescu, M., & Korgaonkar, P. (2011). Viral advertising: Definitional review and synthesis. *Journal of Internet Commerce, 10*, 208–226.

Petronio, S. (2002). *Boundaries of privacy*. Albany, NY: State University of New York Press.

Petronio, S. (2004). Road to developing communication privacy management theory: Narrative in progress, please stand by. *Journal of Family Communication, 4*, 193–207.

Pew Internet & American Life Project. (2012). *Smartphone research* [Infograph]. Retrieved from http://pewinternet.org

Pew Research Internet & American Life Project. (2012). *Trend data (adults)*. Retrieved from http://www.pewinternet.org

Pew Research Internet & American Life Project. (2012). *Trend data (adults). Internet adoptions, 1995–2012*. Retrieved from http://www.pewinternet.org

Pew Internet & American Life Project. (2013). *Device ownership*. Retrieved from http://www.pewinternet.org

Pew Research Center. (2010). *Blogs not neutral on Google*. Retrieved from http://pewresearch.org

Pew Research Center. (2010). *Jesus Christ's return to Earth*. Retrieved from http://pewresearch.org

Pew Research Center. (2010). *World War III and nuclear terrorism*. Retrieved from http://pewresearch.org

Pew Research Center for the People & the Press. (2006). *Online papers modestly boost newspaper readership*. Retrieved from http://www.people-press.org

Pew Research Center for the People & the Press. (2010). *Public sees a future full of promise and peril*. Retrieved from http://people-press.org

Pew Research Center for the People & the Press. (2012). *Partisan polarization surges in Bush, Obama years*. Retrieved from http://www.people-press.org

Pew Research Global Attitudes Project. (2010). *Global publics embrace social networking*. Retrieved from http://www.pewglobal.org

Pew Research Global Attitudes Project. (2011). *Global digital communication: Texting, social networking popular worldwide*. Retrieved from http://www.pewglobal.org

Pfanner, E. (2009, April 17). File-sharing site violated copyright, court says. Retrieved from http://www.nytimes.com/2009/04/18/world/europe/18copy.html?gwh=ED9873A3A3557E4B8B99B25688024B18&gwt=pay

Plambeck, J. (2010, April 27). Newspaper circulation falls nearly 9%. *The New York Times*, p. B2.

Plesser, A. (2010, December 16). The *Huffington Post* hits record 26 million unique visitors in November, now ranked number two "newspaper," comScore. Retrieved from http://www.huffingtonpost.com/andy-plesser/the-huffington-post-hits_b_797947.html

Podlas, K. (2000). Mistresses of their domain: How female entrepreneurs in cyberporn are initiating a gender power shift. *CyberPsychology & Behavior, 3*, 847–854.

Pollard, T. D., Chesebro, J. W., & Studinski, D. P. (2009). The role of the Internet in presidential campaigns. *Communication Studies, 60*, 574–588.

Ponemon Institute (2012, March). 2011 cost of data breach study: United States. Retrieved from http://www.symantec.com/content/en/us/about/media/pdfs/b-ponemon-2011-cost-of-data-breach-us.en-us.pdf

Pontin, J. (2008, August). The future of Web 2.0. *Technology Review, 111*(4), 34.

Pontin, J. (2008, August). The next bubble: Are Web 2.0 companies the unlucky beneficiaries of a speculative mania? *Technology Review, 111*(4), 14.

Porter, L., & Golan, G. J. (2006). From subservient chickens to brawny men: A comparison of viral advertising to television advertising. *Journal of Interactive Advertising, 6*, 26–33.

Postman, N. (1985). *Amusing ourselves to death: Public discourse in the age of show business.* New York, NY: Penguin Books.

Potter, W. J. (2008). *Media literacy* (4th ed.). Thousand Oaks, CA: Sage.

Powell, M. (2001). Knowledge, culture & the Internet in Africa: A challenge for political economists. *Review of African Political Economy, 28*, 241–260.

Pratt, T. (2013, May 2). Las Vegas web site is test for legal online gambling. Retrieved from http://www.nytimes.com/2013/05/03/business/poker-web-site-advances-online-gambling.html?

Prelli, L. J. (1989). *A rhetoric of science: Inventing scientific discourse.* Columbia, SC: University of South Carolina Press.

Purcell, K. (2010). *Seven in ten adult Internet users (69%) have used the internet to watch or download video.* Washington, DC: Pew Research Center.

Purcell, K. (2011). *Search and email still top the list of most popular online activities: Two activities nearly universal among adult internet users.* Washington, DC: Pew Research Center.

Purcell, K., Brenner, J., & Rainie, L. (2012). *Search engine use 2012.* Washington, DC: Pew Research Center.

Purcell, K., Mitchell, A., Rosenstiel, T., & Olmstead, K. (2010). *Understanding the participatory news consumer.* Washington, DC: Pew Research Center.

Radecki, L., Goldman, R., Baker, A., Lindros, J., & Boucher, J. (2013). Are pediatricians "game"? Reducing childhood obesity by training clinicians to use motivational interviewing through role-play simulations with avatars. *Games for Health: Research, Development, and Clinical Applications, 2*(3), 174–178.

Raice, S. (2012, February 2). Facebook sets historic IPO. Retrieved from http://online.wsj.com/news/articles/SB10001424052970204879004577110780078310366

Rainie, L. (2006). *Digital "natives" invade the workplace: Young people may be newcomers to the world of work, but it's their bosses who are immigrants into the digital world.* Washington, DC: Pew Research Center.

Rainie, L. (2009). *The new information ecology.* Washington, DC: Pew Research Center.

Rainie, L. (2012). *Two-thirds of young adults and those with higher income are smartphone owners.* Washington, DC: Pew Research Center.

Rainie, L. (2012). *Social media and voting.* Washington, DC: Pew Research Center.

Rainie, L., Brenner, J., & Purcell, K. (2012). *Photos and videos as social currency online.* Washington, DC: Pew Research Center.

Rainie, L., Cornfield, M., & Horrigan, J. (2005). *The Internet and campaign 2004*. Washington, DC: Pew Research Center.

Rainie, L., & Duggan, M. (2013). *E-book reading jumps; Print book reading declines*. Washington, DC: Pew Research Center.

Rainie, L., Purcell, K., & Smith, A. (2011). *The social side of the Internet*. Washington, DC: Pew Research Center.

Rainie, L., Smith, A., Schlozman, K. L., Brady, H., & Verba, S. (2012). *Social media and political engagement*. Washington, DC: Pew Research Center.

Rainie, L., & Wellman, B. (2012). *Networked: The new social operating system*. Cambridge, MA: The MIT Press.

Rainie, L., Zickuhr, K., Purcell, K., Madden, M., & Brenner, J. (2012). *The rise of e- reading*. Washington, DC: Pew Research Center.

Rauhofer, J. (2008). Privacy is dead, get over it! Information privacy and the dream of a risk-free society. *Information & Communications Technology Law, 17*, 185–197.

Regalado, A., & Leber, J. (2013, May 20). Intel's data economy initiative aims to help people capture the value of personal data. Retrieved from http://www.technologyreview.com/news/514386/intel-fuels-a-rebellion-around-your-data/

Reid, C. (2010, January 30). Amazon halts sales of Macmillan titles. Retrieved from http://www.publishersweekly.com/pw/by-topic/industry-news/publisher-news/article/41879-amazon-halts-sales-of-macmillan-titles.html

Rhoads, C. (2009, October 1). U.S. eases grip over web body: Move addresses criticisms as Internet usages becomes more global. *The Wall Street Journal*, p. B4.

Rice, R., & Katz, J. (2003). Comparing Internet and mobile phone usage: Digital divides of usage, adoption, and dropouts. *Telecommunications Policy, 27*, 597–623.

Rich, M. (2007, July 11). Potter has limited effect on reading habits. *The New York Times*, pp. A1, A15.

Richmond, R. (2011, May 17). VPN for the masses. Retrieved from http://gadgetwise.blogs.nytimes.com/2011/05/17/vpn-for-the-masses/?

Richtel, M. (2005, January 19). Video game industry sales reach record pace in 2004. *The New York Times*, pp. C3.

Richtel, M. (2011, January 28). Egypt cuts off most Internet and cell service. *The New York Times*, p. A13.

Richtel, M. (2012, July 21). He's watching that, in public? Pornography takes next seat. Retrieved from http://www.nytimes.com/2012/07/21/us/tablets-and-phones-lead-to-more-pornography-in-public.html?

Richtel, M. (2013, September 21). Intimacy on the web, with a crowd. Retrieved from http://www.nytimes.com/2013/09/22/technology/intimacy-on-the-web-with-a-crowd.html?

Riegner, C. (2008). Wired China: The power of the world's largest Internet population. *Journal of Advertising Research, 48*, 496–505.

Rieh, S. Y., & Hilligoss, B. (2008). College students' credibility judgments in the information-seeking process. In M. J. Metzger & A. J. Flanagin (Eds.), *Digital media, youth, and credibility* (pp. 49–71). Cambridge, MA: The MIT Press.

Riley, C. (2013, June 10). Booz Allen Hamilton in spotlight over leak. Retrieved from http://mon ey.cnn.com/2013/06/10/news/booz-allen-hamilton-leak/

Risen, J. (2013, July 18). Bipartian backlash grows against domestic surveillance. *The New York Times*, p. A14.

Risen, J., & Lichtblau, E. (2013, June 8). How the U.S. uses technology to mine more data more quickly. The New York Times, p. A1.

Rohter, L. (2013, July 8). Brazil voices "deep concern" over gathering of data by U.S. *The New York Times*, p. A7.

Rosen, J. (2010, July 25). The web means the end of forgetting. *The New York Times Magazine*, 30–37, 44–45.

Rosen, R. J. (2013, June 11). Why should we even care if the government is collecting our data? Retrieved from http://www.theatlantic.com/technology/archive/2013/06/why-should-we-even-care-if-the-government-is-collecting-our-data/276732/

Rosenfeld, M. J., & Thomas, R. J. (2012). Searching for a mate: The rise of the Internet as a social intermediary. *American Sociological Review, 77*, 523–547.

Rosenwald, M. (2012, November). Ducking Google in search engines. Retrieved from http://articles.washingtonpost.com/2012-11-09/business/35505935_1_duckduckgo-search-en gine-search-results

Rule, J. B. (2007). *Privacy in peril.* Oxford, England: Oxford University Press.

Rushe, D. (2013, August 14). Google: Gmail users shouldn't expect email privacy. *The Guardian*, p. 14.

Rushkoff, D. (2013, March 25). Why futurists suck: The real promise of the digital age. Retrieved from http://www.good.is/posts/why-futurists-suck-the-real-promise-of-the-digital-age

Rusli, E. M. (2013, May 16). After IPO, Facebook gets serious about making money. Retrieved from http://online.wsj.com/news/articles/SB10001424127887323582904578487103239166448

Rust, R. T., & Oliver, R. W. (1994). The death of advertising. *Journal of Advertising, 23*, 71–77.

Rutenberg, J. (2008a, October 18). Nearing record, Obama's ad effort swamps McCain. Retrieved from http://www.nytimes.com/2008/10/18/us/politics/18ads.html?

Rutenberg, J. (2008b, October 29). Obama infomercial, a closing argument to the everyman. Retrieved from http://www.nytimes.com/2008/10/29/us/politics/29obama.html?

Ryan, M. (2001). *Narrative as virtual reality: Immersion and interactivity in literature and electronic media.* Baltimore, MD: Johns Hopkins University Press.

Sabino, D. (2012, July 13). Kids age 6–12, parents, and social media. Retrieved from http://www.marketingprofs.com/charts/2012/8424/kids-age-6-12-parents-and-social-media

Sander, D. E. (2013, June 12). N.S.A. chief says phone logs halted terror threats. Retrieved from http://www.nytimes.com/2013/06/13/us/nsa-chief-says-phone-record-logs-halted-terror-threats.html?

Sanger, D. E., Barboza, D., & Perlroth, N. (2013, February 19). China's army seen as tied to hacking against U.S. *The New York Times*, pp. A1, A9.

Sass, E. (2011, January 3). 2000–2010: The decade that killed (newspaper) classifieds. Retrieved from http://www.mediapost.com/publications/article/142129/

Sassen, S. (2007). *Deciphering the global: Its spaces, scales and subjects.* New York, NY: Routledge.

Savage, C. (2013, July 26). Robert's picks reshaping secret surveillance court. *The New York Times*, pp. A1, A3.

Savage, C. (2013, July 27). Defense calls Manning's intention good. *The New York Times*, p. A10.

Savage, C. (2013, July 30). Manning is acquitted of aiding the enemy. Retrieved from http://www.nytimes.com/2013/07/31/us/bradley-manning-verdict.html?

Savage, C. (2013, August 17). N.S.A. calls violations of privacy "minuscule." *The New York Times*, p. A12.

Savage, C. (2013, October 27). Federal prosecutors, in a policy shift, cite warrantless wiretaps as evidence. *The New York Times*, p. 21.

Savage, C., Miller, C. C., & Perlroth, N. (2013, October 31). N.S.A. said to tap Google and Yahoo abroad. *The New York Times*, pp. B1, B6.

Savage, C., & Shane, S. (2013, August 22). Top-secret court castigated N.S.A. on surveillance. *The New York Times*, pp. A1, A14.

Savage, C., & Shear, M. D. (2013, August 10). President moves to ease worries on surveillance. *The New York Times*, pp. A1, A11.

Schatz, A., & Johnson, F. (2009, September 22). Internet providers push back against "net neutrality" proposal. *The Wall Street Journal*, p. A4.

Schencher, S. (2009, September 3). Sawyer to succeed Gibson at ABC News. *The Wall Street Journal*, pp. B1, B2.

Schencher, S. (2011, December 18). Time Warner Cable to offer HBO Go. Retrieved from http://online.wsj.com/news/articles/SB10001424052970204058404577106493650859570

Schmidt, E. (2013, April 15). Internet users in 2020. Retrieved from http://internetstatstoday.com/

Schmidt, E., & Cohen, J. (2013). *The new digital age: Reshaping the future of people, nations and business*. New York, NY: Alfred A. Knopf.

Schmidt, M. S. (2013, October 30). N.S.A. head says European data was collected by allies. *New York Times*, p. A13.

Schwartz, B. (2004). *The paradox of choice: Why more is less*. New York, NY: Ecco.

Schwartz, N. D. (2011). Online banking keeps customers on hook for fees. Retrieved from http://www.nytimes.com/2011/10/16/business/online-banking-keeps-customers-on-hook-for-fees.html?

Scientific American. (2013, January). *The future of science 50, 100, 150 years from now. Scientific American, 308*(1), p. 79.

Scott, J. C. (1990). *Domination and the arts of resistance: Hidden transcripts*. New Haven, CT: Yale University Press.

Segal, D. (2012, February 12). The dirty little secrets of search. Retrieved from http://www.nytimes.com/2011/02/13/business/13search.html?

Segel, D. (2013, June 1). This man is not a cyborg. Yet. Retrieved from http://www.nytimes.com/2013/06/02/business/dmitry-itskov-and-the-avatar-quest.html?

Sengupta, S. (2012, November 26). Courts divided over searches of cellphones. *The New York Times*, pp. A1, A3.

Sengupta, S., & Perlroth, N. (2012, March 5). The bright side of being hacked. *The New York Times*, pp. B1, B8.

Shadbolt, N., & Berners-Lee, T. (October 2008). Web science emerges. *Scientific American, 299*(4), 76–18.

Shane, S. (2011, January 29). Spotlight again falls on web tools and change. *The New York Times*, p. WK1.

Shane, S. (2013, September 11). Court upbraided N.S.A. on its use of call-log data. *The New York Times*, pp. A1, A14.

Shane, S. (2013, November 3). To morsel too minuscule for all-consuming N.S.A. *The New York Times*, pp. 1, 10, 11.

Shane, S., & Sanger, D. E. (2013, July 1). Job title key to inner access held by leaker. *The New York Times*, pp. A1, A8.

Shane, S., & Somaiya, R. (2013, June 13). New leak indicates U.S. and Britain eavesdropped at '09 World Conference. Retrieved from http://www.nytimes.com/2013/06/17/world/europe/new-leak-indicates-us-and-britain-eavesdropped-at-09-world-conferences.html?

Shane, S., & Weisman, J. (2013, June 11). Debate on secret data looks unlikely, partly due to secrecy. *The New York Times*, pp. A1, A13.

Shellenbarger, S. (2012, July 11). Working from home without slacking off. Retrieved from http://online.wsj.com/news/articles/SB10001424052702303684004577508953483021234

Shneiderman, B. (2003). *Leonardo's laptop: Human needs and the new computing technologies.* Cambridge, MA: The MIT Press.

Shneiderman, B., Plaisant, C., Cohen, M. S., & Jacobs, S. M. (2009). *Designing the user interface: Strategies for effective human-computer interactions* (5th ed.). Upper Saddle River, NJ: Prentice-Hall.

Short, M. B., Black, L., Smith, A. H., Wetterneck, C. T., & Wells, D. E. (2012). A review of Internet pornography use research: Methodology and content from the past 10 years. *Cyberpsychology, Behavior, and Social Networking, 15*, 13–23.

Sidel, R. (2013a, March 31). After years of growth, banks are pruning their branches. Retrieved from http://online.wsj.com/news/articles/SB100014241278873236997045783268941463 25274

Sidel, R. (2013b, February 11). Banks make smartphone connection. Retrieved from http://online.wsj.com/news/articles/SB10001424127887323511804578298192585478794

Sigman, S. J. (1991). Handling the discontinuous aspects of continuous social relationships: Toward research on the persistence of social forms. *Communication Theory, 1*, 106–127.

Siibak, A. (2009). Constructing the self through photo selection—Visual impressions management on social networking websites. *Cyberpsychology: Psychosocial Research on Cyberspace, 3*, 1–9.

Silverman, R. E. (2012, September 4). Step into the office-less company. Retrieved from http://online.wsj.com/news/articles/SB10000872396390443571904577631750172652114

Silverman, R. E. (2013, April 17). Telecommuting boosts firms' revenue growth. Retrieved from http://blogs.wsj.com/atwork/2013/04/17/telecommuting-boosts-firms-revenue-growth/

Silverman, R. E., & Fottrell, Q. (2013, February 26). The home office in the spotlight. Retrieved from http://online.wsj.com/news/articles/SB10001424127887323384604578328681101539330

Silverpop. (2013). *2013 email marketing metrics benchmark study*. Retrieved from http://www.sil verpop.com/Documents/Whitepapers/2013/WP_EmailMarketingMetricsBenchmarkStudy 2013.pdf

Singel, R. (2010, October 27). Oct. 27, 1994: Web gives birth to banner ads. Retrieved from http://www.wired.com/thisdayintech/2010/10/1027hotwired-banner-ads/

Singer, N. (2012, December 10). Children's apps fall short on parental disclosure, U.S. says. *The New York Times*, p. B1.

Singer, N. (2013, February 17). If you're collecting our data, you ought to protect it. *The New York Times*, p. 4BU.

Singer, N. (2013, July 15). Wrangling over "do not track." Retrieved from http://bits.blogs.nytimes. com/2013/07/15/wrangling-over-do-not-track/?

Sirota, D. (2013, June 18). How cash secretly rules surveillance policy. Retrieved from http://www. salon.com/2013/06/18/how_cash_secretly_rules_surveillance_policy/

Sisario, B. (2012, January 20). U.S. charges popular site with piracy. *The New York Times*, p. B1.

Slater, D. (1998). Trading sexpics on IRC: Embodiment and authenticity on the Internet. *Body and Society, 4*, 91–117.

Smale, A. (2013, October 25). Indignation over U.S. spying spreads in Europe. *The New York Times*, p. A10.

Smale, A. (2013, November 7). German considers having Snowden testify to inquiry. *New York Times*, p. A8.

Small, G. W., Moody, T. D., Siddarth, P., & Bookheimer, S. Y. (2009). Your brain on Google: Patterns of cerebral activation during Internet searching. *American Journal of Geriatric Psychiatry, 17*, 116–126.

Small, G., & Vorgan, G. (2008, October/November). Meet your ibrain: How the technologies that have become part of our daily lives are changing the way we think. *Scientific American Mind, 19*(5), 42–49.

Small, G., & Vorgan, G. (2009). *iBrain: Surviving the technological alteration of the modern mind*. New York, NY: William Morrow.

Smith, A., & Duggan, M. (2012). *Online political videos and campaign 2012*. Washington, DC: Pew Research Center.

Smith, A., & Rainie, L. (2012). *The future of money: Smart phone swiping in the mobile age*. Washington, DC: Pew Research Center.

Smith, E. (2009, January 2). Music sales decline for seventh time in eight years. Retrieved from http://online.wsj.com/news/articles/SB123075988836646491

Snider, M. (2012, December 3). Even when you gotta go, social media goes too. Retrieved from http://www.usatoday.com/story/tech/2012/12/02/social-media-use-rises/1738009/

Snowden, E. (2013, July 12). Statement by Edward Snowden to human rights groups at Moscow's Sheremetyevo Airport, and Ellen Barry and Andrew Roth. Snowden renews plea for Moscow to grant asylum. Retrieved from http://www.nytimes.com/2013/07/13/world/europe/snowden-russia-asylum.html?

Solove, D. J. (2008). *Understanding privacy*. Cambridge, MA: Harvard University Press.

Sonne, P. (2013, September 4). Data-security expert Kaspersky; There is no more privacy. *Wall Street Journal*, p. 4A.

Souppouris, A. (2012, November 28). South Korea plans to extend "Internet dangers" education to three-year-olds. Retrieved from http://www.theverge.com/2012/11/28/3700928/south-korea-pre-school-internet-danger-education-online-addiction

Souppouris, A. (2013, April 9). Vine is now the number one free app in the US App Store. Retrieved from http://www.theverge.com/2013/4/9/4204396/vine-number-one-us-app-store-free-apps-chart

Spitzberg, B. H., & Hoobler, G. (2002). Cyberstalking and the technologies of interpersonal terrorism. *New Media & Society, 4*, 71–92.

Steel, E. (2001, June 17). TV spots resist flow of dollars to the web. *The Wall Street Journal*, pp. B1, B2.

Stelter, B. (2009, July 06). Rise of web video, beyond 2-minute clips. *The New York Times*, p. B1, B8.

Stelter, B. (2011, May 3). Ownership of TV sets falls in US. *The New York Times*, p. B1.

Stelter, B., & Carter, B. (2010, March 1). Network news at a crossroads. *The New York Times*, pp. B1, B5.

Stelter, B., & Stone, B. (2009, February 5). Digital pirates winning with studios. *The New York Times*, p. A1.

Stelter, B., & Wortham, J. (2010, March 13). F.C.C. plan to widen Internet access in U.S. sets up battle. *The New York Times*, pp. A1, A3.

Stern, L. A., & Taylor, K. (2007). Social networking on Facebook. *Journal of the Communication, Speech & Theatre Association of North Dakota, 20*, 9–20.

Sternstein, A. (2013, April 24). What's the White House policy on neutralizing damaging tweets? Retrieved from http://www.nextgov.com/cybersecurity/cybersecurity-report/2013/04/whats-white-house-policy-neutralizing-damaging-tweets/62742/

Stetzer, A. (2013, May 23). Google Penguin 2013: How to evolve link building into real SEO. Retrieved from http://searchenginewatch.com/article/2269998/Google-Penguin-2013-How-to-Evolve-Link-Building-into-Real-SEO

Steuer, J. (1992). Defining virtual reality: Dimensions determining telepresence. *Journal of Communication, 42*, 73–93.

Steve Jobs. (n.d.). In *Wikipedia*. Retrieved June 29, 2013 from http://en.wikipedia.org/wiki/Steve_Jobs

Stokes, T. (2013). *How to build your brand with branded content*. Retrieved from http://www.forrester.com/How+To+Build+Your+Brand+With+Branded+Content/fulltext/-/E-RES92961

Stone, B. (2009, June, 10). Revenue at Craigslist is said to top $100 Million. *The New York Times*, p. B10.

Strauss, W., & Howe, N. (1991). *Generations: The history of America's future, 1584 to 2069*. New York, NY: William Morrow.

Strauss, W., & Howe, N. (2006). *Millennials and the pop culture: Strategies for a new generation of consumers in music, movies, television, the Internet, and video games*. Great Falls, VA: LifeCourse Associates.

Streitfeld, D. (2012, March 4). In a flood of digital data, an ark full of books. *The New York Times*, p. A1.

Stross, R. (2011, May 1). Opt-in rules are a good start. *The New York Times*, p. B3.

Stross, R. (2012, March 4). The second screen, trying to complement the first. *The New York Times*, p. BU5.

Stroud, N. J. (2008). Media use and political predispositions: Revisiting the concept of selective exposure. *Political Behavior, 30*, 341–366.

Stroud, N. J. (2010). Polarization and partisan selective exposure. *Journal of Communication, 60*, 556–576.

Stuebe, A. (1996, August 15). "Internet alley" brings convention to the Web. Retrieved from http://partners.nytimes.com/library/cyber/week/0815gop.html

Sullivan, L. (2012, May 22). Adobe PrimeTime bridges gap from TV to online video. *Online*. Retrieved from http://www.mediapost.com/publications/article/175195/

Sundar, S. S. (2008). The MAIN model: A heuristic approach to understanding technology effects on credibility. In M. J. Metzger & A. J. Flanagin (Eds.), *Digital media, youth, and credibility* (pp. 73–100). Cambridge, MA: The MIT Press.

Sunde, P. (2012, February 10). The Pirate Bay's Peter Sunde: It's evolution, stupid. Retrieved from http://www.wired.com/threatlevel/2012/02/peter-sunde/

Sunstein, C. (2009). *Republic.com 2.0*. Princeton, NJ: Princeton University Press.

Swire, P. (1999). Financial privacy and the theory of high-tech government surveillance. *Washington University Law Quarterly, 77*, 461.

Swisher, K. (2013, February 25). Survey says: Despite Yahoo ban, most tech companies support work-from-home for employees. Retrieved from http://allthingsd.com/20130225/survey-says-despite-yahoo-ban-most-tech-companies-support-work-from-home-for-employees/

Taleb, N. N. (2013, February 8). Beware the big errors of "big data." Retrieved from http://www.wired.com/opinion/2013/02/big-data-means-big-errors-people/

Tapscott, D. (2012, June). Don Tapscott: Four principles for the open world [Video file]. Retrieved from http://www.ted.com/talks/don_tapscott_four_principles_for_the_open_world_1.html

Taylor, P., & Wang, W. (2010, August 19). *The fading glory of the television and telephone*. Washington, DC: Pew Research Center.

Taylor, R., Funk, C., & Craighill, P. (2006, May 16). *Home "ticket sales" dwarf theater attendance*. Washington, DC: Pew Research Center.

Tedesco, J. C. (2011). The complex Web: Young adults' opinions about online campaign messages. In M. S. McKinney & M. C. Banwart (Eds.), *Communication in the 2008 U.S. election: Digital natives elect a president* (pp. 13–31). New York, NY: Peter Lang.

Telefonica and Financial Times. (2013, June 4). *Today's young adults: The leaders of tomorrow*. Retrieved from http://survey.telefonica.com/

Tess, P. A. (2013). The role of social media in higher education classes (real and virtual)–A literature review. *Computers in Human Behavior, 29*(5), A60–A68.

The Futurist [of the World Future Society]. (2012, September–October). The 22nd century at first light: Envisioning life in the Year 2100. *The Futurist, 46*(5), 33–55.

Theroux, P. (2011, September 5). How Apple revolutionized our world. *Newsweek*, p. 36.

Thompson, J. B. (2005). *Books in the digital age: The transformation of academic and higher education publishing in Britain and the United States*. Malden, MA: Polity Press.

THR poll: 9 out of 10 call social media new form of entertainment; Young people want texting in movies (2012, March 9). Retrieved from http://www.hollywoodreporter.com/news/twit ter-facebook-study-texting-movies-twitter-social-media-302921

Tode, C. (2013, June 11). QR code-enabled virtual stores support merchants' mobile shopping strategies. Retrieved from http://www.mobilecommercedaily.com/qr-code-enabled-virtual-stores-support-merchants-mobile-shopping-strategies

Toffler, A. (1970). *Future shock.* New York, NY: Bantam Books.

Toffler, A. (1990). *Powershift: Knowledge, wealth, and violence at the edge of the 21st century.* New York, NY: Bantam Books.

Toffler, A., & Toffler, H. (2006). *Revolutionary wealth: How it will be created and how it will change our lives.* New York, NY: Alfred A. Knopf.

Tokunaga, R. S. (2011). Friend me or you'll strain us: Understanding negative events that occur over social networking sites. *Cyberpsychology, Behavior, and Social Networking, 14,* 425–432.

Tom Tullis and Bill Albert, *Measuring the User Experience: Collecting, Analyzing, and Presenting Usability Metrics* (Elsevier / Morgan Kaufmann Publishers, 2008).

Toma, C. L., & Hancock, J. T. (2012). What lies beneath: The linguistic traces of deception in online dating profiles. *Journal of Communication, 62,* 78–97.

Toma, C. L., Hancock, J. T., & Ellison, N. B. (2008). Separating fact from fiction: An examination of deceptive self-presentation in online dating profiles. *Personality and Social Psychology Bulletin, 34,* 1023–1036.

Tong, S. T., Van Der Heide, B., Langwell, L., & Walther, J. B. (2008). Too much of a good thing? The relationship between number of friends and interpersonal impressions on Facebook. *Journal of Computer-Mediated Communication, 13,* 531–549.

Tong, S. T., & Walther, J. B. (2010). Just say "no thanks": Romantic rejection in computer-mediated communication. *Journal of Social and Personal Relationships, 28,* 488–506.

Tong, S. T., & Walther, J. B. (2011). Relational maintenance and CMC. In K. B. Wright & L. M. Webb (Eds.), *Computer-mediated communication in personal relationships* (pp. 79–118). New York, NY: Peter Lang.

Trachtenberg, J. A., & Peers, M. (2012, January 6). Barnes & Noble seeks next chapter. *Wall Street Journal,* pp. A1, A10.

Trent, J. S. (2009). The early presidential campaign of 2008: The good, the historical, but rarely the bad. In R. E. Denton (Ed.), *The 2008 presidential campaign. A communication perspective* (pp. 1–17). Lanham, MD: Rowman & Littlefield.

Troup, C., & Rose, J. (2012). Working from home: Do formal or informal telework arrangements provide better work–family outcomes? *Community, Work & Family, 15,* 471–486.

Turkle, S. (1984). *The second self: Computers and the human spirit.* New York, NY: Simon & Schuster.

Turkle, S. (2011). *Alone together: Why we expect more from technology and less from each other.* New York, NY: Basic Books.

United States Census Bureau. (2012). *Home-based workers in the United States: 2010.* Retrieved from http://www.census.gov/prod/2012pubs/p70-132.pdf

United States Census Bureau. (2012). *Voting and registration in the election of November 2008.* Retrieved from http://www.census.gov/prod/2010pubs/p20-562.pdf

United States Postal Service. (2011). *The household diary student: Mail use & attitudes in FY 2011*. Washington, DC.

Utz, S. (2010). Show me your friends and I will tell you what types of person you are: How one's profile, number of friends, and type of friends influence formation on social network sites. *Journal of Computer-Mediated Communication, 15*, 314–335.

Valentino-DeVries, J., & Gorman, S. (2013, July 8). Secret court ruling expanded by spy powers. *The Wall Street Journal*, pp. A1, A4.

Valenzuela, S., Park, N., & Kee, K. F. (2009). Is there social capital in a social network site? Facebook use and college students' life satisfaction, trust, and participation. *Journal of Computer-Mediated Communication, 14*, 875–901.

Valkenburg, P., & Peter, J. (2007). Who visits online dating sites? Exploring some characteristics of online daters. *CyberPsychology & Behavior, 10*, 849–852.

Vargas, J. A. (2008a, August 20). Obama's wide web. Retrieved from http://articles.washingtonpost.com/2008-08-20/news/36858985_1_joe-rospars-obama-supporter-obama-s-internet

Vargas, J. A. (2008b, November 25). Obama raised half a billion online. Retrieved from http://voices.washingtonpost.com/44/2008/11/obama-raised-half-a-billion-on.html

Vass, S. (2006, June 4). What is a newspaper? My hunch is that a purely text-based medium will struggle in future. Retrieved from http://www.redorbit.com/news/technology/529273/what_is_a_newspaper_my_hunch_is_that_a_text_based_medium_will_struggle_in_future

Vega, T. (2010, September 9). Code that tracks users' browsing prompts lawsuits. *The New York Times*, p. B3.

Vega, T. (2012, February, 13). A new tool in protecting online privacy. *The New York Times*, p. B4.

Vega, T. (2012, February, 13). Study finds news sites fail to aim ads at users. *The New York Times*, p. B3.

Vega, T. (2012, February 17). Magazine newsstand sales suffered sharp falloff in second half of 2011. Retrieved from http://mediadecoder.blogs.nytimes.com/2012/02/07/magazine-newsstand-sales-suffered-sharp-falloff-in-second-half-of-2011/?

Vega, T. (2013, April 7). Sponsors now pay for online article, not just ads. Retrieved from http://www.nytimes.com/2013/04/08/business/media/sponsors-now-pay-for-online-articles-not-just-ads.html?

Venetis, M. K., Greene, K., Magsamen-Conrad, K., Banerjee, S. C., Checton, M. G., & Bagdasarov, Z. (2008). "You can't tell anyone but...": Exploring the use of privacy rules and revealing behaviors. *Communication Monographs, 79*, 344–365.

Vranica, S. (2013, June 11). Web display ads often not visible. Retrieved from http://online.wsj.com/news/articles/SB10001424127887324904004578537131312357490

Wahlmann, G. (2009, October 2). Synthetic existence. Retrieved from https://www.adbusters.org/magazine/86/synthetic-existence.html

Wall, A. (2012, April 15). Consumer ad awareness in search results. Retrieved from http://www.seobook.com/consumer-ad-awareness-search-results

Wallop, H. (2011, April 1). Video games sell more than DVDs and albums. Retrieved from http://www.telegraph.co.uk/technology/video-games/8421458/Video-games-sell-more-than-DVDs-and-albums.html

Walsh, M. (2011). Yahoo Mail poised for full rollout, faster time, better ad targeting. Retrieved from http://www.mediapost.com/publications/article/151055/

Walsh, M. (2012, August 14). Most common place to go mobile: In bed. Retrieved from http://www.mediapost.com/publications/article/180768/most-common-place-to-go-mobile-in-bed.html

Walther, J. B. (1996). Computer-mediated communication: Impersonal, interpersonal, and hyperpersonal interaction. *Communication Research, 23*, 3–43.

Walther, J. B., Parks, M. R. (2002). Cues filtered out, cues filtered in: Computer-mediated communication and relationships. In M. L. Knapp & J. A. Daly (Eds.), *Handbook of interpersonal communication* (3rd ed., pp. 529–563). Thousand Oaks, CA: Sage.

Walther, J. B., Van Der Heide, B., Hamel, L. M., & Shulman, H. C. (2009). Self-generated versus other-generated statements and impression in computer-mediated communication. *Communication Research, 36*, 229–253.

Walther, J. B., Van Der Heide, B., Kim, S.-Y., Westerman, D., & Tong, S. T. (2008). The role of friends' appearance and behavior on evaluations of individuals on Facebook: Are we known by the company we keep? *Human Communication Research, 34*, 28–49.

Warf, B. (2010). Uneven geographies of the African Internet: Growth, change, and implications. *African Geographical Review, 29*(2), 41–66.

Warf, B., & Vincent, P. (2007). Multiple geographies of the Arab Internet. *Area, 39*, 83–96.

Wark, M. (2004). *A hacker manifesto.* Cambridge, MA: Harvard University Press.

Wasserman, T. (2012, September 25). What is "native advertising"? Depends who you ask. Retrieved from http://mashable.com/2012/09/25/native-advertising/

Waskull, D. D. (2002). The naked self: Being a body in televideo cybersex. *Symbolic Interaction, 25*, 199–227.

Waskull, D. D., & Radeloff, C. L. (2010). "How do I rate?": Web sites and gendered erotic looking glasses. In F. Attwood (Ed.), *Porn.com: Making sense of online pornography* (pp. 202–216). New York, NY: Peter Lang.

Wayner, P. (2007, August 9). An entire bookshelf, in your hands. *The New York Times*, p. C7.

Web 1.0. In *Wikipedia*. Retrieved from http://en.wikipedia.org/wiki/Web_1.0 *Webster's New Collegiate Dictionary*. (1981). Springfield, MA: Merriam-Webster.

Webber, L. (2013, April 24). How your smartphone could get you a job. Retrieved from http://online.wsj.com/news/articles/SB10001424127887323551004578441130657837720

Weinberger, D. (2011). *Too big to know: Rethinking knowledge now that the facts aren't the facts, experts are everywhere, and the smartest person in the room is the room.* New York, NY: Basic Books.

Weisman, J. (2012, January, 18). In piracy bill fight, new economy rises against old. *The New York Times*, p. A1.

Weisman, J. (2013, July 25). House's attempt to rein in N.S.A. narrowly fails. *The New York Times*, pp. A1, A17.

Westin, A. F. (1967). *Privacy and freedom.* New York, NY: Atheneum.

What We Know About Online Course Outcomes. (2013). *Community College Research Center, Teachers College, Columbia University.* Retrieved from http://ccrc.tc.columbia.edu

White, M. (2013, May/June). We are the players in the game of revolution. Retrieved from https://www.adbusters.org/magazine/107/we-are-players-game-revolution.html

Whiteman, H. (2009). British PM apologizes for treatment of gay code-breaker. Retrieved from http://edition.cnn.com/2009/WORLD/europe/09/11/alan.turing.petition.apology/

Whitty, M. T. (2007a). Revealing the "real" me, searching for the "actual" you: Presentations of self on an Internet dating site. *Computers in Human Behavior, 24*, 1707–1723.

Whitty, M. T. (2007b). Manipulation of self in cyberspace. In B. H. Spitzberg & W. R. Cupach, *The dark side of interpersonal communication* (2nd ed.,) (pp. 93–118). Mahwah, NJ: Lawrence Erlbaum.

Whitty, M. T., & Buchanan, T. (2010). What's in a screen name? Attractiveness of different types of screen names used by online daters. *International Journal of Internet Studies, 5*, 5–19.

Whitty, M. T., & Joinson, A. (2009). *Truth, lies, and trust on the Internet.* London, England: Routledge.

Who was Alan Turing, subject of the latest Google doodle? (2009). Retrieved from http://ibnlive.in.com/news/who-was-alan-turing-subject-of-the-latest-google-doodle/267489-11.html

Wickelgren, I. (2012, December 11). On TV, Ray Kurzweil tells me how to build a brain. Retrieved from http://blogs.scientificamerican.com/streams-of-consciousness/2012/12/11/on-tv-ray-kurzweil-tells-me-how-to-build-a-brain/

Wiener, N. (1950). *The human use of human beings: Cybernetics and society.* New York, NY: Houghton Mifflin.

Wikipedia: Size Comparions. (2013a). In *Wikipedia*. Retrieved from http://en.wikipedia.org/wiki/Wikipedia:Size_comparisons

Wikipedia: Statistics. (2013b). In *Wikipedia*. Retrieved from https://en.wikipedia.org/wiki/Wikipedia:Statistics

Wilford, J. N. (2008, July 31). Discovering how Greeks computed in 100 B.C. *The New York Times*, p. A12.

Williams, A. (2006, July 16). The graying of the record store. *The New York Times*, pp. 1ST, 6ST.

Williams, C. (2012, January 20). Anonymous attacks FBI website over Megaupload raids. Retrieved from http://www.telegraph.co.uk/technology/news/9027246/Anonymous-attacks-FBI-website-over-Megaupload-raids.html

Wilman, J. (2008, April 9). A new kind of Web—don't miss these 11 sites. Retrieved from http://www.computerworld.com/s/article/9072458/Opinion_A_new_kind_of_Web_mdash_don_t_miss_these_11_sites?pageNumber=3

Wolfram, H. J., & Gratton, L. (2012). Spillover between work and home, role importance and life satisfaction. *British Journal of Management*. doi: 10.1111/j.1467- 8551.2012.00833.x

Wood, J. T., & Duck, S. (Eds.). (2006). *Composing relationships: Communication in everyday life.* Belmont, CA: Thomson/Wadsworth.

Woollett, K., & Maguire, E. A. (2011). Acquiring "the knowledge" or London's layout drives: Structural brain changes. *Current Biology, 21*, 2109–2114.

Wortham, J. (2013, January 8). A financial service for people fed up with banks. Retrieved from http://www.nytimes.com/2013/01/09/technology/a-financial-service-for-people-fed-up-with-banks.html?

Wortham, J. (2013, February 9). A growing app lets you see it, then you don't. *The New York Times*, p. A1.

Worthen, B. (2008, May 6). Father of spam marks an infamous anniversary. Retrieved from http://online.wsj.com/news/articles/SB121003279234369267

Wright, A. (2008, January–February). Mastering information through the ages. *The Futurist, 41*(1), 32.

Wyatt, E. (2010, April 6). U.S. court curbs F.C.C. authority on Web traffic. *The New York Times*, pp. A1, B7.

Wyatt, E. (2010, May 26). FCC begins review of hotly debated regulations on media ownership. *The New York Times*, p. B10

Yadron, D., & Ovide, S. (2013, March 12). FTC says tweet ads need some fine print. Retrieved from http://online.wsj.com/news/articles/SB10001424127887324281004578356766374983 9822

Ynalvez, M., & Shrum, W. (2006). International training and the digital divide: Computer and email use in the Philippines. *Perspectives on Global Development & Technology, 5*, 277–302.

Young, D. G., & Caplan, S. E. (2010). Online dating and conjugal bereavement. *Death Studies, 34*, 575–605.

Yousafzai, S., &Yani-de-Soriano, M. (2012). Understanding customer-specific factors underpinning Internet banking adoption. *International Journal of Bank Marketing, 30*, 60–81.

Zhao, J. (2008). A snapshot of Internet regulation in contemporary China: Censorship, profitability and responsibility. *China Media Research, 4*(3), 37–42.

Zhao, S., Grasmuck, S., & Martin, J. (2008). Identity construction on Facebook: Digital empowerment in anchored relationships. *Computers in Human Behavior, 24*, 1816–1836.

Zickuhr, K., & Smith, A. (2012, April 13). *Digital differences*. Washington, DC: Pew Internet & American Life Project.

Zimmer, C. (2010, December 20). Is music for wooing, mothering, bonding—or is it just auditory cheesecake. Retrieved from http://discovermagazine.com/2010/dec/21-music-wooing-moth ering-bonding-enjoyment#.Uq-Kuyes3f0

Zittrain, J. (2008). *The future of the Internet and how to stop it*. New Haven, CT: Yale University Press.

Zweir, S. Arauio, T., Bouaks, M., & Willemsen, L. (2011). Boundaries to the articulation of possible selves through social networking sites: The case of Facebook profilers' social connectedness. *Cyberpsychology, Behavior, and Social Networking, 14*, 571–576.

Index

General Editor: **Steve Jones**

Digital Formations is the best source for critical, well-written books about digital technologies and modern life. Books in the series break new ground by emphasizing multiple methodological and theoretical approaches to deeply probe the formation and reformation of lived experience as it is refracted through digital interaction. Each volume in **Digital Formations** pushes forward our understanding of the intersections, and corresponding implications, between digital technologies and everyday life. The series examines broad issues in realms such as digital culture, electronic commerce, law, politics and governance, gender, the Internet, race, art, health and medicine, and education. The series emphasizes critical studies in the context of emergent and existing digital technologies.

Other recent titles include:

Felicia Wu Song
  Virtual Communities: Bowling Alone, Online Together

Edited by Sharon Kleinman
  The Culture of Efficiency: Technology in Everyday Life

Edward Lee Lamoureux, Steven L. Baron, & Claire Stewart
  Intellectual Property Law and Interactive Media: Free for a Fee

Edited by Adrienne Russell & Nabil Echchaibi
  International Blogging: Identity, Politics and Networked Publics

Edited by Don Heider
  Living Virtually: Researching New Worlds

Edited by Judith Burnett, Peter Senker & Kathy Walker
  The Myths of Technology: Innovation and Inequality

Edited by Knut Lundby
  Digital Storytelling, Mediatized Stories: Self-representations in New Media

Theresa M. Senft
  Camgirls: Celebrity and Community in the Age of Social Networks

Edited by Chris Paterson & David Domingo
  Making Online News: The Ethnography of New Media Production

To order other books in this series please contact our Customer Service Department:

(800) 770-LANG (within the US)

(212) 647-7706 (outside the US)

(212) 647-7707 FAX

To find out more about the series or browse a full list of titles, please visit our website:

WWW.PETERLANG.COM